103
Advances in Biochemical Engineering/Biotechnology

Series Editor: T. Scheper

Advances in Biochemical Engineering/Biotechnology

Series Editor: T. Scheper

Recently Published and Forthcoming Volumes

Tissue Engineering II

Basics of Tissue Engineering and Tissue Applications

Volume Editors: Kyongbum Lee · David Kaplan

With contributions by

J. P. Acker · E. S. Ahn · S. T. Andreadis · F. Berthiaume · S. N. Bhatia
R. J. Fisher · J. A. Garlick · Y. Nahmias · R. A. Peattie · V. L. Tsang
T. J. Webster · M. L. Yarmush

 Springer

76358601

Advances in Biochemical Engineering/Biotechnology reviews actual trends in modern biotechnology. Its aim is to cover all aspects of this interdisciplinary technology where knowledge, methods and expertise are required for chemistry, biochemistry, micro-biology, genetics, chemical engineering and computer science. Special volumes are dedicated to selected topics which focus on new biotechnological products and new processes for their synthesis and purification. They give the state-of-the-art of a topic in a comprehensive way thus being a valuable source for the next 3–5 years. It also discusses new discoveries and applications. Special volumes are edited by well known guest editors who invite reputed authors for the review articles in their volumes.

In references *Advances in Biochemical Engineering/Biotechnology* is abbeviated *Adv Biochem Engin/Biotechnol* and is cited as a journal.

Springer WWW home page: springer.com
Visit the ABE content at springerlink.com

Library of Congress Control Number: 2006929797

ISSN 0724-6145
ISBN-10 3-540-36185-5 Springer Berlin Heidelberg New York
ISBN-13 978-3-540-36185-5 Springer Berlin Heidelberg New York
DOI 10.1007/11749219

Springer is a part of Springer Science+Business Media

springer.com

© Springer-Verlag Berlin Heidelberg 2007

The use of registered names, trademarks, etc. in this publication does not imply, even in the absence of a specific statement, that such names are exempt from the relevant protective laws and regulations and therefore free for general use.

Cover design: WMXDesign GmbH, Heidelberg
Typesetting and Production: LE-TeX Jelonek, Schmidt & Vöckler GbR, Leipzig

Printed on acid-free paper   02/3141 YL – 5 4 3 2 1 0

Advances in Biochemical Engineering/Biotechnology
Also Available Electronically

For all customers who have a standing order to Advances in Biochemical Engineering/Biotechnology, we offer the electronic version via SpringerLink free of charge. Please contact your librarian who can receive a password or free access to the full articles by registering at:

springerlink.com

If you do not have a subscription, you can still view the tables of contents of the volumes and the abstract of each article by going to the SpringerLink Homepage, clicking on "Browse by Online Libraries", then "Chemical Sciences", and finally choose Advances in Biochemical Engineering/Biotechnology.

You will find information about the

– Editorial Board
– Aims and Scope
– Instructions for Authors
– Sample Contribution

at springer.com using the search function.

Attention all Users
of the "Springer Handbook of Enzymes"

Information on this handbook can be found on the internet at
springeronline.com

A complete list of all enzyme entries either as an alphabetical Name Index or
as the EC-Number Index is available at the above mentioned URL. You can
download and print them free of charge.

A complete list of all synonyms (more than 25,000 entries) used for the enzymes
is available in print form (ISBN 3-540-41830-X).

Save 15%

We recommend a standing order for the series to ensure you automatically
receive all volumes and all supplements and save 15% on the list price.

Preface

It is our pleasure to present this special volume on tissue engineering in the series *Advances in Biochemical Engineering and Biotechnology*. This volume reflects the emergence of tissue engineering as a core discipline of modern biomedical engineering, and recognizes the growing synergies between the technological developments in biotechnology and biomedicine. Along this vein, the focus of this volume is to provide a biotechnology driven perspective on cell engineering fundamentals while highlighting their significance in producing functional tissues. Our aim is to present an overview of the state of the art of a selection of these technologies, punctuated with current applications in the research and development of cell-based therapies for human disease.

To prepare this volume, we have solicited contributions from leaders and experts in their respective fields, ranging from biomaterials and bioreactors to gene delivery and metabolic engineering. Particular emphasis was placed on including reviews that discuss various aspects of the biochemical processes underlying cell function, such as signaling, growth, differentiation, and communication. The reviews of research topics cover two main areas: cellular and non-cellular components and assembly; evaluation and optimization of tissue function; and integrated reactor or implant system development for research and clinical applications. Many of the reviews illustrate how biochemical engineering methods are used to produce and characterize novel materials (e.g. genetically engineered natural polymers, synthetic scaffolds with cell-type specific attachment sites or inductive factors), whose unique properties enable increased levels of control over tissue development and architecture. Other reviews discuss the role of dynamic and steady-state models and other informatics tools in designing, evaluating, and optimizing the biochemical functions of engineered tissues. Reviews that illustrate the integration of these methods and models in constructing model, implant (e.g. skin, cartilage), or ex-vivo systems (e.g. bio-artificial liver) are also included.

It is our expectation that the mutual relevance of tissue engineering and biotechnology will only increase in the coming years, as our needs for advanced healthcare products continue to grow. Already, tissue derived cells constitute important production systems for therapeutically and otherwise useful biomolecules that require specialized post-translational processing for their safety and efficacy. Biochemical engineering products, ranging from growth factors to polymer scaffolds, are used as building blocks or signal molecules at virtually every stage of engineered tissue formation. Importantly, the realization of engineered tissues as clinically useful and commercially viable products will at least in part depend on overcoming the same efficiency challenges that the biotechnology industry has been facing. In this light, we see the interface between tissue engineering and various other fields of biochemical engineering as a very exciting area for research and development with enormous potential for cross-disciplinary education. In this regard, we anticipate that this and other similar volumes will also be useful as supplementary text for students.

We extend our special thanks to all of the contributing authors as well as Springer for embarking on this project. We are especially grateful to Dr. Thomas Scheper and Ulrike Kreusel for their incredible patience and hard work as our production editors.

Medford, August 2006   Kyongbum Lee, David Kaplan

Contents

Contents of Volume 102

Tissue Engineering I

Volume Editors: Kyongbum Lee, David Kaplan
ISBN: 3-540-36185-5

Contents of *Advances in Polymer Science, Vol. 203*

Polymers for Regenerative Medicine

Volume Editor: Carsten Werner
ISBN: 3-540-33353-3

Adv Biochem Engin/Biotechnol (2006) 103: 1–73
DOI 10.1007/10_018
© Springer-Verlag Berlin Heidelberg 2006
Published online: 5 July 2006

Controlling Tissue Microenvironments: Biomimetics, Transport Phenomena, and Reacting Systems

Robert J. Fisher[1] · Robert A. Peattie[2] (✉)

[1]Department of Chemical Engineering, Building 66, Room 446,
  Massachusetts Institute of Technology, Cambridge, MA 02139, USA

[2]Department of Chemical Engineering, Oregon State University, 102 Gleeson Hall,
  Corvallis, OR 97331, USA
  peattie@engr.orst.edu

Abstract The reconstruction of tissues ex vivo and production of cells capable of maintaining a stable performance for extended time periods in sufficient quantity for synthetic or therapeutic purposes are primary objectives of tissue engineering. The ability to characterize and manipulate the cellular microenvironment is critical for successful implementation of such cell-based bioengineered systems. As a result, knowledge of fundamental biomimetics, transport phenomena, and reaction engineering concepts is essential to system design and development.

Once the requirements of a specific tissue microenvironment are understood, the biomimetic system specifications can be identified and a design implemented. Utilization of novel membrane systems that are engineered to possess unique transport and reactive features is one successful approach presented here. The limited availability of tissue or cells for these systems dictates the need for microscale reactors. A capstone illustration based on cellular therapy for type 1 diabetes mellitus via encapsulation techniques is presented as a representative example of this approach, to stress the importance of integrated systems.

Keywords Autoimmune and hormone diseases · Biomimetics · Cell/tissue therapy · Cell culture analogs · Diabetes · Encapsulation motifs · Intelligent membranes · Microenvironment · Microreactors · Reaction engineering · Reacting systems · Transport phenomena · Tissue engineering

1
Introduction

1.1
Overview and Motivation

Major thrust areas of research programs in the evolving arena of the engineering biosciences are based upon making significant contributions in the fields of pharmaceutical engineering (drug production, delivery, targeting, and metabolism), molecular engineering (biomaterial design and biomimetics), biomedical reaction engineering (microreactor design, animal surrogate systems, artificial organs, and extracorporeal devices), and metabolic process control (receptor–ligand binding, signal transduction, and trafficking). Since an understanding of the cell/tissue environment can have a major impact on all of these areas, the ability to characterize, control, and ultimately manipulate the cell microenvironment is critical for successful bioengineered system performance. Four major challenges for the tissue engineer, as identified by many sources (e.g., [1]), are: (1) proper reconstruction of the microenvironment for the development of tissue function, (2) scale-up to generate a significant amount of properly functioning microenvironments to be of clinical importance, (3) automation of cellular therapy systems/devices to operate and perform at clinically meaningful scales, and (4) implementation in the clinical setting in concert with all the cell handling and preservation procedures required to administer cellular therapies. The first two of these issues belong primarily in the domain of the research community, whereas the latter two tend to be the responsibility of the commercial sector. Of course, there is significant overlap of emphasis. Without proper coordination of activities, the ultimate goal of a successfully engineered tissue cannot be achieved. A philosophy upheld by many is that a thorough understanding of the fundamentals developed when addressing the first two issues forms the basis for the latter two, and thus implementation of these technologies. Since most tissue engineering groups are concerned with the concepts of how functional tissue can be built, reconstructed, and modified, this chapter is directed toward supporting efforts within that framework. Consequently, the primary objective of this chapter is to introduce the fundamental concepts employed by tissue engineers in reconstructing tissues ex vivo and producing cells of sufficient quantity that maintain a stabilized performance for extended time periods of clinical relevance. The concepts and techniques necessary for the understanding and use of biomimetics, transport phenomena, and reacting systems are presented as focal topics. These concepts are paramount in our ability to understand and control the cell/tissue microenvironment. The delivery of cellular therapies, as a goal, was selected as one representative theme for illustration.

1.2
Background and Approach

Before useful ex vivo and in vitro systems for the numerous applications presented above can be developed, we must have an appreciation of cellular function in vivo. Knowledge of the tissue microenvironment and communication with other organs is essential, since the key questions of tissue engineering are: how can tissue function be built, reconstructed, and/or modified? To answer these questions, we develop a standard approach based on the following axioms [1]: (1) in organogenesis and wound healing, proper cellular communications are of paramount concern since a systematic and regulated response is required from all participating cells; (2) the function of fully formed organs is strongly dependent on the coordinated function of multiple cell types, with tissue function based on multicellular aggregates; (3) the functionality of an individual cell is strongly affected by its microenvironment (within a characteristic length of $100\,\mu m$); and (4) this microenvironment is further characterized by (a) neighboring cells, i.e., cell–cell contact and the presence of molecular signals (soluble growth factors, signal transduction, trafficking, etc.), (b) transport processes and physical interactions with the extracellular matrix (ECM), and (c) the local geometry, in particular its effects on microcirculation.

The importance of the microcirculation is that it connects all the microenvironments in every tissue to their larger whole body environment. Most metabolically active cells in the body are located within a few hundred micrometers from a capillary. This high degree of vascularity is necessary to provide the perfusion environment that connects every cell to a source and sink for respiratory gases, a source of nutrients from the small intestine, hormones from the pancreas, liver, and endocrine system, clearance of waste products via the kidneys and liver, delivery of immune system respondents, and so forth [2]. Further, the three-dimensional arrangement of microvessels in any tissue bed is critical for efficient functioning of the tissue. Microvessel networks develop in vivo in response to physicochemical and molecular clues. Thus, since the steric arrangement of such developing networks cannot be predicted in advance at present, reproduction of the microenvironment with its attendant signal molecule content is an essential feature of engineered tissues.

The engineering of mechanisms to properly replace the role of neighboring cells, the extracellular matrix, cyto-/chemokine and hormone trafficking, microvessel geometry, the dynamics of respiration, and transport of nutrients and metabolic by-products ex vivo is the domain of bioreactor design [3, 4], a topic discussed briefly in this chapter and elsewhere in this volume. Cell culture devices must appropriately simulate and provide these macroenvironmental functions while respecting the need for the formation of microenvironments. Consequently, they must possess perfusion characteristics that allow for uniformity down to the 100 micrometer length scale. These are stringent design requirements that must be addressed with a high priority for each tissue system considered. All these dynamic, chemical, and geometric variables must be duplicated as accurately as possible to achieve proper reconstitution of the microenvironment. Since this is a difficult task, a significant portion of this chapter is devoted to developing quantitative methods to describe the cell-scale microenvironment. Once available, these methods can be used to develop an understanding of key problems, formulate solution strategies, and analyze experimental results. It is important to stress that most useful analyses in tissue engineering are performed with approximate calculations based on physiological and cell biological data; basically, determining tissue "specification sheets". Such calculations are useful for interpreting organ physiology, and provide a starting point for the more extensive experimental and computational programs needed to fully identify the specific needs of a given tissue system (examples are given below). Using the tools obtained from studying subjects such as biomimetics (materials behavior, membrane development, and similitude/simulation techniques), transport phenomena (mass, heat, and momentum transfer), reaction kinetics, and reactor performance/design, systems that control microenvironments for in vivo, ex vivo, or in vitro applications can be developed.

The approach taken in this chapter to achieve the desired tissue microenvironments is through the use of novel membrane systems designed to possess unique features for the specific application of interest, and in many cases to exhibit stimulant/response characteristics. These "intelligent" or "smart" membranes are the result of biomimicry, that is, they have biomimetic features. Through functionalized membranes, typically in concerted assemblies, these systems respond to external physical and chemical stresses to either reduce stress characteristics or modify and/or protect the microenvironment. An example is a microencapsulation motif for beta cell islet clusters, to perform as an artificial pancreas (Sect. 6.2). The microencapsulation motif uses multiple membrane materials, each with its unique characteristics and performance requirements, coupled with nanospheres dispersed throughout the matrix that contain additional materials for enhanced transport and/or barrier properties and respond to specific stimuli. This chapter is structured so as to lead to beta cell microencapsulation as a capstone example of developing understanding and use of the technologies appropriate to design bioengineered systems and to ensure their stable performance.

2
Tissue Microenvironments

Our approach is first to understand the requirements of specific tissue systems and to know how the microenvironment of each meets its particular needs. This requires an understanding of the microenvironment composition, functionality (including communication), cellularity, dynamics, and geometry. Since only a preliminary discussion of these topics is given here, the reader should refer to additional sources [1, 3, 5, 6], as well as other chapters in this volume, for a more in-depth understanding.

2.1
Specifying Performance Criteria

Each tissue or organ undergoes its own unique and complex embryonic developmental program. There are, however, a number of common features of each component of the microenvironment that are discussed in subsequent sections with the idea of establishing general criteria to guide system design. Specific requirements for the application in question are then obtained from the given tissue's "spec sheet". Two representative types of microenvironments (blood and bone) are briefly compared below to illustrate these common features and distinctions. As common examples discussed by many other authors (see [1, 3, 5, 6] for more details), blood and bone are particularly suitable examples for which physiologic and cell biologic data are well understood.

2.2
Estimating Tissue Function

2.2.1
Blood Microenvironment

To fulfill its physiological respiratory functions, blood needs to deliver about 10 mM of oxygen per minute to the body. Given a gross circulation rate of about 5 l/min, the delivery rate to tissues is about 2 mM oxygen per liter during each pass through its circulatory system. The basic requirements that circulating blood must meet to deliver adequate oxygen to tissues are determined by the following: blood leaving the lungs has an O_2 partial pressure between 90 and 100 mm Hg, which falls to 35–40 mm Hg in the venous blood at rest and to about 27 mm Hg during strenuous exercise. Consequently, oxygen delivery to the tissues is driven by a partial pressure drop of about 55 mm Hg on average. Unfortunately, the solubility of oxygen in aqueous media is low. Its solubility is given by a Henry's law relationship, in which the liquid-phase concentration is linearly proportional to its partial pressure with an equilibrium coefficient of about 0.0013 mM/mm Hg. Oxygen delivery from stores directly dissolved in plasma is therefore limited to roughly 0.07 mM, significantly below the required 2 mM.

As a result, the solubility of oxygen in blood must be enhanced by some other mechanism to account for this increase (by a factor of about 30 at rest and 60 during strenuous exercise). Increased storage is of course provided by hemoglobin within red blood cells. However, to see how this came about, let's probe a little further. Although enhancement could be obtained by putting an oxygen binding protein into the perfusion fluid, to stay within the vascular bed this protein would have to be 50–100 kDa in size. With only a single binding site, the required protein concentration is 500–1000 g/L, too concentrated from both an osmolarity and viscosity (10×) standpoint and clearly impractical. Furthermore, circulating proteases will lead to a short plasma half-life for these proteins. By increasing to four sites per oxygen-carrying molecule, the protein concentration is reduced to 2.3 mM and confining it within a protective cell membrane solves the escape, viscosity, and proteolysis problems. Obviously, nature has solved these problems, since these are characteristics of hemoglobin within red blood cells. Furthermore, a more elaborate kinetics study of the binding characteristics of hemoglobin shows that a positive cooperativity exists, and can provide large O_2 transport capabilities both at rest and under strenuous exercise.

These functions of blood present standards for tissue engineers, but are difficult to mimic. When designing systems for in vivo applications, promoting angiogenesis and minimizing diffusion lengths help alleviate oxygen delivery problems. Attempting to mimic respiratory gas transport in perfusion reactors, whether as extracorporeal devices or as production systems, is more complex since a blood substitute (e.g., perfluorocarbons in microemulsions) is typically needed. Performance, functionality, toxicity, and transport phenomena issues must all be addressed. In summary, to maintain tissue viability and function within devices and microcapsules, methods are being developed to enhance mass transfer, especially that of oxygen. These methods include use of vascularizing membranes, in situ oxygen generation, use of thinner encapsulation membranes, and enhancing oxygen carrying capacity in encapsulated materials. All these topics are addressed in subsequent sections throughout this chapter.

2.2.2
Bone Marrow Microenvironment

In human bone marrow cultures, perfusion rates are set by determining how often the medium should be replenished. This can be accomplished through a similarity analysis of the in vivo situation. Blood perfusion through bone marrow in vivo, with a cellularity of about 500 million cells/cc, is about 0.08 ml/cc/min, implying a cell-specific perfusion rate of about 2.3 ml/ten million cells/day. In contrast, cell densities in vitro on the order of one million cells/ml are typical for starting cultures; ten million cells would be placed in 10 ml of culture medium containing about 20% serum (vol/vol). To accomplish a full daily medium exchange would correspond to replacing the serum at 2 ml/ten million cells/day, which is similar to the number calculated previously. These conditions were used in the late 1980s and led to the development of prolific cell cultures of human bone marrow. Subsequent scale-up produced a clinically meaningful number of cells that are currently undergoing clinical trials.

2.3
Communication

Tissue development is regulated by a complex set of events, in which cells of the developing organ interact with each other and with other organs and tissue microenvironments. The vascular system connects all the microenvironments in every tissue to their larger whole body environment. As was discussed above, a high degree of vascularity is necessary to permit transport of signal molecules throughout this communication network.

2.3.1
Cellular Communication Within Tissues

Cells in tissues communicate with each other for a variety of important reasons, such as coordinating metabolic responses, localizing cells within the microenvironment, directing cellular migration, and initiating growth factor mediated developmental programs [6]. The three primary methods of such communication are: (1) secretion of a wide variety of soluble signal and messenger molecules including Ca^{2+}, hormones, paracrine and autocrine agents, catecholamines, growth and inhibitory factors, eicosanoids, chemokines, and many other types of cytokines; (2) communication via direct cell–cell contact; and (3) secretion of proteins that alter the ECM chemical milieu. Since these mechanisms differ in terms of their characteristic time and length scales and in terms of their specificity, each is suitable to convey a particular type of message. In particular, chemically mediated exchanges are characterized by well-defined, highly specific, receptor–ligand interactions that stimulate or control receptor cell activities. For example, the appearance of specific growth factors leads to proliferation of cells expressing receptors for those growth factors.

The multiplicity of tissue–cell interactions, in combination with the large number of signal molecule types and the specificity of ligand–receptor interactions, requires a very large number of highly specialized receptors to mediate transmission of extracellular signals. Broadly, cell receptors can be classified into two types. Lipid-insoluble messengers are bound by cell surface receptors. These are integral transmembrane proteins consisting of an extracellular ligand-binding domain, a hydrophobic membrane spanning region, and one or more segments extending into the cell cytoplasm. The amino acid sequence of these receptors often defines various families of receptors (e.g., immunoglobulin and integrin gene superfamilies). Functionally, surface receptors utilize one of three signal transduction pathways. Either (1) the receptor itself functions as a transmembrane ion channel, (2) the receptor functions as an enzyme, with a ligand-binding site on the extracellular side and a catalytic region on the cytosolic side, or (3) the receptor activates one or more G-proteins, a class of membrane proteins that themselves act on an enzyme or ion channel through a second messenger system.

Lipid-soluble signal molecules and steroid hormones interact with receptors of the steroid hormone receptor superfamily found in the cell cytoplasm or nucleus. Such intracellular receptors, when activated, function as transcription factors to directly initiate expression of specific gene sequences.

2.3.2
Soluble Growth Factors

Growth factors are a critical component of the tissue microenvironment, inducing cell proliferation and differentiation [1, 3, 5–8]. Their role in the signal processing network is particularly important for this chapter. They are small proteins in the size range of 15–50 kDa with a relatively high chemical stability. Initially, growth factors were discovered as active factors that originated in biological fluids, and were known as the colony-stimulating factors. It is now known that growth factors are produced by a signaling cell and secreted to reach target cells through autocrine and paracrine mechanisms. It is also known that in vivo ECM can bind growth factor molecules and thereby provide a storage depot. As polypeptides, growth factors bind to cell membrane receptors, to which they adhere with high affinities. These receptor–ligand complexes are internalized in some cases, with a typical time constant for internalization of the order of 15–30 min. It has been shown that 10 000 to 70 000 growth factor molecules need to be consumed to stimulate cell division in complex cell cultures. Growth factors propagate a maximum distance of about 200 μm from their secreting source. A minimum time constant for growth factor-mediated signaling processes is about 20 min, although far longer times can occur if the growth factor is sequestered after being secreted. The kinetics of these processes are complex, and detailed analyses can be found elsewhere [9] since they are beyond the scope of this chapter.

2.3.3
Direct Cell-to-Cell Contact

Direct contact between adjacent cells is common in epithelially derived tissues, and can also occur with osteocytes and both smooth and cardiac myocytes. Contact is maintained through specialized membrane structures including desmosomes, tight junctions, and gap junctions, each of which incorporates cell adhesion molecules, surface proteins, cadherins, and connexins. Tight junctions and desmosomes are thought to bind adjacent cells cohesively, preventing fluid flow between cells. In vivo they are found, for example, in intestinal mucosal epithelium, where their presence prevents leakage of the intestinal contents through the mucosa. In contrast, gap junctions form direct cytoplasmic bridges between adjacent cells. The functional unit of a gap junction, called a connexon, is approximately 1.5 nm in diameter, and thus will allow molecules below about 1 kDa to pass between cells.

These cell-to-cell connections permit mechanical forces to be transmitted through tissue beds. A rapidly growing body of literature details how fluid mechanical shear forces influence cell and tissue adhesion functions (a topic discussed more thoroughly in other sections), and it is known that signals are transmitted to the nucleus by cell stretching and compression. Thus, the mechanical role of the cytoskeleton in affecting tissue function by transducing and responding to mechanical forces is becoming better understood.

2.3.4
Extracellular Matrix and Cell–Tissue Interactions

The extracellular matrix is the chemical microenvironment that interconnects all the cells in the tissue and their cytoskeletal elements. The multifunctional behavior of the ECM is an important facet of tissue performance, since it provides tissues with mechanical support. The ECM also provides cells with a substrate in which to migrate, as well as serving as an important storage site for signal and communications molecules. A number of adhesion and ECM receptor molecules located on the cell surface play a major role in facilitating cell–ECM communications by transmitting instructions for migration, replication, differentiation, and apoptosis. Consequently, the ECM is composed of a large number of components that have varying mechanical and regulatory capabilities that provide its structural, dynamic, and informational functions. It is constantly being modified. For instance, ECM components are degraded by metalloproteases. About 3% of the matrix in cardiac muscle is turned over daily.

The composition of the ECM determines the nature of the signals being processed and in turn can be governed or modified by the cells comprising the tissue. A summary of the components of the ECM and their functions for various tissues is given in [1].

At present, a major area of tissue engineering investigation is the attempt to construct artificial ECMs. The scaffolding for these matrices has taken the form of polymer materials that can be surface modified for desired functionalities. In some cases, they are designed to be biodegradable, allowing seeded cells to replace this material with its natural counterpart as the cells establish themselves and their tissue function. The major obstacle to successful implementation of a general purpose artificial ECM is that the properties of this matrix are difficult to specify since the properties of natural ECMs are complex and not fully known. Furthermore, two-way communication between cells is difficult to mimic since the information contained within these conversations is also not fully known. At this time, the full spectrum of ECM functionalities can only be provided by the cells themselves.

2.3.5
Communication with the Whole Body Environment

The importance of the vascular system, and in particular the microcirculation, was addressed above, and it was noted that a complex network is needed to connect every cell to a source and sink for respiratory gases, a source of nutrients, a pathway for clearance of waste products to the kidneys and liver, circulating hormones, and immune system components and so forth [2, 3, 9–17]. Transport of mass (and heat) in both normal and pathlogic tissues is driven by convection and diffusion processes occurring throughout the whole circulatory system and ECM [2, 15]. The design of in vivo systems therefore must consider methods to promote this communication process, not just deal with the transport issues of the implanted device itself. The implanted tissue system vasculature must therefore consist of (1) vessels recruited from the preexisting network of the host vasculature and (2) vessels resulting from the angiogenic response of host vessels to implanted cells [2, 18]. Although the implant vessel structure originates from the host vasculature, its organization may be completely different depending on the tissue type, location, and growth rate. Furthermore, the microvessel architecture may be different not only among different tissue types, but also between an implant and any spontaneous tissue outgrowth arising from growth factor stimuli, from the implant [18] or as a whole body response.

A blood-borne molecule or cell that enters the vasculature reaches the tissue microenvironment and individual cells via (1) distribution through the circulatory vascular tree [2, 12], (2) convection and diffusion across the microvascular wall [2, 16, 17], (3) convection and diffusion through the interstitial fluid and ECM [10, 14], and (4) transport across the cell membrane [9, 16]. The rate of transport of molecules through the vasculature is governed by the number, length, diameter, and geometric arrangement of blood vessels through which the molecules pass and the blood flow rate (determining perfusion performance). Transport across vessel walls to interstitial space and across cell membranes depends on the physical properties of the molecules (e.g., size, charge, and configuration), physiologic properties of these barriers, (e.g., transport pathways), and driving force (e.g., concentration and pressure gradients). Furthermore, specific or nonspecific binding to tissue components can alter the transport rate of molecules through a barrier by hindering the species and/or changing the transport parameters [19].

Since the convective component of the transport processes via blood depends primarily on local blood flow in the tissue, coupled with the vascular morphology of the tissue, hydrodynamics must be considered in designing for implanted tissue performance. In addition, perfusion rate requirements must take into account diffusional boundary layers along with the volumes and geometry of normal tissues and implants. In general, implant volume changes as a function of time more rapidly than for normal tissue due to tissue outgrowth, fibrotic tissue formation, and macrophage attachment. All these effects contribute to increased diffusion paths and nutrient consumption.

Notwithstanding these distinctions between different tissues, however, mathematical models of transport in normal, pathologic, and implanted tissues both with and without barriers, whether in vivo, ex vivo, or in vitro, are all identical. Differences between such analyses lie entirely in the selection of physiologic, geometric, and transport parameters. Furthermore, similar transport analyses can also be applied to extracorporeal and novel bioreactor systems and their associated scale-up studies. Examples include artificial organs [8], animal surrogate systems or cell culture analogs (CCAs) for toxicity studies [4], and the coupling of compartmental analysis with CCAs in drug delivery and efficacy testing [20]. Designing appropriate bioreactor systems for these applications is a challenge for tissue engineering teams in collaboration with reaction engineering experts. Many of the required techniques are presented in this chapter and in the voluminous literature for reactor design (see, for example, [21–33]).

2.4
Cellularity

The number of cells found in the tissue microenvironment can be estimated as follows. The packing density of cells is on the order of a billion cells/cc; tissues typically form with a porosity of between 0.5 and 0.7 and therefore have a cell density of approximately 100 to 500 million cells/cc. Thus, an order of magnitude estimate for a cube with a 100 μm edge, the mean intercapillary length scale, is about 500 cells. For comparison, simple multicellular organisms have about 1000 cells. Of course, the cellularity of the tissue microenvironment is dependent on the tissue and the cell types composing it. At the extreme, ligaments, tendons, aponeuroses, and their associated dense connective tissue are acellular. Fibrocartilage is at the low end of cell-containing tissues, with about one million cells/cc or about one cell per characteristic cube. This implies that the microenvironment is simply one cell maintaining its ECM.

In most tissue microenvironments, many cell types are found in addition to the predominant cells which characterize that tissue. Leukocytes and immune system cells, including lymphocytes, monocytes, macrophages, plasma cells, and mast cells, can be demonstrated in nearly all tissues and organs, particularly during periods of inflammation. Precursor cells and residual nondifferentiated cell types are present in most tissues as well, even in adults. Such cell types include mesenchymal cells (connective tissues), satellite cells (skeletal muscle), and pluripotential stem cells (hematopoietic tissues). Endothelial cells make up the wall of capillary microvessels, and thus are present in all perfused tissues.

2.5
Dynamics

In most tissues and organs, the microenvironment is constantly in change due to the transient nature of the multitude of events occurring. Matrix replacement, cell motion, perfusion, oxygenation, metabolism, and cell signaling all contribute to a continuous turnover. Each of these events has its own characteristic time constant. It is the relative magnitude of these time constants that dictates which processes can be considered in a pseudo steady state with respect to the others. Determination of the dynamic parameters of the major events (estimates available in [1, 5–7, 11, 14]) is imperative for successful modeling and design studies.

Time scaling of the systemic differential equations governing the physico-chemical behavior of any tissue is extremely valuable in reducing the number of dependent variables needed to predict responses to selected perturbations and to evaluate system stability. In many cell-based systems, the overall dynamics are controlled by transport and/or reaction rates. Managing selected species transport to and from the system then becomes a major issue since, under certain conditions, transport resistances may be beneficial. For example, when substrate inhibition kinetics are observed, performance is enhanced as the substrate transport rate is restricted.

It is of interest to note that multiple steady states, with subsequent hysteresis problems, have been observed in encapsulated-cell systems as well as in continuous suspension cell cultures [34–36]. Consequently, perturbations in the macroenvironment of an encapsulated cell/tissue system can force the system to a new, less desired steady state in which cellular metabolism, as measured by for example glucose consumption, is altered. Simply returning the macroenvironment to its original state may not be effective in returning the cellular system to its original (desired) metabolic state. The perturbation magnitudes that force a system to seek a new steady state, and subsequent hysteresis lags, are readily estimated from basic kinetics and mass transfer studies [35]. However, incorporation of intelligent behavior into an encapsulation system permits mediation of this behavior. This may be accomplished by controlling the cell/tissue microenvironment through the modification of externally induced chemical, biological, or physical stresses, and through selectively and temporally releasing therapeutic agents or signal compounds to modify cellular metabolism. Various novel bioreactor systems that are currently available as "off-the-shelf" items can be modified to perform these tasks in appropriate hydrodynamic flow fields, with controlled transport and/or contacting patterns, and at a micro scale of relevance.

2.6
Geometry

Geometric similitude is important in attempting to mimic in vivo tissue behavior in engineered devices. The shape and size of any given tissue bed must be known to aid in the design of these devices since geometric parameters help establish constraints for both physical and behavioral criteria. Many microenvironments are effectively two-dimensional surfaces [1]. Bone marrow has a fractal cellular arrangement with an effective dimensionality of 2.7, whereas the brain is a three-dimensional structure. These facts dictate the type of culture technique (high density, as obtained with hollow fiber devices, versus the conventional monolayer culture) to be used for best implant performance. For example, choriocarcinoma cells release more human chorionic gonadotrophin when using high-density culture [3].

2.7
System Interactions

The interactions brought about by communications between tissue microenvironments and the whole body via the vascular network provide a basis for a systems biology approach to understanding the observed performance differences of engineered tissues in vivo versus in vitro. The response of one tissue system to changes in another due to signals generated, metabolic product accumulation, or hormone appearances must be properly mimicked by coupling individual cell culture analog systems through a series of microbioreactors if whole body in vivo responses are to be meaningfully predicted. The need for microscale reactors is obvious given the limited amount of tissue or cells available for many in vitro studies. This is particularly true when dealing with Langerhans islets of the pancreas [8], where intact islets must be used rather than individual beta cells for induced insulin production by glucose stimulation. The supply of islets is extremely limited, and maintaining viability and functionality is quite complex since the islet clusters in vivo are highly vascularized, a feature that is difficult to reproduce in preservation protocols.

An animal surrogate system, primarily for drug toxicity studies, is currently being developed using this CCA concept [4]. A brief discussion of general CCA systems is one of three topics selected to illustrate system interaction concepts in the following subsections. The other two are associated with the use of compartmental analysis in understanding the distribution and fate of molecular species, particularly pharmaceutics, and the need for facilitated transport across the blood–brain barrier, due to its complexities when these species are introduced into the whole body by systemic administration. Compartmental analysis will be discussed first, since it sets the stage for the others.

2.7.1
Compartmental Analysis

Models developed using compartmental analysis techniques are a class of dynamic, i.e., differential equation, models derived from mass balance considerations. These compartmental models are widely used for quantitative analysis of the kinetics of materials in physiologic systems. Materials can be either exogenous, such as a drug or a tracer, or endogenous, such as a reactant (substrate) or a hormone. The kinetics include processes such as production, distribution, transport, utilization, and substrate–hormone control interactions. Compartmental analysis and modeling was first formalized in the context of isotropic tracer kinetics to determine distribution parameters for fluid-borne species in both living and inert systems—particularly useful for determining flow patterns in reactors [4] and tissue uptake parameters [7, 20]. Over time it has evolved and grown as a formal body of theory. Compartmental models have been widely employed for solving a broad spectrum of physiological problems related to the distribution of materials in living systems in research, diagnosis, and therapy at the whole body, organ, and cellular level [2, 4, 7, 10, 14, 16, 17, 20].

The specific goal of compartmental analysis is to represent complicated physiologic systems with relatively simple mathematical models. Once the model is developed, system simulation can be readily accomplished to provide insights into system structure and performance. In compartmental analysis, systems that are continuous and essentially nonhomogeneous are replaced with a series of discrete spatial regions, termed compartments, considered to be homogeneous. Thus, each subsequent compartment is modeled as a lumped parameter system. For example, a physiologic system requiring partial differential equations to describe transient spatial variations in the concentrations of desired components can be simulated using a series of ordinary differential equations using compartmental analysis.

A compartment is an amount of material or spatial region that acts as though it is well mixed and kinetically homogeneous. The concept of well mixed is related to uniformity of information. This means that any samples taken from the compartment at the same time will have identical properties and are equally representative of the system. Kinetic homogeneity means that each particle within a chamber has the same probability of taking any exit pathway. A compartmental model is then defined as a finite number of compartments with specific interconnections among them, each representing a flux of material which physiologically represents transport from one location to another and/or a chemical transformation. When a compartment is a physical space, those parts that are accessible for measurement must be distinguished from those that are inaccessible. The definition of a compartment is actually a theoretical construct which could combine material from several physical spaces within a system. Consequently, the ability to equate a compartment to a physical space depends upon the system being studied and the associated model assumptions.

There are several possible candidates for compartments in specific biological systems. Blood plasma can be considered a compartment as well as a substance such as glucose within the plasma. Zinc in bone and thyroxin in the thyroid can also be compartments. Since experiments can be conducted that follow different substances in plasma, such as glucose, lactate, and alanine, there can be more than one plasma compartment—one for each substance being studied. Extending this concept to other physiologic systems, glucose and glucose-6-phosphate can represent two different compartments within a liver cell. Thus, a physical space may actually represent more than one compartment.

Compartmental analysis then is the combining of material with similar characteristics into entities that are homogeneous, which permits a complex physiologic system to be represented by a finite number of compartments and pathways. The actual number of compartments required depends on both the complexity of these large systems and the robustness of the experimental protocol. The associated model incorporates known and postulated physiology and biochemistry, and thus is unique for each system that is studied. It provides the investigator with invaluable insights into system structure and performance but is only as good as the assumptions that were incorporated into its development.

Identifying the number of compartments and the connections among them that describe the physiologic system under investigation may be the most difficult step in compartmental model building. The structure must reflect a number of facts: (1) there may be some a priori knowledge about the system which can be incorporated in the structure; (2) specific assumptions can be made about the system which are reflected in the structure; and (3) testing via simulation must be conducted on alternate structures to determine what is needed to fit the available data. The result at this stage is a model which has a set of unknown parameters that must be determined using well-established parameter estimation techniques.

One major advantage associated with compartmentalization lies in the ability to reduce the model complexity through the use of lumped versus distributed systems. This permits the use of ordinary differential equations to describe system dynamics instead of more complicated partial differential equations. The following discussion will help clarify these points. Consider the need to remove a toxic compound from a body fluid, such as a xenobiotic drug from blood, by an external device such as an artificial liver. The detoxification may be accomplished by adsorption onto specific receptors bound to solid beads where further reaction can take place. In either event, it is assumed that a first-order process occurs uniformly across a given cross section of flow and that it varies with depth into the packed bed of beads. This system can be modeled as a flow reactor with a time-varying input, for example, as the feed composition of uremic toxins in blood being fed to a dialyzer varies because of the multiple pass requirements. Mass balance considerations for the system, shown schematically in Fig. 1, generate the following partial differential equation:

$$\frac{\partial C}{\partial t} + v\frac{\partial C}{\partial x} = D\frac{\partial^2 C}{\partial x^2} - kC, \tag{1}$$

where C is the concentration of the toxin, v is the linear velocity of blood in this tubular "reactor" (a volumetric flow, F, can be obtained by multiplying v by the cross-sectional area for flow), x is the flow direction, D is the axial diffusion coefficient, and k is the first-order rate constant. This k can represent either a mass transfer coefficient (transport to the bead surface) or biochemical reaction parameter (on the bead surface), dependent upon which mechanism is "controlling". These details will be discussed later when appropriate, but for now the focus will be on how a compartmental system consistent with our model can be structured. Typically in these convective flow situations the axial dispersion term is negligible. Thus, the system becomes:

$$\frac{\partial C}{\partial t} + v\frac{\partial C}{\partial x} + kC = 0. \tag{2}$$

Given an initial and a boundary condition for the situation of interest, an analytical solution for $C(x, t)$ is obtainable. At this time, the distributed par-

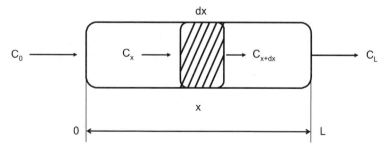

Fig. 1 Schematic diagram of a packed bed (tubular) flow reactor, representing an approach for an artificial liver system

ameter system is represented as a finite number of well-mixed chambers connected in series (Fig. 2).

The inlet concentration to the first chamber is denoted as C_0 and the exit as C_1, which is also the inlet to the second and so forth along the pathway. For algebraic simplicity, the volume of each compartment is taken to be equal and given as $V_j = V/n$, where V is the total volume of the system and n is the number of compartments, which is not known a priori. The actual number required to emulate the "reactor" depends on the accuracy desired from the physical system being simulated.

For chamber 1: $\quad V_1 \dfrac{dC_1}{dt} = FC_0 - FC_1 - k_1 C_1 V_1$

For chamber 2: $\quad V_2 \dfrac{dC_2}{dt} = FC_1 - FC_2 - k_2 C_2 V_2$

For chamber j: $\quad V_j \dfrac{dC_j}{dt} = FC_{j-1} - FC_j - k_j C_j V_j \,.$ $\hfill (3)$

The simplest way to observe the equivalence of these two approaches is to study the system at steady state. The dynamic response comparison yields the same conclusion. For ease of illustration, the rate constant k will be considered to be independent of concentration, and the state variables, such as pressure and temperature, which affect its value are held constant. This allows the subscript on k to be dropped. Since all the volumes are equal, they can be referred to as Vn. The corresponding residence time for each chamber is $\theta_n = Vn/F$, while that for the tubular reactor is $\tau = L/v$, which is also equal to $n\theta_n$. The steady-state solution to Eq. 2 is

$$C(x) = C_0 \exp(-kx/v) \,, \hfill (4)$$

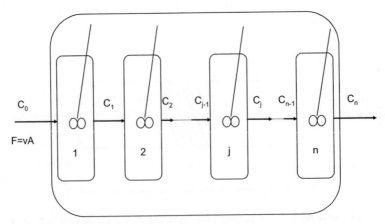

Fig. 2 Schematic diagram of a finite number of well-mixed chambers in series

where C_0 is the steady input to the system. This could be viewed as a "single pass" analysis for the reactor. The compartment model equations simplify to an algebraic form. Represented as transfer functions, they become the following for the jth element:

$$\frac{C_j}{C_{j-1}} = \frac{1}{k\theta_n + 1} .$$

(5)

Writing this for the entire series of chambers, i.e., from C_0 to C_n gives:

$$\frac{C_n}{C_0} = \left(\frac{C_n}{C_{n-1}}\right)\left(\frac{C_{n-1}}{C_{n-2}}\right) \cdots \left(\frac{C_2}{C_1}\right)\left(\frac{C_1}{C_0}\right) = \prod_{j=0}^{n} \frac{C_{j+1}}{C_j} .$$

(6)

Thus

$$\frac{C_n}{C_0} = \left[\frac{1}{k\theta_n + 1}\right]^n$$

(7)

or

$$\theta_n = \left(\frac{1}{k}\right)\left[\left(\frac{C_0}{C_n}\right)^{\frac{1}{n}} - 1\right] .$$

(8)

The total residence time for the full length of the system must be considered for comparison to the distributed system. Given that $\tau = \frac{L}{v} = n\theta_n$ and noting that $C(L) = C_n$ gives

$$\frac{C_n}{C_0} = \exp(-k\tau) \quad \text{or} \quad \tau = \frac{1}{k}\ln\left(\frac{C_0}{C_n}\right) .$$

(9)

To demonstrate the equivalence between the approaches, Eq. 8 is multiplied by n. Then the limit is taken as n gets large, i.e., $n \to \infty$.

$$\lim_{n\to\infty} (n\theta_n) = \lim_{n\to\infty} \left(\frac{n}{k}\right)\left[\left(\frac{C_0}{C_n}\right)^{\frac{1}{n}} - 1\right] .$$

(10)

Application of l'Hopitals' rule allows clarification of the limit process, i.e.,

$$\lim_{n\to\infty} \frac{\frac{d}{dn}\left[\left(\frac{C_0}{C_n}\right)^{\frac{1}{n}} - 1\right]}{\frac{d}{dn}\left(\frac{k}{n}\right)} = \lim_{n\to\infty}\left[\frac{\left(\frac{C_0}{C_n}\right)^{\frac{1}{n}}\ln\left(\frac{C_0}{C_n}\right)\frac{d\frac{1}{n}}{dn}}{k\frac{d\frac{1}{n}}{dn}}\right] = \frac{1}{k}\ln\left(\frac{C_0}{C_n}\right) .$$

(11)

This, of course, is identical to the result from the distributed system. Systems originally modeled by a complicated partial differential equation can be represented by a "large" number of simpler, lumped parameter, ordinary differential equations. How large is large needs to be evaluated for each particular system studied, while with the model objectives are kept in focus.

To further illustrate the usefulness of this compartmentalization approach, a complex physical situation can be taken to show how lumping due to significant differences in system response times (i.e., characteristic times, also referred to as system time constants) simplifies both the overall view of the system and its analysis. An example from pharmacokinetics has been selected that is concerned with the study and characterization of the time course of drug absorption, distribution, metabolism, and excretion. The purpose is to determine the relationship of these processes to the intensity and time course of therapeutic and toxicological effects of the substance in question. A schematic of the various steps in the transfer of a drug from its absorption site (e.g., the gastrointestinal tract) to the blood and its subsequent distribution and elimination in the body is given in Fig. 3.

Here, k_{ij} is the rate constant of the species (drug in this case) from compartment i to compartment j. The reversible step is obviously characterized by k_{ji}. Once the drug is absorbed into the blood, it quickly distributes itself between the plasma and erythrocytes. Within the plasma, it distributes between the water phase and the plasma proteins, particularly albumin (sometimes to α_1-acid glycoproteins but rarely to globulin). Most drugs are relatively small molecules that are readily transported through capillary walls and reach the extracellular fluids of essentially every organ in the body. They are also sufficiently lipid soluble to be distributed into the intracellular fluids of various tissues. In every location, the drug is partitioned between body water and proteins (or other macromolecules) dispersed in the body fluids or as components of the cells. The body can therefore be thought of as a collection of individual compartments, each containing a portion of the administered

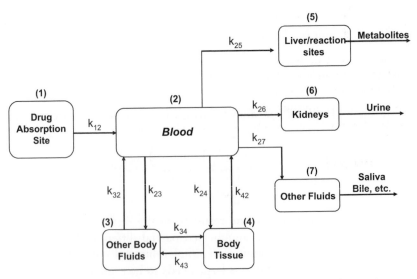

Fig. 3 Schematic representation of drug absorption, distribution, and elimination

drug. The communication between the compartments is, as previously discussed, through a transport rate constant (k_{ij}). The transfer of the drug from the blood to the other body fluids and tissues is termed distribution. Typically, this process is extremely rapid and reversible so that a state of distribution equilibrium is assumed to exist between the plasma, erythrocytes, other body fluids, and tissue components. This concept permits the interpretation of the variations in drug concentration in the plasma as an indication of changes in drug levels at the other sites, including those of pharmacological interest.

The elimination process consists of three major components (or routes): (1) from the blood to urine via the kidneys; (2) into other excretory fluids, such as bile and saliva; and (3) enzymatic or biochemical transformation (i.e., metabolism) in the tissues/organs (e.g., the liver) or in the plasma itself. These processes are typically characterized as irreversible and are responsible for the physical and biochemical removal of the drug from the body.

The distribution and elimination processes occur concurrently, although at different rates. The instant a drug reaches the blood its distribution generally occurs more rapidly than elimination, primarily due to the difference in the rate constants. If the difference is significantly large, distribution equilibrium can be assumed, i.e., the drug can be distributed before any appreciable amount is eliminated. Under these conditions, the body can be characterized by a single compartment (Fig. 4). The compartments for blood, other body fluids, and tissues are "lumped" into one chamber with one inlet and three outlets, all considered irreversible. Since all three elimination rate processes are considered to be linear, they can be combined into one single transport rate constant by simply summing them. This is now an extremely simple representation for a complex system, i.e., a single compartment with one time variant input and one time variant output. It is apparent that the amount of drug at the absorption site decreases with time as the drug is distributed and eliminated. These dynamic processes are the object of investigation. After administration, there is a continual increase in the amount of drug converted to metabolites and/or physically eliminated.

As a result, the amount of drug in the body at any time is the net transient response to these input and output processes. Compartmental analysis allows

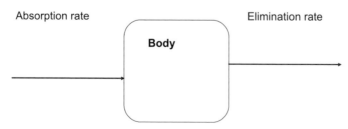

Fig. 4 Simplified version of Fig. 3 for drug dynamics

the system to be represented by three differential equations as follows:

for absorption (input)
$$\frac{dA}{dt} = -k_0 A \tag{12}$$

for body transients
$$\frac{dB}{dt} = k_0 A - k_1 B \tag{13}$$

for elimination
$$\frac{dE}{dt} = k_1 B, \tag{14}$$

where A is the drug concentration from the absorption site, B is in the body, and E is the appearance of eliminated drug. The appropriate initial conditions are $E(0) = B(0) = 0$ and $A(0) = A_0$, the initial dose of drug at the absorption site as a consequence of either oral or intramuscular administration. The general form for the solution is represented in Fig. 5.

The analytic solutions to Eqs. 12–14 are obtained straightforwardly by use of Laplace transforms or by conventional techniques and are:

$$A(t) = A_0 \exp(-k_0 t) \tag{15}$$

$$B(t) = \left(\frac{k_0 A_0}{k_1 - k_0} \right) \left[\exp(-k_0 t) - \exp(-k_1 t) \right] \tag{16}$$

$$E(t) = A_0 - A(t) - B(t)$$
$$= A_0 \left\{ 1 - \left(\frac{1}{k_1 - k_0} \right) \left[k_1 \exp(-k_0 t) - k_0 \exp(-k_1 t) \right] \right\}. \tag{17}$$

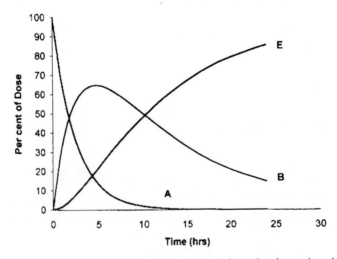

Fig. 5 *Curve A* is the time course of drug disappearance from the absorption site. *Curve E* is the appearance of eliminated drug in all forms. *Curve B* is the net result indicating drug transients in the body

The time when the maximum in B occurs is

$$t_{\max} = \frac{\ln(k_1/k_0)}{(k_1 - k_0)} ,\qquad(18)$$

and the corresponding value for B at $t_{\max}$ is

$$B(t_{\max}) = B_{\max} = A \left(\frac{k_1}{k_0}\right)^{\frac{k_1}{k_0 - k_1}} .\qquad(19)$$

This model and variations of it provide excellent illustrations of the usefulness of single-compartment models. Applications other than for pharmacokinetics are discussed in subsequent sections. However, this current model or modifications to it is useful in providing a quantitative index of the persistence of a drug in the body. It will be used to determine the duration of clinical effect and a dose administration schedule. From Eqs. 18 and 19 it is apparent that as the absorption rate increases compared to elimination processes, i.e., the ratio of $(k_1/k_0) \ll 1$, the maximum in drug plasma concentration increases and the time to reach this maximum is shortened. This maximum needs to be determined, since any drug can be toxic at high concentrations. The absorption rate can be controlled by administration procedures. The current model reflects either an oral or intramuscular injection with a known release/absorption rate.

A delta (δ) function can be used for the initial condition for the drug concentration in the body to simulate a rapid intravenous injection (referred to as bolus). The model is simplified since only the change in the amount of drug in the body (or plasma) with time must be described. This is represented by

$$\frac{dB}{dt} = - k_1 B \quad \text{with} \quad B_0 = A_0 \qquad(20)$$

and

$$\frac{dE}{dt} = - \frac{dB}{dt} .\qquad(21)$$

Therefore, only the time course for B must be monitored. This is given by

$$B(t) = B_0 \exp(- k_1 t) .\qquad(22)$$

Due to the possible harmful effects related to either excessive dosages or lack of level maintenance from missed timed drug administration, many drugs are encapsulated for a long-term controlled release schedule. This can lead to a constant infusion rate (i.e., zero-order process) and thus Eq. 12 for the absorption rate is replaced by the following:

$$\frac{dA}{dt} = k_0 .\qquad(23)$$

The rate of change of the amount of drug in the body (B) during infusion is given by

$$\frac{dB}{dt} = k_0 - k_1 B \,, \tag{24}$$

and the solution is obtained straightforwardly.

$$B(t) = \left(\frac{k_0}{k_1}\right)\left[1 - \exp(-k_1 t)\right] \,. \tag{25}$$

The concentration in the body (and thus plasma concentration) increases during infusion and is at its maximum when infusion is terminated at time T. After this, it will decline according to

$$B(t) = B_{max} \exp\left[-k_1(t - T)\right] \,, \tag{26}$$

where B_{max} is obtained from Eq. 25 with $t = T$. Two important points are to be observed. First, the maximum drug plasma concentration after intravenous infusion is always lower than after bolus injection of the same dose (i.e., $k_0 T$). Second, since B_{max} is linearly proportional to k_0, doubling the dose, (i.e., doubling the infusion rate over the same time period) doubles the maximum concentration.

It is apparent that various administration strategies can be simulated using Eqs. 12 and 13 and subsequent modifications. Some of these were described above. This simulation capability is imperative, since many drugs will not produce a pharmacological effect or the desired response unless a minimum concentration is transported to the site of action. This requires a threshold level be exceeded in the plasma due to the distribution equilibrium that is established. A therapeutic plasma concentration range can be predicted and experimentally validated. By prescribing a drug in an appropriate dosing regime, the physician expects to elicit the desired clinical response promptly.

An introduction to the single-compartment concept was given above. It is the simplest model and may be considered as a special case of a two-compartment model because the substance in question is transported by one mechanism or another from one compartment and can enter another. A dilute suspension of red blood cells is one of the simplest examples of this concept, where the erythrocytes themselves form the compartment. A tracer such as radioactive potassium can be put into the medium, and changes in radioactivity within the cells can be monitored. It is also possible to start with prelabeled erythrocytes and measure their decay rate. Since the potassium transport rate is quite slow, measurements can be performed at moderate rates.

The most troublesome problem in applying compartmental analysis to this system is the issue of linearity. This requires, among other things, that all unidirectional fluxes from a compartment are linearly proportional to the concentration therein. Of course, zero-order behavior is an acceptable alternative. To meet this requirement, a Taylor series expansion could be performed about a "steady-state value" (e.g., a stable isotope). Discarding all higher-order terms in this perturbation variable will provide an appropriate linearized form with an apparent first-order rate constant. To illustrate this point, the efflux is considered to be proportional to the square of the internal potassium concentration (c). This flux can be symbolically represented as

$$F_1 = k_1 c^2 . \tag{27}$$

If a small amount of labeled tracer (δc^*) is added and the (δc^*)^2 terms and higher are discarded,

$$F_1 = F_{1S} + F_1^* \cong k_1 c_S^2 + 2k_1 c_S \delta c^* , \tag{28}$$

where the subscript S refers to a stable isotope. From Eqs. 8–29 the apparent rate constant k_{1a} is defined as

$$k_{1a} = 2k_1 c_S \tag{29}$$

instead of k_1 itself. It should be noted that there are several mechanisms for transport of potassium ions between the red blood cells and the surrounding medium, and that some are actually zero- or first-order naturally.

It is appropriate at this point to establish a basis for zero-order and first-order kinetic behavior in biological systems. This is easily demonstrated for both chemical kinetics and diffusion processes. This will be illustrated for the chemical kinetics case since a similar analysis is applied for the other processes and nothing new is revealed. Not only do biochemical reactions govern metabolism, but also they are coupled to many diffusion and adsorption phenomena. Most of these reactions are catalyzed by specific enzymes, and thus are often represented by the Michaelis–Menton reaction network scheme:

$$E + S \underset{k_{-1}}{\overset{k_1}{\rightleftharpoons}} ES^* \overset{k_2}{\longrightarrow} E + P , \tag{30}$$

where the enzyme E is viewed as reacting with a specific substrate/reactant S in a reversible step to form an activated enzyme–substrate complex ES*, which subsequently is converted to the product P and the original form of the enzyme. If the kinetics are assumed to be controlled by the rate of complex decomposition to product, then the reversible steps can be considered to be in equilibrium, or at least at a quasi-steady state. The functional form obtained is similar in either situation so the equilibrium assumption can be selected due to the simpler algebra and ease of physical interpretation of the model parameters. Thus,

$$\frac{[E][S]}{[ES^*]} = K_{eq} = \text{an equilibrium constant.} \tag{31}$$

The total amount of enzyme E is invariant so

$$[E_0] = [E] + [ES^*] = \text{initial amount of enzyme.} \tag{32}$$

It is easily shown using Eqs. 31 and 32 that

$$[ES^*] = \frac{[E_0][S]}{([S] + k_{eq})} . \tag{33}$$

This is difficult to measure directly, so through use of Eq. 33 the rate of product formation can be written as

$$r = \frac{k_2[E_0][S]}{([S] + K_{eq})} . \tag{34}$$

From this form, it becomes clear that both zero-order and first-order kinetics can be observed in various substrate concentration regimes. For example, when $[S] \gg K_{eq}$, Eq. 34 becomes $r = k_2[E_0]$, i.e., a constant and thus zero-order kinetics. In this concentration regime, the rate of reaction is limited by the quantity of enzyme present. Consequently, the rate is at its maximum value. When $K_{eq} \gg [S]$, Eq. 34 becomes $r = k_2[E_0][S]/K_{eq}$ which yields first-order kinetics with an apparent rate constant $k_a = k_2[E_0]/K_{eq}$.

Use of these analysis techniques by the tissue engineer in developing simplified models is relatively straightforward. They can be useful for system identification, parameter estimation from experimental results, and predictive and/or design capabilities.

2.7.2
Blood–Brain Barrier

A substantial challenge for distribution of therapeutic agents is the development of a method for delivering drugs to the brain, since systemic administration is inadequate when targeting to the central nervous system (CNS) is desired [16]. Many drugs, particularly water-soluble or high molecular rate compounds, do not enter the brain following traditional administration because their permeation rate through blood capillaries is very slow. This blood–brain barrier (BBB) severely limits the number of drugs that are candidates for treating brain disease. New strategies for increasing the permeability of brain capillaries to drugs are therefore frequently proposed. For example, transient increases in BBB permeability can be accomplished by intra-arterial injection of hyperosmolar solutions that disrupt endothelial plate junctions. Unfortunately, osmotically induced BBB disruptions affect capillary permeability throughout the CNS, enhancing permeability to all compounds not just the agent of interest. Other methods take advantage of the fact that the BBB is generally permeable to lipid-soluble compounds that can diffuse through endothelial cell membranes. One such approach involves chemical modification of therapeutic compounds to improve their lipid solubility, although lipidization approaches are not useful for drugs of molecular weight larger than 1000 Da. Another approach is to entrap the drugs in liposomes, but delivery may be limited by liposome stability in blood plasma as well as nonspecific uptake by other tissues. A potentially more effective technique, with the promise of confining localized drug transport specifically to brain capillaries, is based on nutrient transport mechanisms. Since some metabolic precursors are transported across CNS endothelial cells by the neutral amino acid transport system, it has been suggested that analog compounds could be used as both chaperones and targeting species. In addition, there has been some success with protein cationization and anionization, as well as with linkage of the drugs themselves to an anti-transferrin receptor antibody.

Methods for direct administration of drugs into the brain parenchyma have also attracted attention. Direct infusion, implantation of a drug-releasing matrix, and transplantation of drug-secreting cells are all techniques under consideration. These approaches provide sustained drug delivery that can be confined to specific sites, localizing therapy to a given brain region. Because they provide a localized and continuous source of active drug molecules, the total drug dosage can be less than that with systemic administration. Infusion systems, however, require periodic refilling. In turn, this usually requires the drug to be stored in a liquid reservoir at body temperature, at which many drugs are unstable. In contrast, polymeric implants can be designed to protect unreleased drug from degradation, while permitting extremely high doses to be delivered at precisely defined locations in the brain.

Coupling of these approaches through the use of nanosphere technologies to entrap the drug, after modification as necessary for stability in the encapsulated state, offers an intriguing possibility of successfully enhanced transport into the brain. Surface modification of the spheres for specific targeting can further augment localized delivery. Proof of concept experiments using inulin (5200 Da) were conducted recently (R Nicolosi, D Bobilya, RJ Fisher, A Watterson, 2004, personal communication) to verify this premise. A brief description of the approach and results obtained follows.

An in vitro coculture system comprising porcine brain capillary endothelial cells (BCECs) with porcine astrocytes, a widely accepted BBB model, was established with the astrocytes seeded on the bottom of permeable membrane filters and BCECs seated on the top. This configuration permitted communication between the two cell lines without disruption of the endothelial cell monolayer. The filters were suspended in a chamber of fluid such that an upper chamber was formed analogous to the lumen of a brain capillary blood vessel. BBB permeability was determined from measurement of transendothelial electrical resistance (TEER), a standard technique that measures the cell layer's ability to resist passage of a low electrical current. TEER essentially represents the passage of small ions and is the most sensitive measure of BBB integrity. The nanospheres used to encapsulate the radioactive-labeled inulin were tested for their toxicity to the BBB using TEER, and no loss of barrier properties was observed. Inulin was selected as a marker species to represent potential pharmaceutical drugs. When used without nanosphere encapsulation, it provided a reasonable control with slow transport across the BBB, less than 2% after 4 h of exposure. For the same time period and drug concentrations, more than 16% of the inulin within the nanospheres crossed the BBB. Although the mechanism of nanosphere-enhanced transport has not yet been elucidated, this greater than eightfold increase in rate represents a dramatic increase, supporting the premise that nanosphere encapsulation can facilitate drug delivery across the BBB (R Nicolosi, D Bobilya, RJ Fisher, A Watterson, 2004, personal communication).

2.7.3
Cell Culture Analog (CCA): Animal Surrogate System

Recent research in biomedical engineering has contributed significantly to advances in the pharmaceutical and biotechnology areas. The majority of these efforts have been focused on drug discovery, design, and delivery, with less emphasis given to drug metabolism and interactions. In order to optimize patient treatment regimes, minimizing dosages and side effects, the kinetics of drug degradation and accumulation of drug metabolites need to be characterized. These studies will also provide valuable insight into the toxicity of the drug and its metabolites over time. Assessment of the effects on viability, signal transduction, and metabolic pathways is critical when designing optimal drug delivery protocols.

Characterization studies for potential toxicity and the kinetics of degradation of a drug and its metabolites traditionally utilize whole animal studies. However, for ethical and scientific reasons, alternative testing systems are desirable, and some prototypes have been developed [37]. In vitro methods using isolated cells and cell cultures are inexpensive and can provide rapid results; however, they often do not accurately represent in vivo responses and therefore are inherently unreliable. Some successes have been reported in the development of an in vitro CCA system [38, 39] to physically mimic the multicompartmental nature of animal physiology (e.g., metabolizing liver cells in one compartment, lung cells in another compartment, cell culture media serving as a blood surrogate, etc.). However, these systems are not optimized and are often transport-limited. In order to accurately mimic biotransformation processes, efficient transfer of substrates, nutrients, metabolites, and other materials to and from the cells must be achieved, and the tissue microenvironment integrity maintained. Innovative intelligent encapsulation motifs, combined with novel reactor designs/configurations, can eliminate and/or control transport limitations. These "reaction engineering" designed systems will better mimic physiological conditions.

Ultimately, CCA systems will integrate separate compartments that can mimic dose-release kinetics, conversion of the drug into specific metabolites from each organ, and the interchange of these metabolites among organs. By permitting dose-exposure scenarios that can better replicate those of whole animal studies, CCA systems will work in conjunction with physiologically based pharmacokinetic models as a tool to evaluate and modify proposed reaction/degradation mechanisms.

Numerous reactor configurations have been implemented in attempts to establish relevant hydrodynamic conditions suitable for engineered CCA applications [4]. Transport limitations and shear stress effects are primary concerns due to their effects on cellular processes [22, 40, 41]. Operation as chemostats (i.e., with well-mixed and uniform extracellular environments) is desired, but is generally in conflict with the need for low shear fields. Continued efforts to couple reaction engineering technologies [23] with encapsulation techniques will lead to innovative systems that approach ideal in vivo emulation. Moderate successes with novel hollow fiber modules [41] and modified heterogeneous (three-phase) contacting systems [29, 42] have led to applications as extracorporeal devices. Prototypes have been characterized hydrodynamically, and preliminary transport studies indicate that incorporation into existing research programs is probable.

The analysis and design of reactors is highly dependent upon knowing the nonideal flow patterns that exist within the reaction vessel. In principle, if a complete velocity distribution map is known for a given vessel, a local Peclet number can be used to characterize the hydrodynamics and mass transfer occurring at a fluid/membrane interface, and to assess the magnitude of each resistance [43, 44]. Computational fluid dynamics (CFD) is a very powerful and versatile tool for analyses of this type, and is finding many diverse applications in the emerging areas of tissue engineering and biomimetics [12, 45].

Further research and development efforts into CCA modeling of whole body systems and their response to external stimuli need to focus on the following areas:

1. Development of cell encapsulation motifs to better mimic tissue functionality for relevant cell lines. The majority of current in vitro data are obtained using monolayer cultures, which often do not effectively mimic functional tissue. Encapsulated cells within hydrogel-based matrices should provide a more realistic three-dimensional representation of a microenvironment and thus better mimic in vivo conditions, i.e., tissue morphology (including the extracellular matrix), diffusion profiles, and vascularization.

2. Development of reactor systems that permit control of transport processes and minimize shear stress effects. Chemostat operation is desirable to obtain near-ideal in vivo emulation, particularly with system improvements

using biomimetic membranes, hollow fiber modules, and other innovative reactor redesigns.

3. Characterization of specific xenobiotic metabolism and the effects of species and degradation products on cellular functions, such as metabolism and signaling features, when using encapsulated cultures in combination with novel reactor configurations.

4. Testing the cell encapsulation/reactor systems as extracorporeal devices, a bridge to organ transplantation in clinical settings.

3
Biomimetics

Phenomenological events, physical and biological in nature, revealed by our environment have always been a source of fascination. Designing processes to mimic their beneficial aspects, in combination with analysis of our observations, allows us to understand and utilize the fundamental mechanisms involved. Experience has demonstrated the complexity and durability of natural processes, and has shown that adaptability with multifunctionality is a fundamental prerequisite for biological systems to survive. The ability to mimic these processes permits understanding them and utilizing their fundamental aspects for research, developmental, and clinical purposes. A particularly attractive feature of living systems is their unique ability to diagnose and repair localized damage through a continuously distributed sensor network. Mimicry of these networks couples transformation and separation technology with detection and control systems. A major research emphasis has been toward the development of intelligent membranes, specifically receptor/reporter technology.

The concept of intelligent barriers and membranes arises from combining such technology with controlled chemistry and reaction engineering. These intelligent membranes may take the form of polymer films, composite materials, ceramics, supported liquid membranes, or laminates. Their important feature is specific chemical functionality, engineered to provide selective transport and structural integrity, to control stability and release, and to permit sensor/reporter capabilities. Application of intelligent membranes as active transport mimics will create valuable insight into cellular mechanisms.

3.1
Fundamentals of Biomimicry

Understanding of the fundamental concepts inherent in natural processes has led to a broad spectrum of new processes and materials modeled on mimetic systems [46]. Natural processes, such as active transport systems functioning in living systems, have been successfully mimicked and useful applications in other fields, such as pollution prevention, have been demonstrated [47]. Key properties on which living systems rely include multifunctionality, hierarchical organization, adaptability, reliability, self-regulation, and repairability. All of these properties can be incorporated into engineered "intelligent" systems that mimic living processes. For example, molecular design has permitted the fabrication of liquid membranes that mimic active transport of ions. Similarly, noninvasive sensors have been designed to monitor in vivo glucose concentration, and carrier molecules have been developed for controlled release of pharmaceuticals. All these successes were accomplished through interdisciplinary approaches, following three major organizational themes: morphology and properties development, molecular engineering of thin films, and biotechnology and engineering biosciences.

3.1.1
Morphology and Properties Development

Polymer blends are receiving a great deal of attention in this thrust area of polymer research and development. New materials have been created in recent years at a pace that exceeds the rate at which detailed understanding of the phase behavior, phase architecture, and morphology and interfacial properties of polymer blends can be achieved. Successful material design has been related to the wide range of mechanical properties, transformation processes and shapes, and low production costs that can be implemented through appropriate polymer selection. Medical areas in which such novel materials have found use include artificial organs, the cardiovascular system, orthopedics, dentistry, and drug delivery systems. However, even wider deployment of copolymers has been limited by difficulties related to the interaction of synthetic materials with living tissue. Hence, to overcome the biological deficiencies of synthetic polymers and enhance mechanical characteristics, a class of bioartificial polymeric materials has been introduced based on interpenetrating polymer networks of both synthetic and biological polymers. Preparations fabricated from copolymerization of biopolymers such as fibrin, collagen, and hyaluronic acid with synthetic polymers such as polyurethane, poly(acrylic acid) and poly(vinyl alcohol) are now available.

3.1.2
Molecular Engineering

The focus of thin film design is to develop the fundamental understanding of how morphology can be controlled in (1) organic thin film composites prepared by Langmuir–Blodgett (LB) monolayer and multilayer techniques and (2) molecular design of membrane systems using ionomers and selected supported liquids. The ability to control structures of this nature will find immediate application in several aspects of smart materials development, particularly microsensor design.

Surfaces, interfaces, and microstructures play important roles in many research frontiers. Exploration of structural property relationships at the atomic and molecular level, applying modern theoretical methods for predicting chemical dynamics at surfaces, is within the realm of surface and interfacial engineering. The control of surface functionality by appropriate selection of LB film composition can mimic many functions of a biologically active membrane. An informative comparison is that between inverted erythrocyte ghosts and their synthetic mimics, when environmental stresses are imposed on both systems. Such model systems can assist mechanistic studies that seek to understand the functional alterations caused by ultrasound, electromagnetic fields, and UV radiation. The behavior of carrier molecules and receptor site functionality must be mimicked properly along with simulating disturbances in the proton motive force of viable cells. Use of electron/ion transport ionomers in membrane–catalyst preparations is beneficial for programs such as electro-enzymatic synthesis [48]. New membranes for artificial organs and advances in micelle reaction systems are some of the expected outcomes from these efforts.

3.1.3
Biotechnology and Engineering Biosciences

This area of biomimetics has focused on (1) sensor/receptor reporter systems and detection methods, (2) transport processes in biological and synthetic membranes, (3) biomedical and bioconversion process development, and (4) smart film development and applications for intelligent barrier systems. These topics can be coupled with biochemical reaction engineering techniques, as well as with metabolic engineering, the modification of the metabolism of organisms to produce useful products. Extensive research in bioconversion processes is currently being directed toward producing important pharmaceuticals. Future efforts also need to be directed toward the manipulation or reconstruction of cell and tissue function using molecular approaches.

3.2
Biomimetic Membranes: Ion Transport

A cell must take nutrients from its extracellular environment if it is to grow or maintain its normal metabolic activity. Thus, the selectivity and rate of uptake of specific molecular species can regulate cell metabolic processes. The mechanisms by which molecules are transported across cell membranes depend upon the size, shape, and charge distribution of the molecules to be transported. Cell membranes consist of a continuous double layer of phospholipid molecules in which various membrane proteins are embedded. Individual phospholipid molecules are able to diffuse rapidly within their own monolayer. However, they rarely "flip-flop" spontaneously between the two monolayers. These molecules are amphoteric and assemble spontaneously into bilayers when placed in water, forming sealed components that reseal if torn.

The topic of membrane transfer mechanisms is discussed in many sources [2–4, 7–17], and therefore will not be discussed here. It is, however, essential to realize that simple, protein-free, synthetic lipid bilayers can mimic only passive diffusion processes, since they are impermeable to ions. Thermodynamically, virtually any molecule can diffuse across a lipid bilayer down its concentration gradient. However, the rate of diffusion is highly dependent on the size of the molecule and its relative solubility in the hydrophobic interior of the bilayer. Consequently, small nonpolar molecules such as oxygen readily diffuse across lipid bilayers, as do uncharged polar molecules such as carbon dioxide, ethanol, and urea if small enough. In contrast, glucose is essentially excluded. Charged particles such as sodium and potassium ions, no matter how small, are essentially excluded since their charge and high degree of hydration prevent them from entering the hydrophobic phase. Thus, only nonpolar molecules and small uncharged polar molecules can cross a cellular lipid membrane directly by simple passive diffusion. Other molecules and ions require specific membrane transport proteins as either carriers or channels. Synthetic membranes can be designed for specific biomedical applications that can mimic the above discussed transport processes. Membrane selectivity and transport are enhanced with the aid of highly selective complexing agents, positioned as either fixed-site or mobile carriers. To use these membranes to their full potential, the transport mechanisms need to be thoroughly understood.

3.2.1
Active Transport Biomimetics

Extensive theoretical and experimental work has previously been reported for supported liquid membrane systems (SLMs) as effective mimics of active transport of ions [43, 47, 49]. This was successfully demonstrated with an organic acid as a carrier in an organic liquid supported in a hydrophobic, microporous matrix, using copper and nickel ions as the transported species. A pH differential between the aqueous feed and strip streams, separated by the SLMs, mimicked the PMF required for the active transport process in living systems. A local Peclet number (which determines the relative magnitude of species transport by bulk flow to that by diffusion) was used to characterize the hydrodynamics and mass transfer occurring at the fluid/SLM interface.

The SLM itself was modeled as a heterogeneous surface with mass transfer and reaction occurring only at active sites on the surface; in this case, the transverse pores. At present, long-term stability and toxicity problems limit SLM applications in biomedical settings. However, their use in combination with fixed-site carrier membranes as entrapping barriers has great potential and is an active research area.

3.2.2
Facilitated Transport via Fixed Carriers

A theoretical analysis of the mechanism of diffusion through a membrane, using a fixed carrier covalently bound to the solid matrix, has been described elsewhere [7, 50]. The conceptual basis of this analysis is that the solute molecule jumps from one carrier to the next in sequence. A carrier at the fluid/membrane interface reacts with the species to be transported and subsequently comes into contact with an unoccupied carrier, reacting in turn with it. This transfer process is then repeated across the entire width of the membrane. Facilitated diffusion can occur only if these "chained" carriers are reasonably close to each other and have some limited mobility. Although the carrier–solute complex does not diffuse as in a classical random walk concept, it can "jiggle" around its equilibrium position. This movement can bring it into contact range with an uncomplexed carrier also "jiggling" and result in a reversible interaction. As compared to a mobile carrier in a liquid membrane, a fixed-chain carrier in a solid matrix improves stability, avoids the potential of losses from the system, and may enhance transport.

3.2.3
Facilitated Transport via Mobile Carriers

The concept of using a mobile carrier to deliver a drug across a nonreceptive barrier was discussed above in the context of the blood–brain barrier. It was shown that using a chaperone to deliver the agent to a selective site enhances efficacy, due to the protective nature of the nanosphere encapsulating the active ingredient within the vascular network, and its ability to bind to the BBB at a specific site to release the transported species. In that case, the nanospheres mimicked the performance of blood proteins as carriers/binding agents. Since the surface of these spheres can be functionalized with relative ease, they can be tailored to perform numerous tasks. For example, inside an encapsulating membrane motif fabricated to be biocompatible with a given tissue system, nanospheres could facilitate transport operations or provide a reactive site for elimination of unwanted species. These capabilities will be discussed more specifically in the last section of this chapter with the capstone illustration.

3.3
Biomimetic Reactors

3.3.1
Uncoupling Mass Transfer Resistances

Characterization of mass transfer limitations in biomimetic reactors is essential when designing and evaluating reactor performance. When used in CCA systems, the proper mimicry of the role of intrinsic kinetics and transport phenomena cannot be overemphasized. Since lack of similitude will negate the credibility of the phenomenological observations of toxicity and/or pharmaceutical efficacy, these systems must be designed to allow the manipulation and control of all interfacial events. The majority of material transfer studies for gaseous substrates are based on the assumption that the primary resistance to transfer occurs at the gas/liquid interface. Examination of the use of hollow fiber membranes to enhance gas/liquid transfer has indicated that the liquid/cell interface resistance can successfully be uncoupled from that of the gas/liquid interface, so that their potential impacts can be evaluated separately [21, 22]. A reduction in the mean velocity gradient, while maintaining a constant substrate flux into the liquid, resulted in a shift in the limiting resistance from the gas/liquid to liquid/cell interface. This shift manifested itself as an increase in the Monad apparent half-saturation constant for the chemo-autotrophic methanogenic microbial system selected as a convenient analog. The result of these studies significantly influences the design and/or evaluation of reactors used in the tissue engineering research area, especially for animal surrogate or CCA systems. Although a reactor can be considered as well-mixed if it develops spatially uniform cell density, it was demonstrated that significant mass transfer limitations may remain at the liquid/cellular boundary layer in spite of thorough mixing.

Three major points can be stressed. First, the liquid/cellular interface may contribute significantly to mass transfer limitations. Second, when mass transfer limitations exist, intrinsic biokinetics parameters that are essential for accurate modeling of system performance cannot necessarily be determined. Finally, without an understanding of the intrinsic biokinetics, transport mechanisms across biological membranes cannot be correctly evaluated. The determination of passive or active transport across membranes is strongly affected by the extent of the liquid/cellular interfacial resistance.

3.3.2
Pharmacokinetics and CCA Systems

The potential actions and toxicity of a pharmaceutical are tested primarily by using animal studies. Since this technique can be problematic on both a scientific and ethical basis, alternatives have been sought. In vitro methods using isolated cells are inexpensive, rapid, and generally present no ethical issues. However, the use of isolated cell cultures does not duplicate the full range of biochemical activity as in the whole organism. Tissue slices and engineered tissues have also been studied, but are not without inherent problems of their own, such as a lack of interchange of metabolites among organs and the time-dependent exposure within the animal. An alternative to both in vitro and animal studies is the use of computer models based on physiologically based pharmacokinetics (PBPK) analysis. These models mimic the integrated multicompartment nature of animals, and thus can predict time-dependent changes in the blood and tissue concentrations of the parent chemical and its metabolites. Their obvious limitations lie in that a response is based on assumed mechanisms; therefore, secondary sequelae and unexpected events are not included, and parameter estimation is difficult. Consequently, there remains the need for an animal surrogate or CCA system suitable for accurate evaluation of the effects of administration of a variety of pharmaceutical agents. Pioneering work from many research groups has led to the following CCA approach (summarized in [4]).

CCA systems are physical representations of the PBPK structure when cells or engineered tissues are used in organ compartments. The fluid medium that circulates between compartments acts as a blood surrogate. Small-scale bioreactors housing the appropriate cell types provide the physical compartments that represent organs or tissues. This concept combines attributes of both PBPK and in vitro systems. Furthermore, these are integrated systems that can mimic dose-release kinetics, conversion into specific metabolites from each organ, and interchange of these metabolites between compartments. Since CCA systems permit dose-exposure scenarios that can replicate those of animal studies, they work in conjunction with PBPK as a tool for the evaluation and modification of proposed mechanisms. Thus, bioreactor design and performance evaluation testing is crucial to the success of this animal surrogate concept.

Efficient transfer of substrates, nutrients, stimulants, and other solutes from the gas phase across all interfaces is critical for the efficacy of most biotransformation processes, and for improving the blood compatibility of biosensors monitoring the compartments. Gas/liquid mass transfer theories are well-established for microbial processes [51]; however, biotransformation processes also involve liquid/cellular interfacial transport. In these bioreactor systems, a gaseous species is transported across two interfaces. Each could be a rate-determining step and can mask intrinsic kinetics modeling studies associated with cellular growth and/or substrate conversion and product formation.

CCA systems have been studied through evaluation of representative methanogenic, chemo-autotrophic processes, selected for study because of their relative simplicity and strong dependence on gaseous nutrient transport, thus establishing a firm quantitative base case [21]. The primary objective was to compare the effect of altering hydrodynamic conditions on mass transfer across the liquid/cellular interface of planktonic cells, and the subsequent impact upon growth kinetics. A standard experimental protocol was employed to measure the gas/liquid resistance [21, 51]. However, determination of the liquid/cellular resistance is more complex. Thus, boundary layer thicknesses were calculated under various hydrodynamic conditions and combined with molecular diffusion and mass action kinetics to obtain the transfer resistance. Microbial growth kinetics associated with these hydrodynamic conditions can also be examined. Since Monad models are commonly applied to describe chemo-autotrophic growth kinetics, the half-saturation constant can be an indicator of mass transfer limitations. The measured apparent half-saturation constant will then be greater than its intrinsic value, as demonstrated in these earlier studies.

3.4
Electron Transfer Chain Biomimetics

The biosynthesis of lactate from pyruvate has been used as a model system to demonstrate the applicability of electro-enzymatic membranes as electron transport chain biomimetics [48, 52]. A multipass, dynamic input operating scheme permitted optimization studies to be conducted on system parameters, including concentrations of all components in the free solution, flow rates, and electrode compositions, as well as their transport characteristics. By varying the system hydrodynamics in membrane reactor systems, operating regimes that determine the controlling mechanism for process synthesis, i.e., mass transfer versus kinetics limitations, were readily identified. Procedures for developing operational maps were thus established.

3.4.1
Mimicry of In Vivo Coenzyme Regeneration

In many biosynthesis processes, a coenzyme is required for the base and enzymes to function as high-efficiency biocatalysts. A regeneration system is then needed to repeatedly recycle the coenzyme, both to reduce operating costs in continuous in vitro synthesis processes and to mimic the in vivo regeneration process. Multiple reaction sequences can be initiated, as in metabolic cycles, and these typically involve an electron transfer chain system.

Nicotinamide adenine dinucleotide (NADH) is one such coenzyme. Much effort has been focused on improving the NADH regeneration process because of its high cost, with electrochemical methods receiving increased attention and some success. However, direct regeneration on an electrode has proven to be extremely difficult. Accordingly, either acceleration of protonation or inhibition of intermolecular coupling of NAD^+ is required. Redox mediators have permitted the coupling of enzymatic and electrochemical reactions to accomplish regeneration or recycling of the coenzyme during a biosynthesis reaction. A mediator accepts electrons from the electrode and transfers them to the coenzyme via an enzymatic reaction. Immobilization of mediator and enzyme on electrodes can reduce the separation procedure, increase the selectivity, and stabilize enzymatic activity. The mechanism and kinetics of various viologen mediators and electrodes have been investigated for the NADH system in batch configurations by cyclic voltammetry, rotating disk electrode, and impedance measurement techniques.

3.4.2
Electro-Enzymatic Membrane Bioreactors

Flow experiments were conducted using a modified flow-by porous reactor configured with two electrodes and a membrane barrier. A key feature of this reaction system was the in situ regeneration of the coenzyme NADH. The electro-enzymatic reaction process utilized an immobilized enzyme system (lipoamide dehydrogenase (LipDH) and methyl viologen as a mediator) within porous graphite cathodes, encapsulated by a cation-exchange membrane (Nafion® 124, DuPont). The cathodic electrolyte contained the pyruvate, lactate, and coproducts reaction mixture, the enzyme lactate dehydrogenase (LDH), and the coenzyme NADH/NAD$^+$ system.

The flow reactor system consisted of a multipass plug-flow reactor and a well-mixed feed/recycle tank. Lactate synthesis and NADH regeneration occurred simultaneously in the packed-bed reactor. Lactate yields of up to 70% were obtained when the reactor system was operated in a semibatch (i.e., recirculation) mode for 24 h.

These proof of concept experiments established the biomimetic character of this flow-through membrane bioreactor system. Batch reactor and cyclic voltammetry results verified the reaction sequences as proposed. The reaction scheme operative in this system consisted of a three-reaction sequence: an NADH-dependent should be enzymatic (LDH) synthesis of lactate from pyruvate, the regeneration of NADH from NAD$^+$ via an enzymatic (LipDH) reaction using the mediator (methyl viologen), and the electrochemical (electrode) reaction. Methyl viologen (MV^{2+}) accepted electrons from the cathode and donated them to NAD$^+$ via the LipDH reaction. The regenerated NADH in solution was converted to NAD$^+$ in the enzymatic (LDH) conversion of pyruvate to lactate.

3.5
Biomimicry and the Vascular System

Medium supply and gas exchange can become limiting in high cell density situations, as in tissue emulation in vitro, due to lack of an effective transport system mimicking microvessel perfusion conditions in vivo. Many different system configurations and designs have been considered to overcome this deficiency, including cellulose and gel-foam sponge matrix materials, filter-well inserts, and mimetic membranes in novel bioreactor systems (see material in this volume and [3]). Of particular interest here is the use of plastic capillary fibers (hollow fibers) in perfusion chambers. These fibers can also be used in nonreactive systems such as blood oxygenation and continuous dialysis systems. Some applications, for both reactive and nonreactive systems, are discussed in the following subsection along with further discussions of their characteristics that are beneficial for these applications.

3.5.1
Hollow Fiber Systems

The most useful characteristic of these fibers (Fig. 6) for the tissue engineer is that they can support cell growth on their outer surfaces and are gas and nutrient permeable, and thus have been successfully used in perfusion chambers. They are also useful for many other bioreactor systems and contacting devices that are of more general applicability in the broad area of the engineering biosciences.

In most hollow fiber perfusion chambers, medium saturated with 5% CO_2 in air is pumped through the centers (lumen) of a bundle of capillary fibers, while cells are added to the outer region surrounding the fiber bundle. The cells attach and grow on the outer surface of the fibers, fed by diffusion from the perfusate, and can reach tissue-like cell densities. Ultrafiltration properties can be selected by the choice of fiber polymer to provide molecular cutoffs at 10–100 kDa, thus regulating macromolecule diffusion. It is then possible for the cells to behave as they would in vivo. For example, in such cultures, choriocarcinoma cells release more human chorionic gonadotrophin than they would in conventional monolayer culture, and colonic carcinoma cells produce elevated levels of carcinoembryonic antigen [3].

Unfortunately, sampling from these commercially available chambers to determine cell concentrations is difficult because of the cartridge configu-

Fig. 6 Schematic diagram of a hollow fiber cartridge

ration in which the chambers are manufactured. This may be a significant limitation in research or clinical settings when knowledge of cell density is desired, although it need not inhibit use otherwise. In particular, extracorporeal devices that bridge transplantation, such as the use of encapsulated islet clusters to treat pancreatic failure or hepatocytes for liver failure, are well-suited to the perfusion chamber approach. Furthermore, these systems may have commercialization potential in the other bio-based fields. Systems for studying the synthesis and release of biopharmaceuticals seem to be progressing well, as briefly stated below and illustrated in Sect. 5.3.

Applications in the membrane bioreactor field as gas delivery systems have great potential [21, 22, 27]. Their use in blood oxygenation systems during surgery and/or as blood detoxification systems (when coupled, for example, with immunocapture techniques) are also conceivable [28]. Hollow fiber membrane systems promise higher transport efficiencies compared to conventional equipment. To capitalize on this potential, customized designs need to be implemented for the numerous unique applications arising in engineering science. The development of novel, spiral-wound techniques that yield customized configurations with flexibility in fiber placement for optimum pitch layout and favorable flow characteristics is ongoing. Multiple designs can be implemented, and characterization studies have been performed for systems applicable in blood oxygenation [28]. These devices use hydrophobic microporous membranes for the exchange of O_2 and CO_2 between two segregated fluids. Hydrodynamics play a major role in both the transport processes involved and the protection of fragile blood cells. Uneven spacing in commercial modules results in channeling—apparently the major reason for the observed compromised performance. These modules, with their uniform spacing and optimal pitch layout, have high transport characteristics and small hold-up volumes. Results from preliminary mass transport characterization studies indicate a 25-fold enhancement compared to a commercial, well-packed fiber bundle cartridge.

3.5.2
Pulsatile Flow in Biomimetic Blood Vessels

Because of the elasticity of vascular tissue, the pulsatile blood flow of the arterial circulation generated by the cardiac cycle causes deformation of artery walls. Figure 7 shows the flow waveform of the abdominal aorta. Blood vessel compliance thus changes the nature of the resulting flow fields, creating a dynamic coupling between the fluid motion and wall deformation. Better modeling of the hemodynamic effects of this coupling will assist in understanding a variety of vascular pathologies, including atherosclerotic plaque formation, aneurysms, and stenoses. Models currently available treat the vessel wall as a fixed boundary, which suppresses the governing dynamics of this coupled system. However, a number of research groups have identified this

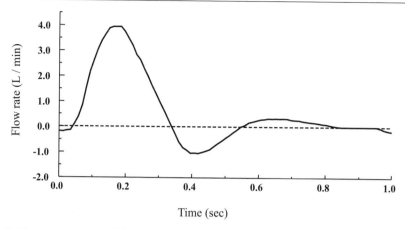

Fig. 7 Flow waveform of the abdominal aorta. This is considered a triphasic wave, as it consists of a large systolic surge of forward directed flow, followed by a period of net retrograde motion that is itself followed by the slow flow of diastole

limitation and are presently developing computer models of pulsatile flows bounded by compliant walls. A biomimetic flow-through system is also being developed and will be used for model validation studies. The initial system is being developed with biomimetic materials; however, actual tissue systems will be incorporated when appropriate.

Furthermore, arterial wall elasticity can create slow-moving wave propagation in long vessels in response to pulsatile flow and wall pressure. Wall deformation may also either damp or stimulate flow instabilities of various frequencies. These phenomena are beyond the predictive capability of current computational fluid dynamics models. Future work therefore needs to accurately represent fluid–structure interactions and produce greater understanding of the coupled system.

The two major required performance specifications of this innovative new model [53] will be (1) development of a suitable constitutive relation and biomimetic material to adequately represent the vessel walls and (2) implementation of appropriate fluid–wall field equations and coupled boundary conditions [54]. Unsteady blood flow must be predicted in three dimensions through simultaneous calculation of the flow field and wall deformation. It is both desired and anticipated that the model implementation will be based on representation of realistic in vivo vessel geometry reconstructed from CT scans of actual patients. Successful models should not only calculate all pertinent details of the flow fields, but should also provide all the requisite data for future studies of endothelial response. Thus, such models must determine fluid shear stress on the walls and tensile stresses in the tissue, as well as time rate-of-change of these quantities as predicted variables.

At present, efforts to produce a high-quality biomimetic material for representing the vessel wall have not yet produced a material capable of simulating wall properties accurately. Candidate materials must possess mechanical characteristics similar to those of vascular tissue, not only in global behavior, but also in their ability to create areas of local stiffness representing the behavior of diseased tissue. These biomimetic materials can then be used to construct flow-through devices for validating computed wall deformations and strains.

Fluid flow simulations thus must apply advanced technologies to biofluid mechanics. One promising approach to solving the three-dimensional, unsteady equations governing arterial hemodynamics is through the use of the embedded-boundary method, a technique that has previously been used successfully for a very different application, namely, two-phase flow in a polymer processing stability study [45]. The embedded-boundary technique is especially suited for blood flow studies in that it is inherently designed to accurately represent flow in deforming blood vessels of complex geometry. More traditional finite volume methods suffer from the difficulties of meshing the numerous features of real vascular shapes and the problems of using a moving mesh.

In the embedded-boundary method, the walls are represented to the fluid as a set of forces. Rather than impose wall position as a boundary condition, the walls become sources and sinks of momentum, sufficient to represent the no-slip boundary (or, as appropriate, a slip condition). One may then employ a fixed, regular Cartesian mesh. As the walls move, so do the location and magnitude of these forces in the fixed mesh. Thus, the difficulties of re-meshing the fluid flow domain after each increment of wall motion are avoided, and accurate flow solutions are generated.

Validation of computed deformation profiles will be conducted experimentally in biomimetic flow-through systems with flexible walls. The current focus has been on wall materials that are known to mimic the mechanical properties of healthy vascular tissue, such as collagen gels, poly(L-lactic acid)–poly(glycolide) systems, poly(glycolic acid)–poly(4-hydroxybutyrate) blends, and poly(acrylic acid)-based networks. These systems all possess a compliance, tensile strength, and failure strength that are similar to those of healthy tissue, and many have been used for tissue engineering of vascular grafts. None of these materials exhibits the complete spectrum of mechanical properties or the composite nature of healthy tissue. However, this is not a limitation for the objective of providing experimental validation of computational results, rather than an exact match for physiological conditions. Thus, only elastic wall materials with well-defined mechanical properties are required, not a perfect representation of actual vascular tissue.

Many of the materials listed above also offer the possibility of rigidification or stiffening via glycation reactions or chemical cross-linkers (e.g., glutaraldehyde). The resulting mechanical properties are directly related to the degree of cross-linking. Thus, by controlling the amount of cross-linking agent in different regions of the wall, it will be possible to create tubes with tunable, well-defined "pockets" of rigidity. Using this method, vessel walls can be designed to simulate tissue that has been damaged by plaque formation in certain regions; that is, the wall properties will be inhomogeneous with particular areas of decreased elasticity. It will also be possible to create walls with varying degrees of thickness to mimic vascular pathologies with that feature.

Salient mechanical properties, such as tensile strength, elastic modulus, loss modulus, recoverable compliance, and Poisson's ratio, of homogeneous materials with different degrees of cross-linking must be retained by model walls through conditions of repeated cyclical strain. A flow-through system with well-defined wall properties is thus created, with either homogeneous properties, spatially varying elasticity, or varying wall thickness, as desired. The experimentally measured wall properties will be used as initial inputs into the CFD calculations described above to predict wall deformation and aneurysm formation. The experimental system will be subjected to pulsatile flow, and wall profiles will be monitored and compared to predictions from CFD calculations. Efforts like these will provide requisite information for future studies of cellular-level response, particle deposition, and cellular attachment that will be useful in a clinical setting, by producing the technology to provide detailed calculations of flow and deformation for the actual patient with a particular physiological condition. These topics are addressed briefly in the next section and in more detail elsewhere in this volume [54].

3.5.3
Abdominal Aortic Aneurysm Emulation

Ultimately the evaluation of time-varying fluid/solid interface conditions, as influenced by platelet, cellular, and particulate deposition and removal, is needed. Steady and unsteady flow patterns, and consequent particle transport and deposition fields, need to be analyzed computationally and experimentally in realistic geometries derived from actual patients. As one example of this process, ongoing research efforts in our laboratories have demonstrated the capability to discern aneurysms from patient CT scans and build rigid flow chambers for emulation (Fig. 8) [53, 55]. The data obtained to generate these systems are readily passed on to CFD simulation packages. Consequently, valuable insight can be obtained into the processes of thrombus deposition and mechanisms by which abdominal aortic aneurysms (AAAs) expand and rupture.

Predicted time-variant stress profiles from computations in such models can be used to design flow field validation devices. Initially, rigid wall systems were studied to minimize degrees of freedom for simplicity. The rigid systems

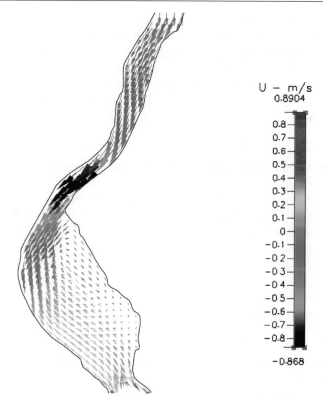

Fig. 8 Patient-based aortic aneurysm model, showing flow pattern within the aneurysmal bulge. Forward directed flow (*top to bottom*) is indicated by negative values of velocity

provide the basis for the more advanced studies in biomimetic, deformable systems, as discussed above. Multiple vessel geometries can be fabricated and investigated experimentally. A cellular detachment technique can be used to measure variations of the stress field with respect to time and position in these three-dimensional chambers. A cell line of known attachment strength will permit the identification of the stress variations by simply measuring cell densities throughout the system. This approach is similar to the radial flow detachment assay technique currently used to measure cell-to-surface binding strengths. However, these new devices possess more flexibility and realism in generating flow fields that accurately reproduce specific patient lesions.

The biocompatible materials mentioned above, such as collagen and poly(lactic acid), allow coating of flow-through models with endothelial cells and thereby make possible examination of gene expression under shear stress. Preliminary results from other projects using simple rigid systems indicate that the use of complex geometric models will not be prohibitive. At this time promising results have been obtained with human arterial endothelial cells, an SV-40 virus-transformed immortalized line that grows rapidly and provides markers for stress-induced responses. These include nitric oxide synthase (a vasodilator), interleukin-8 cytokine (which promotes release of adhesion molecules and acts as a chemoattractant for monocytes and neutrophils), and C-fos (an indicator of cell growth and intermediate gene for chemotactic proteins).

3.5.4
Stimulation of Angiogenesis with Biomimetic Implants

The transport of mass to and within a tissue is determined primarily by convection and diffusion processes occurring throughout the whole body system [2, 15]. The design of interventional therapeutic in vivo systems must consider methods to promote this communication process, not only deal with the transport issues of the device itself. For example, an encapsulated tissue system implant must develop an enhanced localized vasculature. Since an efficient geometric arrangement of microvessels currently cannot be planned in advance, the development of local microvasculature may best be accomplished by (1) recruiting vessels from the preexisting network of the host vasculature and/or (2) stimulating new vessel growth resulting from an angiogenic response of host vessels to the implant [2, 18]. Therefore, when considering implantation of encapsulated tissue or cells, it is prudent to design the implant to elicit an angiogenic response from the ECM. For example, it is known that hyaluronic acid (HA)-based hydrogels can be synthesized to be biodegradable and that these degradation products stimulate microvessel growth. Also, any biocompatible matrix could be preloaded prior to implantation with cytokines that would diffuse out on their own and/or be released via degradation mechanisms.

A recent study by a multi-university collaborative team (including the authors and their colleagues from the University of Utah) demonstrated these facts and identified synergistic behaviors (Fig. 9). Cross-linked HA-based hydrogels were evaluated for their ability to elicit new microvessel growth in vivo when loaded with one of two cytokines, vascular endothelial growth factor (VEGF) or basic fibroblast growth factor (bFGF). HA film samples were surgically implanted in the ear pinnas of mice, and the ears retrieved 7 or 14 days post implantation. Histologic analysis showed that all groups receiving an implant demonstrated significantly more microvessel density than control ears undergoing surgery but receiving no implant. Moreover, aqueous administration of either growth factor produced substantially more vessel growth than an HA implant with no cytokine. However, the most striking result obtained was a dramatic synergistic interaction between HA and VEGF. New vessel growth was quantified by a metric developed during that study, i.e., a dimensionless neovascularization index (NI). This index is de-

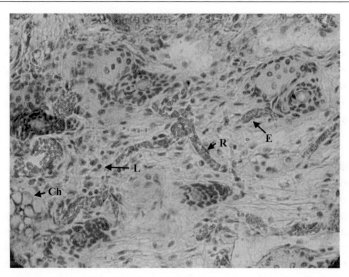

Fig. 9 Mouse ear tissue section, day 14 post-implant of a VEGF-loaded hydrogel film sample, 400×, hematoxylin and eosin staining. *E*: endothelial cell; *R*: erythrocyte; *Ch*: chondrocyte; *L*: polymorphonuclear leukocyte

fined to represent the number of additional vessels present post-implant in a treatment group, minus the additional number due to the surgical procedure alone, normalized by the contralateral count. Presentation of VEGF in cross-linked HA generated a vessel density of NI = 6.7 at day 14. This was more than twice the effect of the sum of HA alone (NI = 1.8) plus VEGF alone (NI = 1.3), and was twice the vessel density generated by coaddition of HA and bFGF (NI = 3.4). New therapeutic approaches for numerous pathologies could be notably enhanced by this localized, synergistic angiogenic response produced by release of VEGF from cross-linked HA films.

4
Transport Phenomena

In all living organisms, but especially the higher animals, diffusion and flow limitations are of critical importance. Moreover, we live in a very delicate state of balance with respect to these two processes. In adapting problem-solving strategies for mathematical analysis and modeling of specific physiologic transport problems, it thus becomes particularly important to establish orders of magnitude and to make realistic limiting calculations. Especially attractive for these purposes are dimensional analysis and pharmacokinetic modeling; it seems, in fact, that these may permit unification of the whole of biological mass transport. Distributed-parameter transport modeling is also helpful. To obtain useful models, one must set very specific goals and work toward them by systematically comparing theory and experiment. It is also important to put results in a form that is readily accessible to potential users. Particularly important are the estimation of transport properties and the development of specialized conservation equations, as for ultrathin membranes, about which only limited information is available. Thus, analysis of physiologic transport stands in strong contrast to classical areas of transport phenomena [56–60].

4.1
Mass Transfer

Mass transfer lies at the heart of cardiovascular, respiratory, urinary, and gastrointestinal physiology and provides major constraints on the metabolic rates and anatomy of living organisms, from the organization of organ networks to molecular intracellular structures. Limitations on mass transfer rates are major constraints for nutrient supply, waste elimination, and information transmission. The primary functional units of metabolically active muscle and brain, liver lobules, and kidney nephrons have evolved to just eliminate significant mass transfer limitations in the physiological state [7]. Turnover rates of highly regulated enzymes are just less than diffusion limitations. Signal transmission rates are frequently mass transport limited [9] and very ingenious mechanisms have evolved to speed these processes [5]. Consequently, understanding tissue mass transport is important for both the engineer and scientist. Examples of engineering interests include the design of extracorporeal devices and biosensors. For the scientist, it is important to understand the often complex interactions of transport and reaction to accurately interpret transport-based experiments.

Since one emphasis in this chapter is to describe the issues relevant to the development of tissue and cellular encapsulation motifs that help control their microenvironments, the following analysis focuses on transport within these systems. The objective here is to identify mechanisms by which molecular transport occurs through complex media. Experimental protocols focus on the selective transport of solute molecules to evaluate proposed mechanisms and establish performance criteria. Numerous models have been postulated to explain these phenomenological observations and to develop methodologies to predict performance, thereby facilitating the design of successful encapsulation systems. Issues that need to be incorporated into these models include (1) interfacial phenomena between the bulk fluid and the outer membrane surfaces and/or along a pore wall; (2) sorption into the membrane matrix itself, with diffusion possibly affected by immobilization at specific interactive sites; (3) free and/or fixed site diffusion within the matrix and, if appropriate, through porous regions, whether as distinct pores, microchannels, or other nonhomogeneous/discrete areas; and (4) any chemical reactions that could alter the nature of the diffusing species or the medium itself.

Two models developed previously for analysis of transport through hydrogel membranes [61,62] address these issues. Both pore and sorption mechanisms may be active, but our classification and characterization will be based on which, if either, dominates. When considering a pore mechanism for solute transport, the solute is envisioned as passing through fluid-filled micropores or channels in the membrane. For molecules with a mean free path much smaller than these openings, a simple Fickian diffusion model suffices to accurately describe net transport. Knudsen diffusion is considered when the pore size is smaller than the mean free path yet still larger than the molecular diameter. As the pore and molecular sizes approach each other, hindered diffusion occurs, with physical and chemical interactions between solute molecules and the pore wall, not simply elastic collisions, dominating transport rates.

In this latter case, interactions at the molecular level, such as surface adsorption (physical or chemisorption), absorption into the matrix (solubility), and molecular transformation, become critically important. In modeling solute transport, consideration must then be given to rate events, extent of reaction, equilibrium partitioning, irreversibility, site proximity, degree of saturation, and desorption. Solute migration rates along a fiber or pore surface are dependent on energetics as well as reactive site proximity. In some materials, shunt pathway formation can enhance transport if close enough. If such alternative paths do not develop, random adsorption events can hinder the transport process, as occurs during diffusion with immobilization [63]. Once reaction sites are fully saturated, however, a normal diffusive zone can be established. The influence of chemical reactive sites can be determined following prior analyses [51] when appropriate, as in functionalized surfaces and/or sites distributed throughout a membrane. In contrast, in porous materials possessing microchannels but lacking a well-defined pore structure, the dominant phase is the fluid that fills these voids. This situation is analogous to that in a swollen fibrous medium, in which solute can diffuse readily but encounters fibers in its sojourn through the membrane. These encounters have a random aspect due to the nonhomogeneous nature of the membrane. They could be passive, as in elastic collisions, or active through affinity interactions, as in the pore concept described above.

Transport through nonporous materials requires solute to be absorbed (solubilized) in the matrix material. Solute molecules are thus subject to thermodynamic equilibrium factors at the fluid/solid interfaces, as well as the nature of the fluid phase itself. These include ion strength, degree of solute hydration, and other interactive forces. Once the solute is within the membrane, a simple Fickian diffusion mechanism can take place. Furthermore, all the other affinity events discussed for the pore model could occur at interactive sites, thus establishing a more complex transport process.

Combinations of these mechanisms may be observed in any membrane system that has distinct fluid, amorphous, crystalline, and/or functionalized regions, whether classified as porous or nonporous. Membranes may be characterized with respect to these mechanistic events, as modeled based on experimental transport measurements. The analysis tools used to interpret these results are developed and presented in the following sections.

4.1.1
Membrane Physical Parameters

A variety of physical parameters may be measured or calculated as necessary for analysis and interpretation of transport measurements. The hydrated encapsulation matrix or membrane volume (V) is obtained using a water displacement technique. In a membrane system, the membrane area (A) may be measured directly, and using the volume the width of the membrane (l) calculated. The porosity can be estimated through a simple mass balance. A matrix or membrane sample is removed from the storage solution, its surface wiped dry, and then it is weighed. The sample is then thoroughly dried and reweighed. The equilibrium weight swelling ratio, q, is then determined from:

$$q = {W_s}/{W_d} \tag{35}$$

with W_s the equilibrium weight of the swollen sample and W_d its dry weight [64].

4.1.2
Permeability

Encapsulation motif permeability may be determined using a pseudo steady-state analysis based on Fick's law, equilibrium partitioning to the membrane surface, and the observed concentration profile in the membrane. Three key assumptions are necessary for the applicability of this analysis: (1) that the concentration or penetration wave has had sufficient time to traverse the membrane and establish a uniform flux throughout the membrane, beyond lag and induction times; (2) a new uniform flux is established more rapidly than the concentration changes in the chamber; and (3) the membrane surface concentration is in equilibrium with the bulk solution. These assumptions are usually valid, since the motif sample volumes used for experimental characterizations are typically small compared to the volume of the chambers in which transport properties are measured. Thus, transport is relatively fast once past the induction period. Under these pseudo steady-state conditions, the instantaneous flux, j, through the matrix or membrane is then given by:

$$-j = \frac{[P]}{l} \left(C_D - C_R \right) ,$$ (36)

where C_D and C_R are the concentration within the donor cell and the receptor cell, respectively [65]. The parameter P is the membrane permeability and is defined as:

$$P = D \cdot H ,$$ (37)

where the partition coefficient, H, is the ratio of solute at/on the membrane surface to that free in solution, and D is the effective diffusion coefficient in the membrane. Combining Eqs. 36 and 37 with a time variant mass balance for each compartment and subjecting the resulting differential equation to the initial condition, $(C_{0,D} - C_{0,R}) = (C_{1,D} - C_{1,R})$ at $t = 0$, the solution obtained provides a method to interpret transport measurements [65] and calculate solute permeability:

$$P = \frac{1}{\beta t} \cdot \ln \frac{(C_{0,D} - C_{0,R})}{(C_{1,D} - C_{1,R})} ,$$ (38)

with β a physical constant containing the dimensions of the diffusion cell and the membrane.

4.1.3
Dextran Diffusivity

Measurements of dextran transport can be used to validate the effectiveness of a pore model for a given motif. Investigation of the relation between molecular size and rate of transport can establish whether diffusion is hindered due to pore walls or simply due to collisions with fibers in the diffusion path. To determine the unhindered diffusion rate of dextran molecules in water, the diffusivity at infinite dilution [66–71] is first calculated using the Stokes–Einstein equation:

$$D_0 = kT / (6 \pi \eta_W r_H) ,$$ (39)

where r_H, the hydrodynamic radius of the dextran molecules, is obtained using the Stokes–Einstein correlation based on molecular weight, k is Boltzmann's constant, T is the absolute temperature, and η_W is the viscosity of water at that temperature. Calculated diffusion coefficients of various molecular weight dextrans in water at dilute and semidilute concentrations (5–80 mg/ml) are reported in the literature [67]. The effective diffusivities of dextran solutions through any membrane tested may then be calculated using the results of these permeability experiments, if the partition coefficient for that particular system is known. The specific case $H = 1$ then implies that diffusion occurs only through the water phase within the membrane. A ratio of D_{eff}/D_0 less than one indicates that a hindered diffusion process is present. If this ratio is dependent on molecular size, then pore (or microchannel) dimensions dominate versus collisions with individual fibers in a relatively open structure.

4.1.4
Marker Molecule Diffusivity

The diffusivity of a solute through a particular membrane can only be determined directly, from experimental transport measurements obtained with a horizontal diffusion cell under select conditions, if the partition coefficient and all physical parameters of the membrane and apparatus are known. Although membrane external dimensions can often be measured with reasonable accuracy, pore characteristics are typically lumped into an apparatus parameter that must be determined by calibration experiments and is accordingly subject to experimental error. Consequently, permeability determination is not a fundamental process. Its usefulness is restricted to applications within a data collection regime. Extensions to predict performance and provide better design protocols for novel applications require that the fundamental parameters, diffusivity and partition coefficients, be known. Both can be obtained from desorption experiments, but data analysis from such experiments is more complicated, particularly when the motif geometry is nonspherical, a not uncommon situation. As described below, we have used this technique to estimate both parameters, and then have used adsorption tests to provide a direct measure of the partition coefficients, thereby providing redundancy checks for all three parameters P, D, and H. Even with this more extensive data analysis program, one can only obtain effective diffusivities since the internal membrane structure is quite complex.

The marker species vitamin B12, bovine serum albumin (BSA), and lysozyme are generally representative selections that provide a reasonable range in size and properties for the solutes in a desorption experiment. Membranes are initially saturated with one solute, then immersed in a buffer solution of known volume [72]. Mathematical analysis of the resulting desorption is based on an infinite sheet of uniform thickness $2l$ placed in a solution, allowing solute to diffuse from the sheet. Since membrane diameters are more than 100 times greater than the thickness, assumption of an infinite sheet is appropriate. The sheet occupies the space $-l \leq x \geq l$, with an initial concentration C_0 while the bath is free of solute. Solute transport in this system is governed by a diffusion equation:

$$\frac{\partial C}{\partial t} = D \frac{\partial^2 C}{\partial x^2},$$

$$(40)$$

with initial condition:

$$C = C_0, \quad -l \le x \ge l, \quad t = 0 . \tag{41}$$

This boundary condition expresses the fact that the rate at which solute leaves the material is always equal to that at which it enters the sheet over the two surfaces $x = \pm l$. Noting that the total transfer area is therefore $2A$, we obtain:

$$\frac{V_{\text{solution}}}{2A} \cdot \frac{\partial C}{\partial t} = \mp D \frac{\partial C}{\partial x} , \quad x = \pm l, \quad t = 0 , \tag{42}$$

where V_{solution} is the volume of the buffer bath into which diffusion occurs. The solution to this differential equation was developed by Crank [73] in a form expressing the total amount of solute, M_t, in the solution at time t as a fraction of the amount after infinite time, M_∞:

$$\frac{M_t}{M_\infty} = 1 - \sum_{n=1}^{\infty} \frac{2\alpha(1 + \alpha)}{1 + \alpha + \alpha^2 q_n^2} e^{-Dq_n^2 t/l} , \tag{43}$$

where the q_n terms are the nonzero positive roots of the transcendental equation:

$$\tan q_n = - \alpha q_n , \tag{44}$$

and α is the ratio of volumes of solution and sheet, including the partition factor H:

$$\alpha = V_{\text{solution}}/2AHl . \tag{45}$$

At equilibrium, the mass balance on the sheet and solution is:

$$lC_0 = aC_\infty + C_\infty lH , \tag{46}$$

where a is the length of solution, being equal to $V_{\text{solution}}/2A$, and C_∞ is the uniform concentration in the membrane at infinite time. M_∞ is then given by:

$$M_\infty = 2aC_0 = \frac{2lC_0}{1 + 1/\alpha} , \tag{47}$$

and the fractional uptake of the solution is:

$$\frac{M_\infty}{2lC_0} = \frac{1}{1 + 1/\alpha} . \tag{48}$$

It should be noted that the solution for the differential equation system (Eqs. 40–42) is not valid when $\alpha = \infty$, as is the situation with BSA desorption. For this case, the solution is:

$$\frac{M_t}{M_\infty} = 1 - \sum_{n=0}^{\infty} \frac{8}{(2n + 1)^2 \pi^2} e^{-D(n+1/2)^2 \pi^2 t/l^2} , \tag{49}$$

where the roots of Eq. 44 are $q_n = (n + 1/2)\pi$. The analysis proceeds using Eq. 48 to obtain α, then H from Eq. 45. Once the q_n values are determined from Eq. 44, either Eq. 43 or Eq. 49 is used to recover a value for D from a nonlinear fitting routine. The number of terms retained in the summation is dependent on the magnitude of time and the relative spacing of the roots.

4.1.5
Interpreting Experimental Results

For investigations of this type, three different experimental protocols are executed to quantify the transport and thermodynamic properties of permeability, diffusivity, solubility, and adsorption onto the membranes. The permeability experiments are performed using a horizontal flow cell. They provide a first estimate of the transport process and enable comparisons with other membranes. Desorption (diffusion) experiments allow effective diffusivity and solubility to be obtained independently, resulting in a more fundamental understanding of transport in the membranes. The calculated diffusion coefficients are then compared to those in pure solvent to establish a basis for identifying a hindered diffusion mechanism. An analysis of the adsorption behavior of the marker molecules in the membranes assists in the investigation of the mass transport phenomena by identifying if solute–matrix fiber interactions are significant. Membrane morphology studies are usually conducted using scanning electron microscopy (SEM)—an available, simple, and straightforward method to determine the physical characteristics of the membranes. Along with the equilibrium weight swelling ratio porosity measurement technique described above, surface morphology can be examined by SEM to determine the type and structure of the void space [64, 74].

4.2
Heat Transfer

Models for microvascular heat transfer are useful for understanding tissue thermoregulation, for optimizing thermal therapies such as hypothermia treatment, for modeling thermal regulatory response at the tissue level, for assessing microenvironment hazards that involve tissue heating, for using thermal means of diagnosing vascular pathologies, and for relating blood flow to heat clearance in thermal methods of blood perfusion measurement [2, 15]. For example, the effect of local hypothermia treatment is determined by the length of time that the tissue is held at an elevated temperature, normally 43 °C or higher. Since the tissue temperature depends on the balance between the heat added by artificial means and the tissue's ability to clear that heat, an understanding of the means by which the blood transports heat is essential for assessing the tissue's response to any thermal perturbation. This section outlines the general problems associated with such processes, while more extensive reviews and tutorials on microvascular heat transfer can be found elsewhere [2, 15, 75]. In general, under normal physiological conditions, each of the following heat transfer effects can be expected to contribute to the thermal balance of any tissue, and therefore to be requisite for the accuracy of either a lumped or distributed parameter model: metabolic heat generation; conduction and convection within the body; heat exchange with the environment by radiation, conduction, and convection; heat loss from the skin by evaporation of sweat and water diffused across the skin; and respiratory heat loss. Quantitative expressions for each of these terms, along with important material parameter values, are given elsewhere [75].

The temperature range of interest for most physiologic applications is intermediate between freezing and boiling, making only sensible heat exchange by conduction and convection as important mechanisms of heat transfer. At much higher and lower temperatures, such as those present during laser ablation or electrocautery and cryopreservation or cryosurgery, phase changes and the accompanying mass transport present problems that are beyond the scope of this section. The general field equations that govern heat transfer are mathematically similar to those that govern passive mass transport processes. However, the physical processes involved differ from microvascular solute transport and exchange in fundamental ways. First, the thermal diffusivity of most tissues is roughly two orders of magnitude greater than the diffusivity for mass transport of most mobile species. In addition, mass exchange with the surrounding ECM is largely restricted to the smallest blood vessels, the capillaries, arterioles, and venules, whereas in principle, heat exchange can occur across the wall of any blood vessel. In turn, this leads to differences in the mechanisms of heat transfer, since thermally significant vessels are often found as countercurrent pairs in which the artery and vein may be separated by one vessel diameter or less. Moreover, while the details of the vascular architecture for particular organs have been well characterized in individual cases, variability among individuals makes use of such published data valid only in a statistical sense. Finally, an additional challenge arises from the spatial and temporal variability of the blood flow in tissue. The thermal regulatory system and the metabolic needs of tissues can change the blood perfusion rates in some tissues by a factor of 25 [15].

4.2.1
Models of Perfused Tissues: Continuum Approach

The intention of a continuum model of microvascular heat transfer is to average the effects of many vessels so that details of the blood velocity field in any one are not required. Such models are usually in the form of a modified heat diffusion equation in which the effects of blood perfusion are accounted for by one or more additional terms. These equations can then be solved to yield a sort of moving average of local temperature that, while not elucidating the local temperature distribution around each individual blood vessel or capillary bed, still effectively characterizes the thermal response of a larger tissue sample. Much of the confusion concerning the proper form of this bioheat equation stems from the difficulty in precisely defining appropriate length scales. Unlike a typical porous medium, such as water percolating through sand for which the grains fall into a relatively narrow range of sizes, blood vessels form branching structures with length scales spanning many orders of magnitude. However, some useful forms have been identified, limited to specific applications only.

The Pennes heat sink model, first described in 1948 [15], was initially developed to model the temperature profile in the human forearm. Its major assumptions were that the primary site of equilibrium is the capillary bed, and that each volume of tissue is supplied with a source of arterial blood at the core body temperature. Key advantages of the Pennes model are that it is readily solvable for constant parameter values, requires no anatomical data, and in the absence of independent measurement of the actual blood and heat generation rates, only two adjustable parameters are needed to fit the majority of experimental results available. Conversely, major drawbacks are that the model gives no predictions of the actual details of the vascular temperatures, the actual blood perfusion rate is usually unknown and not exactly equal to the "best fit" parameter from the thermal data, the assumption of constant arterial temperature is not generally valid, and thermal equilibrium in fact occurs proximal to a capillary bed. However, despite these weaknesses the Pennes model remains the primary choice of investigators seeking to analyze tissue thermal behavior.

Several specific shortcomings of the Pennes model can be addressed in a formulation that is essentially analogous to that used for common porous media, referred to as the directed perfusion approach. These and other useful forms are summarized in [15], as are the following alternative approaches.

4.2.2
Alternative Approaches

One important feature of continuum models of tissue heat transfer is that they do not require a separate treatment of the blood subvolume. In each of these formulations, thermal transport of blood and blood vessels is handled by introducing assumptions about flow and temperature profiles, thus requiring solution of only a single differential equation. Several investigators have sought to improve the resolution of temperature distribution within any tissue by introducing multiequation analyses that consider the tissue, arteries, and veins as three separate but interacting subvolumes. Unfortunately, as with the other non-Pennes formulations, these models are difficult to apply to particular clinical applications of practical importance. They do, however, provide valuable theoretical insights into microvascular heat transfer.

Even more complex formulations [76] can be based on more complete reconstructions of the vasculature in the tissue under consideration, coupled with a scheme for solving the resulting flow, conduction, and advection equations. In addition to the mean temperatures predicted by coarser continuum analyses, such models provide insight into the mechanisms of heat transport, the sites of thermal interaction, and degree of thermal perturbations produced by vessels at a given size. Unfortunately, however, since the reconstructed vasculature is similar to the actual vascular tree only in a statistical sense, they cannot provide the actual details of the temperature field in any given living tissue. Further, solution techniques in these approaches are computationally intensive due to the high spatial resolution needed to properly account for all of the thermally significant blood vessels. Typically, CFD software packages are used to obtain a more realistic solution [23], since these techniques significantly reduce the degree of approximation needed when compared with the other modeling efforts.

4.3
Momentum Transfer

Biological processes within living systems are significantly influenced by the flow of liquids and gases. The objective of this section is to summarize the role that momentum transport plays in these systems. More extensive coverage of the theory and implementation of these concepts is given in the chapter on perfusion and hydrodynamics in this volume, and elsewhere [17, 45, 56–60, 75]. Examples of its role in arterial diseases, such as AAA formation and atherosclerosis [17], were discussed above (Sect. 3.5). A more general discussion of its importance, and the use of computational techniques for quantification, is presented here.

Quantifying the impact of fluid flow requires invasive experimentation and/or extensive computational analysis with physiologically based models. Fluids with nonideal rheological properties add a significant level of complexity to the computations, necessitating intricate numerical methods for accurate analysis. As a result, CFD techniques can become powerful and versatile tools for the investigation of many diverse problems in tissue engineering and biomimetics. Examples include predicting the strengths of adhesion and dynamics of detachment of mammalian cells, analysis of material processing systems (useful in encapsulation technology and designing functional surfaces), and predicting transport properties for nonhomogeneous materials and nonideal interfaces. Furthermore, the analysis and design of reactors is highly dependent on knowing the nonideal flow patterns that exist within the vessel. In principle, if we have a complete velocity distribution map for the fluid, then we are able to predict the behavior of any given vessel as a reactor. Once considered an impractical approach, this is now obtainable by computing the velocity distribution using CFD-based procedures in which the full conservation equations are solved for the reactor geometry of interest. Both the nonlinear nature of these equations themselves and the nonlinear constitutive relationships between a flux and a gradient of an appropriate state property (e.g., a non-Newtonian fluid) are taken into consideration [77, 78].

Much attention in biofluid mechanics is directed toward studying blood flow in the vasculature and air flow in the tracheobronchial tree because of their major physiological significance, although important issues related to momentum transport occur with urine, sweat, tears, and the synovial fluid of the joints as well. For mathematical simplicity, analyses of the motion of these fluids generally assume the fluids to be Newtonian. However, the synovial fluid and blood at low shear rates exhibit substantial non-Newtonian behavior. Since blood is a suspension it has interesting properties; it behaves as a Newtonian fluid for large shear rates, and is highly non-Newtonian for low shear. Synovial fluid exhibits viscoelastic characteristics that are critical in the joints, where elasticity is required along with lubrication. Further complicating the analysis is the fact that physiological fluids are subjected to many types of three-dimensional, usually distensible, passageways in which mixed laminar and turbulent flow regimes are developed. The pulsatile flow of blood is mostly laminar in a healthy circulatory system; however, peak Reynolds numbers approaching 10 000 can be developed during exercise conditions, when cardiac output is a maximum. Even at rest, small "bursts" of turbulence are detected in the aorta during a small fraction of each cardiac cycle. Occlusions or stenoses in the circulatory system promote such turbulence. Airflow in the lung is normally laminar during inspiration, but rapid expiration, coughing, or an obstruction can produce turbulent flow, with Reynolds numbers reaching as much as 20 000.

Elasticity combined with strength is a particularly important characteristic of arterial blood vessel walls, since the cyclic contraction of the heart leads to systolic pressure pulsations that deform the large arteries. The ability of the walls to bulge slightly with each heartbeat minimizes the amplitude of this pressure pulse, preventing a "water hammer" effect that would otherwise be highly damaging were the arteries rigid. It is believed that the combination of strength and elasticity is the principal factor permitting arterial walls to withstand systolic pressure and flow surges throughout life (over 3×10^9 heartbeats for an 80 year old).

Understanding the basic pressure and flow mechanisms involved in biofluid processes is essential for our ability to design biomimetic systems. Simulations using CFD to develop the necessary similitude and performance predictions, and the experimental evaluations of prototype devices, are crucial for successful in vivo implantation applications. Even for systems intended to remain in vitro, the effects of physiological fluids must be anticipated. Other complex devices, such as the development of bioreactors as surrogate systems, also need design guidance from CFD methods. The interaction of fluids and supported tissue is paramount in tissue engineering. The strength of adhesion and dynamics of detachment of mammalian cells from engineered biomaterials scaffolds [79] is important ongoing research, as is the effect of shear on receptor–ligand binding at cell/fluid interfaces. In addition to altering transport across the cell membrane, more importantly receptor location, binding affinity, and signal generation with subsequent trafficking within the cell [9] can also be changed.

5
Reacting Systems

Living systems are dynamic in nature. Chemical reactions, usually closely coupled with transport phenomena, sustain and support life processes. Combining these with thermodynamics identifies the area of reaction engineering. Knowledge of reaction fundamentals is essential not only for understanding reacting systems, but also to design and control engineering devices such as the novel flow reactor systems needed by tissue engineers.

The coupling of encapsulation technologies with cell culture techniques permits extended use of bioreactors with complex tissue systems such as islet clusters. Existing systems are also useful in obtaining the transport and reaction parameters necessary to design extracorporeal devices that bridge transplantation. In summary, the relevance of this work to the biotechnology and health care based industries is in the development of artificial organ systems, extracorporeal devices to bridge transplantation, biosensors, and drug design, discovery, development, and controlled delivery systems. Furthermore, the control and stabilization of metabolic processes has a major impact on many research programs, including the development of animal surrogate systems for toxicity testing and biotechnological processes for the production of pharmaceuticals.

5.1
Metabolic Pathway Studies: Emulating Enzymatic Reactions

Many experimental programs of this type have been conducted using novel reactor systems and their relevance has been discussed previously [21–30, 48, 52, 80, 81]. The objective of this section is to understand and control the microenvironment of enzymatic reacting systems wherever they occur, whether in free solution, on a supporting scaffold or surface, or either intra- or extracellularly as in tissue engineering applications. It is well established that the apparent activity of an immobilized enzyme is affected by factors such as steric hindrance, support–enzyme interactions, and diffusive and electrostatic phenomena. Prior studies have described the combined effects of diffusive resistance and electrostatic fields on the rate of reaction catalyzed by an enzyme immobilized on a nonporous surface or within a porous matrix [82]. Of paramount concern is the effect of surface charge density and bulk substrate concentration on the effectiveness factor, quantifying the importance of variation in species concentration at the enzyme site from that in the bulk solution.

Understanding solute concentration inhomogeneities is essential for both basic biochemistry studies evaluating intrinsic kinetics of metabolic pathways and also process studies utilizing chemical reaction engineering principles. The kinetics of compartmentalized enzyme systems that are sensitive to hydrogen ion concentration and produce a fully ionized acidic product, such as enzymatic hydrolysis of esters to yield a carboxylic acid and an alcohol, are good examples. When an enzyme that in free solution exhibits a typical hyperbolic dependence of rate on substrate concentration and a bell-shaped dependence of rate on pH is compartmentalized by a membrane exhibiting a mass transport limitation, multiple steady states can be observed. To understand such nonlinear behavior by systems like these that are far from equilibrium, we can study them both experimentally in biomimetic reactor systems and from a theoretical perspective using dynamical systems theory [83].

Transport limitations leading a dynamic system to demonstrate multiple steady states are also associated with the possibility of limit cycles, hysteresis, and chaos [83]. Multiple steady-state behavior has been observed experimentally in bacterial, yeast, and mammalian cell culture systems [34], as well as for enzymatic reactions. In such systems, the particular pathway of a given reaction sequence depends on the environmental history and the route taken to the specific reaction operating conditions. When hysteresis can occur, the subsequent tracking to an alternate steady state requires significant perturbations in operating parameters. Furthermore, the stability of each steady state needs to be evaluated individually since rhythmic chemical oscillators can be observed, as in biological clocks.

These phenomenological observations led our group to establish the following overall objectives for a global investigation of sequestered enzyme behavior: (1) to demonstrate the use of electro-enzymatic membrane reactors as electron transfer chain biomimetics, and thus to accomplish metabolic pathway emulation via isolating key steps in the reaction sequence; (2) to control reaction microenvironments by identifying the regimes where transport and/or kinetics limitations exist; (3) to conduct associated stability studies and apply chaos theory as related to chemical oscillators; and (4) to develop the requisite operational maps and perform process alternative and optimization studies.

Objectives (1) and (2) were met in a series of studies [48, 81] using the biosynthesis of lactate from pyruvate as a model system to study the mechanisms present (see Sect. 3.4). Novel reactor systems coupling electron transfer chain membranes, enzyme immobilization/encapsulation, and electrode technologies were used to isolate key steps of the pyruvate–lactate redox cycle. Varying system hydrodynamics and state variables was shown to control whether equilibria were determined by transport or reaction kinetics limitations. Operating regimes determining controlling mechanisms were thereby identified, and procedures for developing operational maps established.

To accomplish objectives (3) and (4), efforts need to focus on studying irreversible time evolution in nonlinear dissipative systems that are far from equilibrium, by applying dynamical systems theory, deterministic chaos, and self-organization mathematics to these processes. Prior attempts to apply these theories by mathematical biologists have been somewhat successful, albeit considered as speculative by critics. As examples, rhythmic biochemistry has been found to play a role in signal transmission both inside and outside cells and possibly in cell differentiation, while dissipative structures have been discovered in solutions of microtubules, the principal component molecules of the cytoarchitecture [84].

Glycolytic rhythms became the first confirmed example of a biological dissipated structure when a self-organizing pattern in time was demonstrated both experimentally and theoretically using an excellent quantitative model predicting limit cycles, hysteresis, and chaos [85]. Under appropriate concentration conditions, the [ATP]/[ADP] ratio was shown to vary in a repetitive cycle (i.e., a limit cycle) with a period of approximately 1 min. The enzyme phosphofructokinase (PFK) is inhibited by high concentrations of ATP and activated by high concentrations of ADP and/or AMP. PFK itself is a phosphorylating agent that attaches a phosphate group from ATP to the monosaccharide fructose 6-phosphate, in the process converting ATP to ADP. The rate of that phosphate group transfer is decelerated more than 20-fold by the presence of ATP, which attaches to PFK at a regulatory, nonactive site. It is this inhibition, and the subsequent reaction reacceleration as ATP is consumed by cellular activities, that provide the autocatalytic nonlinear step necessary for self-organization, along with the possibility for limit cycles, hysteresis, and chaos.

An understanding of enzyme control mechanisms can provide important features needed to address more complicated issues, such as cell metabolic regulation presently being studied by many groups [4, 5, 8, 12, 86]. Furthermore, the stability of critical cellular metabolic pathways is crucial to the in vivo behavior of cell-containing implants. Electro-enzymatic emulators can be employed to demonstrate the role of transport properties on the multiple steady-state behavior of glucose conversion [34]. In addition, they provide the requisite data to aid in the design of chemostats for studying overall system dynamical behavior, to validate use as extracorporeal devices that bridge transplantation and to obtain required performance data on a microencapsulated system prior to its use in clinical trials.

5.2
Bioreactors

An entire chapter in this volume is devoted to bioreactor analysis. Consequently, only a brief summary of the topics pertinent to the present chapter will be discussed and some references given that provide simplified overviews.

5.2.1
Reactor Types

Characterization of the mass transfer processes in bioreactors, as used in cell/tissue culture systems, is essential when designing and evaluating their performance. Flow reactor systems bring together various reactants in a continuous fashion while simultaneously withdrawing products and excess reactants (Fig. 10). These reactors generally provide optimum productivity and performance. They are classified as either tank- or tube-type reactors [12, 23]. Each represents extremes in the behavior of the gross fluid motion within the vessel. Tank-type reactors are characterized by instant and complete mixing of the contents and are therefore called perfectly mixed, or back-mixed, reactors. Tube-type reactors are characterized by lack of mixing in the flow direction and are called plug flow or tubular reactors. The performance of some actual reactors, though not fully represented by these idealized flow patterns, may match them so closely that they can be modeled as such with negligible error. Others can be modeled as combinations of tank and tube types over various regions.

Also in use are batch systems, in which there is no flow in or out. The feed (initial charge) is placed in the reactor at the start of the process and the products are withdrawn all at once some time later. The reaction time is thus readily determined. This is significant, since conversion is a function of time and can be obtained from a knowledge of the reaction rate and its dependence upon concentration and process variables. Since operating conditions such as temperature, pressure, and concentration are typically controlled at a constant value, the time for reaction is the key parameter in the reactor design process. In a batch reactor, the concentration of reactants changes with time and, therefore, the rate of reaction does as well.

Most actual reactors deviate from these idealized systems primarily because of nonuniform velocity profiles, channeling and bypassing of fluids, and the presence of stagnant regions caused by reactor shape and internal components such as baffles, heat transfer coils, and measurement probes. Disruptions to the flow path are common when dealing with heterogeneous systems, particularly when solids are present, as when using encapsulated systems. To model these actual reactors, various regions are compartmentalized and represented as combinations of plug flow and back-mixed elements.

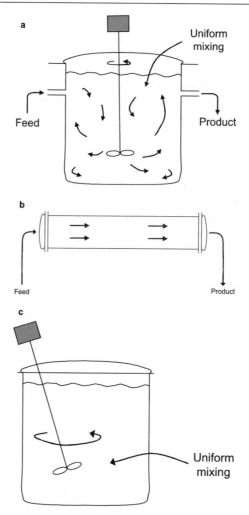

Fig. 10 Schematic diagrams of reactor types. **a** Ideal continuous flow stirred tank reactor; **b** ideal plug flow reactor; **c** ideal batch reactor

The design, analysis, and simulation of bioreactors thus becomes an integral part of the bioengineering profession. The study of biochemical kinetics, particularly when coupled with complex physical phenomena such as the transport of heat, mass, and momentum, is required to determine or predict reactor performance. For accurate modeling and analysis, it is imperative to uncouple and unmask fundamental phenomenological events in reactors and to subsequently incorporate them in a concerted manner to meet the objectives of specific applications. This need further emphasizes the role played by the physical aspects of reactor operation in the stability and controllability of the reaction process.

5.2.2
Design of Microreactors

Microscale reaction systems are often desirable for biomedical and biotechnological applications, in which the component substrates may be expensive or available only in very small quantities. This is particularly true when it is desired to conduct reactions with living cells. The advantages of microreactors are briefly summarized below along with some of the salient features of two representative devices. An important point to bear in mind is that these microscale systems, by their very nature, are differential reactors, that is, very low single pass conversions. Single pass analysis is quite beneficial for intermediate species kinetics studies, and multiple passes and subsequent reservoir dynamics can be readily followed to obtain significant concentrations for analysis [28, 51].

A few of the intrinsic qualities of microreactors are as follows:

- Minimize reactant (serum) requirements.
- Minimize the quantity of cells (beads or otherwise).
- Short contact time reactions can be studied more reliably at reduced flow rates.
- Extend duration of experiments, i.e., longer continuous operation with small resource consumption. "Aging" studies, both accelerated and long term, are more easily conducted.
- Minimize transport effects. Coping with both mixing and flow distribution problems is less formidable and the data are more amenable to analysis.
- Compatibility with spectroscopic systems. In situ studies utilizing advanced spectroscopic techniques are easier to conduct since size restrictions are reduced.
- Control of aseptic environment. Operation in a biohazard/laminar flow hood is easily implemented, as is the optimum placement of ancillary equipment.

Two systems currently in use in our laboratories are:

1. Microchemostat: In this device, mixing is accomplished without mechanical stirring, by superimposing a reciprocating flow with the main flow. The oscillatory motion component is generated by two membrane compressors set up in opposition and located on the inlet and exit ports of the reactor. Feeds can be gases and/or liquids, easily delivered at 1.0 to 600 cc/h (HPLC pumps are this reliable) or as pulses injected into a steady carrier stream. The device performance has been characterized both hydrodynamically and by chemical rate processes [25, 31–33]. With current reactor volumes of 1 to 10 cc, residence times in the range from milliseconds to hours are easily obtained. Thus, the advantages of this reactor compared to the conventional differential reactor with external

recycle and/or the mechanically stirred internal recycle reactors are numerous [23]. A few representative examples are: smaller hold-up volumes, smaller "dead" zones, stability of flow rates, ease of operation, and greater reliability. Applications for transport phenomena studies are apparent; local Peclet numbers are in a sufficiently controllable range to measure external mass transport resistance and assess dominant mechanisms.

2. Microparallel plate flow chamber: These systems have typically been used to study the effects of wall stress on anchorage-dependent cultured cells or tissues and to evaluate protective barrier properties. The system can easily be modified and integrated into flow loops, as used with the microchemostat, with or without the reciprocating motion. Thus, it can be used as a microscale plug flow reactor and/or transport phenomena device. Reactor volumes of 0.006 to 0.05 cc are available, with volume flows up to 2 cc/s permissible, limited by device geometry and/or the maximal desired shear rate. Nominal ranges for wall shear stress, shear rates, and residence times are $1-100$ dynes/cm^2, $0-70\,000$ s^{-1}, and 3 ms, respectively.

5.2.3
Scale-up

This is a complex topic requiring a thorough background in chemical reaction engineering. An extensive literature has been published on its art and technology. Here our objective is only to direct readers to some of the sources most pertinent to bioreactors. In addition to the bioreactor analysis chapter in this volume, additional works worthy of attention are sections of some textbooks on biomedical engineering [1, 59, 77] and cell culture methods [3].

5.2.4
Performance and Operational Maps

The biosynthesis of lactate from pyruvate (Sect. 5.1) is an illustrative example of the use of reactor hydrodynamic characterization and mass transport studies to improve performance through identification of operating regimes in which selected mechanisms are dominant. Exploitation of this knowledge can then enhance the contribution from the desired phenomena. The operational protocols to enter these regimes, once they are documented, are termed operational maps [48, 52]. The multipass, dynamic-input operating scheme used in that study permitted system parameter optimization studies to be conducted, including concentrations of all components in the free solution, flow rates, and electrode compositions, as well as their transport characteristics. Consequently, by altering these system variables and operating conditions, operating regimes that determined the controlling mechanism for process synthesis (mass transfer vs kinetics limitations) were readily identified. Procedures for developing operational maps were thus established.

5.3
Integrated Systems

Hollow fiber membrane systems (Sect. 3.5.1) promise higher transport efficiencies compared to those of conventional equipment. To capitalize on this potential, customized designs need to be implemented for the numerous unique applications in which their use can be beneficial. For example, a novel spiral wound technique, which provides flexibility in fiber placement for optimum pitch layout and good flow characteristics, has been developed [28]. Multiple designs were implemented and characterization studies performed for systems applicable in blood oxygenation. These devices use hydrophobic microporous membranes for the exchange of O_2 and CO_2 between two segregated fluids. Hydrodynamics play a major role, both in the transport processes involved and the protection of the fragile blood cells. These modules, with their uniform spacing and optimal pitch layout, have high transport characteristics and small hold-up volumes. Results from preliminary mass transport characterization studies indicate that a 25-fold enhancement compared to a commercial well-packed fiber bundle cartridge can be achieved. Integration with reactors has shown improved system performance [21, 22, 28] with respect to conversion efficiency, operability, and control.

The use of various membrane configurations coupled with bioreactors has led to multiple functionality improvements and innovations. Implementation as guard beds, recycle conditioning vessels with solids separation capabilities, in situ extraction systems, and slip-stream and/or bypass reactors for catalyst activity maintenance are but a few important examples.

As an illustration of the use of these innovative membrane systems, continuous production of pharmaceutical agents provides a successful example [27]. These products and their derivatives include antibiotics, antioxidants, carriers for and sources of medicinal agents, and so-called green solvents. Representative compounds are hyaluronic, levulinic, and succinic acids, alcohols, angelica lactones, and proteins such as recombinant human interleukin-2 (rhIL-2). The sugars obtained from acid or enzymatic hydrolysis of lignocellulosic biomass are used as feedstocks for both bacterial and yeast fermentation processes to produce these value-added products. Biomimetic membrane systems are presently extending the online capabilities of continuous fermenters (chemostats). Prior studies had shown that the use of recycle streams became problematic after about 40 days of continuous, aseptic operation. Proof of concept experiments were conducted for ethanol production from a mixed sugar feed (hydrolyzate) using *Saccharomyces cerevisiae* in a benchtop chemostat. Recycling to uncouple the hydraulic detention time from the solids retention time (i.e., cell age/mean residence time) gives better control and performance. Immobilized hollow fiber reactors and supported liquid membrane (SLM) systems were evaluated for recycle conditioning and in situ product recovery. Removal of metal ions such as Cu^{2+}, Fe^{2+}, Mg^{2+}, and Zn^{2+}

and conversion of the "unfermentable" tri- and tetrasaccharides into glucose was accomplished using slipstream and bypass bioreactors. A proprietary "smart" resin, formulated using existing encapsulation materials techniques, accomplished acid–sugar separation in the feed conditioning phase. An SLM system was used as a guard bed to remove heavy metal ion contaminants in the virgin feed.

6
Capstone Illustration: Control of Hormone Diseases via Tissue Therapy

The treatment of autoimmune disorders with cell/tissue therapy has shown significant promise. A successful implant comprises cells or tissue surrounded by a biocompatible matrix permitting the entry of small molecules such as oxygen, nutrients, and electrolytes and the exit of toxic metabolites, hormones, and other small bioactive compounds, while excluding antibodies and T cells, thus protecting the encapsulated cells/tissue. Systems of this type are currently being evaluated for the treatment of a variety of disorders, including type I diabetes mellitus, Hashimoto's disease (thyroiditis), and kidney failure. Several key issues need to be addressed before the clinical use of this technology can be realized: tissue supply, maintenance of cell viability and functionality, and protection from immune rejection. Viability and metabolic functionality are controlled by the transport of essential biochemical signal molecules, nutrients, and respiratory gases. In particular, the maintenance of sufficient oxygen levels in the encapsulation device is critical to avoid local domains of necrotic and/or hypometabolic cells. Oxygen transport can be enhanced through several means [72, 87], including the selection of optimal encapsulation configurations, promotion of vascularization at the implantation site, and seeding at an optimal cell density. Despite research efforts directed in these areas, O_2 availability remains one of the major limitations in maintaining cell viability and functionality. To improve this situation, recent efforts have focused on the design of a novel nanotechnology-based encapsulation motif containing specific oxygen carriers. Such a motif will ensure complete metabolic functionality of the encapsulated cells and allow the cells to retain that functionality over extended time, i.e., months as opposed to weeks.

6.1
Selection of Diabetes as Representative Case Study

The National Institute of Diabetes and Digestive and Kidney Diseases (NIDDKD) estimates that there are 16 million people (nearly one in 17) that have diabetes in the USA alone. It is one of the most common and widespread diseases, with as many as 6% of the world population suffering from diabetes. In addition to the primary symptom, loss of blood glucose regulation, the complications and sequelae of diabetes include blindness, cardiovascular disease, and loss of peripheral nerve function. When including these complications, diabetes is the fourth most important cause of mortality in the USA and the main cause of permanent blindness. The American Diabetes Association estimates that diabetics consume 15% of US healthcare costs (more than twice their percentage of the population), with the total cost of diabetic morbidity and mortality exceeding $90 billion per year. At least 50% of that figure is attributed to direct medical costs for the care of diabetic patients. Although most of the affected individuals are not dependent upon interventional insulin replacement, NIDDKD estimates that 800 000 diabetics in the USA are insulin-dependent, and can only manage blood glucose regulation through insulin therapy.

Insulin delivery is not a cure, however. Restoration of normal glucose regulation by improved insulin therapy techniques that regulate insulin delivery offers the hope of circumventing the need for injection treatments and eliminating the serious debilitating secondary complications. Consequently, many research paths are being followed to determine how normal pancreatic functions can be returned to the body. These include whole pancreas transplants, human and animal islet transplantation, fetal tissue exchange, and creation of artificial beta cells, each with its pros and cons. The two major problems are the lack of sufficient organs or cells to transplant and the rejection of transplants. Since there is a severe shortage of adult pancreases, 1000 patients per available organ, alternatives such as islet cell transplantation are being sought. Cell transplantation, if it could be successfully achieved, would help with both of these major problems. Use of either "artificial" islets, potentially grown from stem cells or beta cells themselves, or xenogenic islet clusters, in combination with designed materials for immunoisolation functionality, could lead to restoration of normoperative glycemic control without the need for insulin therapy.

6.2
Encapsulation Motif: Specifications and Design

The main goal of these ongoing research programs [42] is to improve the success rate of pancreatic islet cell transplantation and provide a better means to regulate glucose levels for diabetic patients. This is being accomplished through immunoisolation and immunoalteration technologies implemented using intelligent membrane encapsulation systems. These systems exist as multilayered microcapsules that utilize semipermeable membranes designed to permit transport of nutrients, insulin, and metabolic waste products while excluding antibody and T cell transport.

The immunoisolative capabilities of the encapsulation motif are based on a size-exclusion principle, whereby antibodies (primarily IgM and IgG) and complement proteins of the immune system are unable to reach the implanted cells. In order to activate the complement pathway, in which antibodies bind specific complement proteins and ultimately destroy the implanted cells through lysis, one IgM molecule (MW = 800 kD; $\sim$ 30 nm diameter) and one molecule of complement protein C1q (MW = 410 kD, $\sim$ 30 nm diameter) must bind together. Alternatively, two IgG molecules (MW = 150 kD each, $\sim$ 20 nm total diameter) bind in concert with C1q to destroy the implant cells. Encapsulating cells in materials with a molecular weight cutoff of roughly 200 kD therefore shields cell surface antigens from exposure to these antibodies, which thus cannot destroy the encapsulated cells.

In addition, it is important to consider the ability of implanted cells to shed antigens into the surrounding host tissues, triggering another type of immune response. Many such shed antigens are composed of major histocompatibility complex (MHC) antigens, which are too small (57–61 kD) to be retained by the encapsulation motif. Activation of the immune system in this manner recruits macrophages, which release reactive oxygen species in an attempt to destroy the implanted cells. The inclusion of reactive sites within the matrix that function as free radical scavengers is thus desirable to protect the cells from these toxic compounds. This may be possible using nanosphere technologies to disperse them, whether active ingredient-loaded or by their own behavior, throughout the microsized beads. Simultaneously, nanosphere technology may also be designed to augment respiratory gas exchange. The solubility of O_2 in water or blood plasma is approximately 2 ml O_2 per 100 ml of solution at standard temperature and pressure, which is at least an order of magnitude too low for encapsulation purposes. Thus, selected O_2 carriers, such as perfluorocarbons, could be incorporated into the nanospheres or as microemulsions into the matrix of the encapsulating motif.

Transport to and from encapsulated cells of nutrients, respiratory gases, and similar small molecules is not affected by the pore sizes commonly found in encapsulation motifs. However, secretion of desired proteins synthesized by the cells out of the encapsulation matrix can be limited by the pore size of the material. Large secreted proteins (e.g., MW $\sim$ 660 kDa) will be blocked by the MW cutoffs used to shield implanted cells from antibodies. In contrast, the diffusion of MW = 28 kDa and smaller will not be significantly influenced by the encapsulation matrix due to their small sizes. Consequently, the molecular weight cutoff of the encapsulation material must be carefully chosen when transport of a secreted natural compound (specific to each cell type) is desired.

The hypothesis underlying this effort is that macromolecular biomaterial encapsulation materials can be engineered to promote islet cluster viability while simultaneously facilitating desirable biological responses. For encapsulation materials to be physiologically functional at metabolite transport and hormone secretion, their most important properties are biocompatibility and selective semipermeability. However, to promote implant longevity as well as augment tissue interstitium transport and exchange characteristics, an equally important subcharacteristic is the ability to stimulate neovascularization and vascular in-growth in situ. Approaches such as biodegradation of a sacrificial outer layer, which either releases or "generates" growth factors to promote angiogenesis, are currently being studied as an extension of that work [18]. There is also a concern that the implant should not induce acute inflammation secondary to its presence alone.

In general, a single encapsulating material does not possess the spectrum of properties required for successful implantation. An alternative approach is to design a multilayer motif in which each layer is selected to contribute specific functions to the motif. Major considerations that must be derived from the individual layers include biocompatibility, mechanical strength, selective transport, and stability. The properties of each potential individual lamina can be characterized using established techniques with diffusion cells, rotating disks, hollow fiber cartridges, and accumulation dynamics. When appropriate, a composite assemblage can be similarly evaluated.

The dimensions of these encapsulating systems must be chosen considering many opposing factors; the membranes must be of sufficient thickness to protect cells from environmental stress, shear, and immunogenic responses without limiting the transport of necessary compounds. However, in certain situations, transport resistances can be beneficial. For example, when substrate inhibition kinetics behavior is observed, performance is improved as the substrate transport rate is restricted. As described above, mammalian cells can establish multiple steady states, with subsequent hysteresis effects, while in continuous culture at the same dilution rate and feed medium. Consequently, perturbations in the macroenvironment of an encapsulated cell/tissue system can force the system to a new, less desired, steady state with altered cellular metabolism. Simply returning the macroenvironment to its original state may not be effective in returning the cellular system to its original metabolic state, and hence the predicted performance is unfavorable. The magnitudes of the macroenvironment perturbations forcing the system to seek a new steady state may be estimated from models developed by simulating experimentally measured metabolism. Such behavior can be mediated by the encapsulation motif through its control of the cell/tissue microenvironment. The intelligent behavior of these proposed encapsulation systems will, in principle, allow them to maintain the desired microenvironment through the modification of stresses and release of necessary compounds. Characterization of the required materials in this motif and subsequent testing of its efficacy with appropriate cell/tissue systems is an integral component of future efforts. Candidate materials for the various encapsulation layers in the motif that may meet the objectives/desired functionality are mostly available at this time.

Examples include:

1. Immobilizing hydrogel (inner core): A pH-sensitive, smart hydrogel [chitosan–poly(vinyl alcohol)(PVA)] is currently a prime candidate, as is a novel hydrophobically modified hydrogel as an effective alternative.
2. Alginate buffer layer: A highly purified alginate covering for the hydrogel can be used to buffer cellular components from the outer encapsulating membrane. We have used this type of material, typically obtained from seaweed, to form immunomagnetic beads [25] to capture waterborne pathogens.
3. Outer membrane layer: Various fixed site carrier membranes are being tested as the entrapping outer layer. Materials that have great potential, with selective transport characteristics, are cation-exchange membranes (e.g., Nafion® 124), ionomers (e.g., ethylene–acrylic acid films), and "developmental" membranes (e.g., cytokine-loaded hyaluronic acid hydrogels).

The ability to execute a research program to obtain the requisite data to evaluate and implement designed encapsulation motifs, e.g., develop a prototype from experimental data for clinical testing, is dependent upon coordinating all the efforts described above. This includes using the various novel bioreactor systems discussed above to perform these tasks in appropriate flow fields, with controlled transport and/or contacting patterns, and at a microscale of relevance. Concerted programs will help in attaining the goal of understanding the microenvironment of encapsulated systems to control and optimize tissue function.

References

1. Palsson B (2000) Tissue engineering. In: Enderle J, Blanchard S, Bronzino JD (eds) Introduction to biomedical engineering. Academic, Orlando, chap 12
2. Jain RK (1994) Transport phenomena in tumors. In: Advances in chemical engineering, chap. 19. Academic, Orlando, pp 129–194
3. Freshney RI (2000) Culture of animal cells: a manual of basic technique, 4th edn. Wiley-Liss, New York
4. Shuler MJ (2000) Animal surrogate systems. In: Bronzino JD (ed) The biomedical engineering handbook, 2nd edn. CRC, Boca Raton, chap 97
5. Palsson BO, Hubbell JA (2000) Tissue engineering. In: Bronzino JD (ed) The biomedical engineering handbook, 2nd edn, sect XII. CRC, Boca Raton
6. Long MW (2000) Tissue microenvironments. In: Bronzino JD (ed) The biomedical engineering handbook, 2nd edn. CRC, Boca Raton, chap 118
7. Lightfoot EN (1974) Transport phenomena and living systems. Wiley, New York
8. Galletti PM, Colton CK, Jaffrin M, Reach G (2000) Artificial pancreas. In: Bronzino JD (ed) The biomedical engineering handbook, 2nd edn. CRC, Boca Raton, chap 134
9. Lauffenburger DA, Linderman JJ (1993) Receptors: models for binding, trafficking, and signaling. Oxford University Press, New York

10. Lightfoot EN, Duca KA (2000) The roles of mass transfer in tissue function. In: Bronzino JD (ed) The biomedical engineering handbook, 2nd edn. CRC, Boca Raton, chap 115
11. Fisher RJ (2000) Biomimetic systems. In: Bronzino JD (ed) The biomedical engineering handbook, 2nd edn. CRC, Boca Raton, chap 95
12. Fisher RJ (2000) Transport phenomena and biomimetic systems. In: Bronzino JD (ed) The biomedical engineering handbook, 2nd edn, sect XII. CRC, Boca Raton
13. Davidson MG, Deen WM (1988) Macromol 21:3474
14. Lightfoot EN (2000) Diffusional processes and engineering design. In: Bronzino JD (ed) The biomedical engineering handbook, 2nd edn. CRC, Boca Raton, chap 96
15. Baish JW (2000) Microvascular heat transfer. In: Bronzino JD (ed) The biomedical engineering handbook, 2nd edn. CRC, Boca Raton, chap 98
16. Saltzman WM (2000) Interstitial transport in the brain: principles for local drug delivery. In: Bronzino JD (ed) The biomedical engineering handbook, 2nd edn. CRC, Boca Raton, chap 99
17. Tarbell JM, Qui Y (2000) Arterial wall mass transport: the possible role of blood phase resistance in the localization of arterial disease. In: Bronzino JD (ed) The biomedical engineering handbook, 2nd edn. CRC, Boca Raton, chap 100
18. Peattie RA, Nayate AP, Firpo MA, Shelby J, Fisher RJ, Prestwich GD (2004) Biomaterials 25:2789
19. Fisher RJ (1989) Diffusion with immobilization in membranes: transport and failure mechanisms; Part II—transport mechanisms. In: Butterfield A (ed) Biological and synthetic membranes. Liss, New York
20. Fisher RJ (2000) Compartmental analysis. In: Enderle J, Blanchard S, Bronzino JD (eds) Introduction to biomedical engineering. Academic, Orlando, chap 8
21. Grasso DK, Strevett KA, Fisher RJ (1995) Chem Eng J 59:195
22. Sweeney LM, Shuler MJ, Babish JG, Ghanem A (1995) Toxicol In Vitro 9:307
23. Fisher RJ (1993) Appl Sci 2:987
24. Sneider HG (1998) MS thesis, University of Connecticut
25. Koufas D, Fisher RJ (1998) Proc 24th NEBC/IEEE Trans 24:12
26. Gunasekaran S, Fisher RJ (1991) Performance characteristics of a counter-current diffusion reactor. IFT meeting, Dallas
27. Fisher RJ, Gunasekaran S, Magliocco CA (1999) Proc 25th NEBC/IEEE Trans 25:3
28. Fisher RJ, Zhang Y, Magliocco CA (1999) Proc 25th NEBC/IEEE Trans 25:11
29. Yang MC, Cussler EL (1986) AIChE J 32:1910
30. Shuler ML, Ghanem A, Quick D, Wang MC, Miller P (1996) Biotech Bioeng 52:45
31. Pirard JP, L'Homme GA (1978) J Catal 51:422
32. Mencier B, Figueras F, de Mourgues L (1969) J Phys 66:1950
33. DeMaria F, Longfield JE, Butler G (1961) Ind Eng 53:259
34. Europa AF, Grambhir A, Fu PC, Hu WS (2000) Biotech Bioeng 67:25
35. Bruns D, Bailey J, Luss D (1973) Biotech Bioeng 15:1131
36. Fisher RJ, Roberts SC, Peattie RA (2000) Ann Biomed Eng 28(S1):39
37. Gura T (1997) Science 273:1041
38. Sweeney L, Shuler ML, Quick D, Babish J (1996) Ann Biomed Eng 24:305
39. Shuler ML, Ghanem A, Quick D, Wang M, Miller P (1996) Biotech Bioeng 42:45
40. Converti A, Perego P, Lodi A, Fiorito G, Del Borghi M, Ferraiolo G (1991) Bioproc Eng 7:3
41. Wickramasinghe SR, Semmens MJ, Cussler EL (1993) J Memb Sci 84:14
42. Bronzino JD, Fisher RJ (1999) Funded research proposal: encapsulating islet cells in intelligent membranes for successful transplantation in the treatment of diabetes. Connecticut Innovations Inc.

43. Thoresen K, Fisher RJ (1995) Biomimetics 3:31
44. Juhasz NM, Deen WM (1991) Ind Eng Res 30:556
45. Ovaici H, Mackley MR, McKinley GH, Crook SJ (1998) J Rheol 42:125
46. Scinivasan AV, Haritos GK, Hedberg FL (1991) Appl Mech Rev 44:463
47. Thoresen K, Fisher RJ (1995) Biomimetics 3:57
48. Fisher RJ, Fenton JM, Iranmahboob J (2000) J Memb Sci 177:17
49. Kalachev AA, Kardivarenko LM, Plate NA, Bargreev VV (1992) J Memb Sci 75:1
50. Cussler EL, Aris R, Brown A (1989) J Memb Sci 43:149
51. Cussler EL (1984) Diffusion: mass transfer in fluid systems. Cambridge University Press, New York
52. Fry AJ, Soblov SB, Leonida MD, Viovodov KI (1994) Denki Kagaku 62:1260
53. Atkinson SJ, Feller KJ, Peattie RA (2001) Trans ASME-BED 50:729
54. Peattie RA, Fischer RJ (2006) Adv Biochem Engin/Biotechnol 103:75–156
55. Feller KJ, Atkinson SJ, Peattie RA (2001) Trans ASME-BED 50:753
56. Bird RB, Stewart WE, Lightfoot EN (2002) Transport phenomena, 2nd edn. Wiley, New York
57. Rosner DE (1986) Transport processes in chemically reacting flow systems. Butterworth, Boston
58. Deen WM (1996) Analysis of transport phenomena. Oxford University Press, New York
59. Fournier RL (1999) Basic transport phenomena in biomedical engineering. Taylor and Francis, Philadelphia
60. Brodkey RS, Hershey HC (1988) Transport phenomena: a unified approach. McGraw-Hill, New York
61. Fang YE, Cheng Q, Lu XB (1998) J Appl Polym Sci 68:1751
62. Yasuda H, Lamaze CE (1971) J Macromol Sci Phys B5:111
63. Fisher RJ (1989) Diffusion with immobilization in membranes: transport and failure mechanisms. Part II—Transport mechanisms. In: Butterfield IA (ed) Biological and synthetic membranes. Liss, New York
64. Bell CL, Peppas NA (1996) Biomaterials 17:1203
65. Cussler EL (1997) Diffusion: mass transfer in fluid systems. Cambridge University Press, Cambridge
66. Deen WM, Bohrer MP (1981) AIChE J 27:952
67. Callaghan PT, Pinder DN (1983) Macromolecules 16:968
68. Davidson MG, Deen WM (1988) Macromolecules 21:3474
69. Lebraun L, Junter GA (1994) J Memb Sci 88:253
70. Bu Z, Russo BS (1994) Macromolecules 27:1187
71. Furukawa RJ, Arauz-Lara B, Ware BR (1991) Macromolecules 24:599
72. Colton CK, Stroeve P, Zahka J (1975) J Appl Phys 35:307
73. Crank J (1956) The mathematics of diffusion. Oxford University Press, London
74. DeRossi D, Kajiwara K, Osaka Y, Yamauchi A (1991) Polymer gels. Plenum, New York
75. Cooney DO (1976) Biomedical engineering principles. Dekker, New York
76. Anderson JC, Babb AL, Hlastala MP (2003) Ann Biomed Eng 31:1402
77. Mueller TJ (1978) Computational Fluid Dynamics Modeling. In: Wirz HJ, Smolderen JJ (eds) Numerical Methods in Fluid Dynamics. McGraw-Hill, New York
78. Kulpers JAM, Van Swaaij WPM (1998) Adv Chem Eng 24:227
79. Goldstein AS, DiMilla PA (1997) Biotech Bioeng 55:616
80. Roberts SC, Fisher RJ, Roberts L (2000) Ann Biomed Eng 28(S1):19
81. Chen X, Fenton JM, Fisher RJ, Peattie RA (2004) J Electrochem Soc 151:E56
82. Hamilton BK, Stockmeyer LJ, Colton CK (1973) J Theor Biol 41:547

83. Strogatz SH (1994) Nonlinear dynamics and chaos. Addison-Wesley, Reading
84. Hess B (1985) Trends Biol Sci 2:37
85. Marcus M, Hess B (1984) Proc Natl Acad Sci USA 81:4394
86. Fisher RJ, Bronzino JD, Peattie RA (2001) Report TC/CCI-99.01; from BEACON Partners to Connecticut Innovations Inc.
87. Lewis AS, Colton CK (2004) Tissue engineering for insulin replacement in diabetes. In: Ma PX, Elisseeff J (eds) Scaffolding in tissue engineering. Marcel Dekker, New York (in press)

Adv Biochem Engin/Biotechnol (2006) 103: 75–156
DOI 10.1007/10_019
© Springer-Verlag Berlin Heidelberg 2006
Published online: 5 July 2006

Perfusion Effects and Hydrodynamics

Robert A. Peattie[1] (✉) · Robert J. Fisher[2]

[1]Department of Chemical Engineering, 102 Gleeson Hall, Oregon State University,
Corvallis, OR 97331, USA
peattie@engr.orst.edu

[2]Department of Chemical Engineering, Building 66, Room 446,
Massachusetts Institute of Technology, Cambridge, MA 02139, USA

Abstract Biological processes within living systems are significantly influenced by the motion of the liquids and gases to which those tissues are exposed. Accordingly, tissue engineers must not only understand hydrodynamic phenomena, but also appreciate the vital role of those phenomena in cellular and physiologic processes both in vitro and in vivo. In particular, understanding the fundamental principles of fluid flow underlying perfusion effects in the organ-level internal environment and their relation to the cellular microenvironment is essential to successfully mimicking tissue behavior.

In this work, the major principles of hemodynamic flow and transport are summarized, to provide readers with a physical understanding of these important issues. In particular, since quantifying hemodynamic events through experiments can require expensive and invasive techniques, the benefits that can be derived from the use of computational fluid dynamics (CFD) packages and neural networking (NN) models are stressed. A capstone illustration based on analysis of the hemodynamics of aortic aneurysms is presented as a representative example of this approach, to stress the importance of tissue responses to flow-induced events.

Keywords Fluid flow · Perfusion · Biomimetics · Steady flow · Pulsatile flow · Laminar flow · Turbulence · Computational fluid dynamics · Transport phenomena · Cell/tissue therapy · Tissue engineering

1
Introduction

1.1
Overview and Motivation

Biological processes within living systems are significantly influenced by the flow of liquids and gases. Biomedical engineers must therefore have an understanding of hydrodynamic phenomena [1] and their vital role in the biological processes that occur within the body [2]. In particular, the tissue engineer is concerned with perfusion effects in the cellular microenvironment, and the ability of the circulatory and respiratory systems to provide a whole body communication network with dynamic response capabilities. Understanding the fundamental principles of fluid flow involved in these processes is essential to accomplishing the primary objective of mimicking tissue behavior, whether in extra-corporeal devices or for in vivo cellular therapy, i.e., to know how tissue function can be built, reconstructed and/or modified to be clinically relevant. From a geometric and flow standpoint, the body may be considered a network of highly specialized and interconnected organs. The key elements of this network for transport and communication are its pathway (the circulatory system) and its fluid (blood). Of most interest for tissue engineering purposes is the ability of the circulatory system to transport oxygen and carbon dioxide, glucose, other nutrients and metabolites and signal molecules to and from the tissues, and to provide an avenue for stress-response agents from the immune system, including cytokines, antibodies, leukocytes, and macrophages, and system repair agents such as stem cells and platelets. The bulk transport capability provided by convective flow helps to overcome the large diffusional resistance that would otherwise be offered by such a large entity as the human body. At rest, the mean blood circulation time is of the order of one minute. Therefore, given that the total amount of blood circulating is about 76–80 ml/kg (5.3–5.6 liters for a 70 kg "standard male"), the flow from the heart to this branching network is about 95 ml/sec. This and other order of magnitude estimates for the human body are available elsewhere, for example [2–4].

Although the main fluids considered in this work are blood and air, other fluids such as urine, perspiration, tears, ocular aqueous and vitreous fluids and the synovial fluid in the joints can also be important in evaluating tissue system behavioral responses to induced chemical and physical stresses. For purposes of analysis, these fluids are normally assumed to exhibit Newtonian behavior, although the synovial fluid and blood under certain conditions can be non-Newtonian. Since blood is a suspension it has interesting properties; it behaves as a Newtonian fluid for large shear rates, but is highly non-Newtonian for low shear rates. The synovial fluid exhibits visco-elastic characteristics that are particularly suited to its function of joint lubrication, for which elasticity is beneficial. These visco-elastic characteristics must be accounted for when considering tissue therapy for joint injuries.

Further complicating physiologic fluid flow analysis is the fact that these fluids travel through three dimensional, usually distensible, passageways, in which highly disturbed or turbulent flow regimes may be mixed with stable, laminar regions. For example, pulsatile blood flow is laminar in many parts of a healthy circulatory system in spite of the potential for peak Reynolds numbers (defined below) of the order of 10 000. However, small "bursts" of turbulence are detected in the aorta during a fraction of each cardiac cycle. An occlusion or stenosis in the circulatory system, such as the stenosis of a heart valve, will promote such turbulence. Airflow in the lung is normally stable and laminar during inspiration, but less so during expiration, and heavy breathing, coughing or an obstruction can result in turbulent flow, with Reynolds numbers of 50 000 a possibility.

Elasticity of vessel walls is an important characteristic necessary for control of homeostasis. For example, pulsatile blood flow induces accompanying flow channel expansions and contractions in healthy elastic-wall vessels. These wall displacements then influence the flow fields. Elastic behavior maintains the norm of laminar flow that minimizes wall stress, lowers flow resistance and thus energy dissipation and fosters maximum life of the vessel. In combination with pulsatile flow, distensibility permits strain-relaxation of the wall tissue with each cardiac cycle, which provides an exercise routine promoting extended "on-line" use.

The term *perfusion* is used in engineering biosciences to identify the rate of blood supplied to a unit quantity of an organ/tissue system. It is clear that perfusion of in vitro tissue systems is necessary to maintain cell viability along with functionality to mimic in vivo behavior. Furthermore, it is highly likely that cell viability and normoperative metabolism are dependent on the three-dimensional structure of the microvessels distributed through any tissue bed, which establishes an appropriate microenvironment through both biochemical and biophysical mechanisms. This includes transmitting both intracellular and long range signals along the scaffolding of the extra-cellular matrix.

Developing an appreciation for the complexities of fluid dynamics is the primary objective of this work, as hydrodynamic and hemodynamic principles have many important applications to tissue engineering. In fact, the interaction of fluids and supported tissue is of paramount importance to successful tissue development. The strength of adhesion and dynamics of detachment of mammalian cells from engineered biomaterials and scaffolds is important ongoing research [5], as is the effects of shear on receptor–ligand binding at the cell-fluid interface. The consequences of shear stress are far more reaching than simply altered transport across the cell membrane. Receptor location, binding affinity and signal generation with subsequent trafficking within the cell [6] can also be changed. In addition, design and use of perfusion systems such as membrane biomimetic reactors and hollow fibers is most effective when careful attention is given to issues of hydrodynamic similitude. Similarly, understanding the role of fluid mechanical phenomena in arterial disease and subsequent therapeutic applications is clearly dependent on appreciation of hemodynamics. For that reason, working with team members with expertise in fluid mechanics can often be of substantial benefit towards completion of tissue engineering projects. However, understanding their approach and having a fundamental grasp of the technology and its terminology is a prerequisite for effective communication.

A thorough treatment of the mathematics needed for model development and analysis is beyond the scope of this review, and is presented in numerous sources [1,2]. Herein, the major principles of hemodynamic flow and transport are summarized, with the goal of providing the reader with a physical understanding of the important issues. Benefits derived from the use of computational fluid dynamics (CFD) packages and neural networking (NN) models are stressed. In particular, quantifying hemodynamic events can require invasive experimentation and/or extensive computational analysis. Tissue engineers will therefore often find computational fluid dynamics (CFD) a very powerful and versatile tool for analysis and solution of many complex but crucial problems. Notable examples include simulation of prototype devices to develop similitude and performance predictions. Analysis and design of bioreactors, in particular, is highly dependent on understanding the nonideal flow patterns that exist within the reactor vessel. In principle, if the complete velocity distribution of the reactor is known, mixing and transport characteristics and the pressure and stress profiles affecting cellular processes can be fully predicted. Once considered an impractical approach, this is now obtainable by computing the velocity distribution using the CFD-based procedures [7, 8].

1.2
Background and Approach

Although hydrodynamics as a field is too complex to be fully discussed within the scope of this text, it is important to provide at least an introduction to the subject so that the interested reader will be able to pursue more in depth studies when needed. Consequently, the approach taken in this work is to present a brief overview of the pertinent background and associated tools for a series of concepts rather than a full treatment. Each topic has been selected for its relevance to problems encountered by the tissue engineer. The review includes: (1) mathematical tools, model development, computational techniques, and the role of nonlinear analysis in stability and flow regime development; (2) perfusion, design specifications, and devise performance; (3) mixing and transport phenomena as related to (i) nano-encapsulation and (ii) extra-corporeal systems; (4) stress effects on cellular viability and behavior; (5) characteristics and analysis of pulsatile flow; and (6) the role of hydrodynamics in arterial diseases; diagnosis, risk assessment and therapy. Most of these topics are discussed with respect to a representative application, to illustrate their importance but not to delve deeply into the details. The review closes with a capstone illustration of hemodynamic analysis techniques based on the pathophysiology of abdominal aortic aneurysms (AAAs).

2
Elements of Theoretical Hydrodynamics

It is essential that tissue engineers understand both the advantages and the limitations of mathematical theories and models of biological phenomena, since those models often involve approximations that are not always fully justified in biological systems. Accordingly, making choices between theoretical models so as to keep erroneous conclusions to a minimum requires an understanding of the assumptions underlying the models.

In this work, an overview of the basic ideas of continuum mechanics and their mathematical formulation is presented, to provide tissue engineers with a framework for appreciating continuum mechanics analysis. Our discussion of fluid dynamics begins with the Navier–Stokes equations. In brief, these equations provide an expression for Newton's second law ($F = ma$) when applied to continuous distributions of Newtonian fluids; that is, fluids such as air and water for which the rate of motion is linearly proportional to the applied stress producing the motion. To illustrate the usefulness of the Navier–Stokes equations when applied to realistic problems, the basic concepts from which these equations have been developed are summarized along with a few general ideas about boundary layers and turbulence. Application of the Navier–Stokes equations to the vascular system as developed by Womersley and others is treated in a later section on pulsatile flow. It is hoped that this very generalized approach will facilitate understanding of the sections that follow. Furthermore, it is anticipated that the reader will appreciate the complexities involved in an analytic solution to pulsatile phenomena; a necessity for properly analyzing the vascular system and describing its behavior for clinical evaluations.

2.1
Mathematical Preliminaries

The objective of this section is to briefly describe the most important concepts of vector and tensor calculus needed for hydrodynamics modeling. Vector field theory is introduced, and vector differentiation is used to determine field properties such as the divergence, gradient and curl. Since the dependent variables typically encountered in continuum mechanics, notably pressure, flow, and tension, typically are functions of several independent variables, these operations make extensive use of partial differentiation. For example, if the scalar quantity pressure, P, is a function of the three spatial coordinates x, y, and z, as well as of time, t, than the total change in P with respect to all variables, that is its total differential dP, would be

$$dP = \frac{\partial P}{\partial x}\,dx + \frac{\partial P}{\partial y}\,dy + \frac{\partial P}{\partial z}\,dz + \frac{\partial P}{\partial t}\,dt\,. \tag{1}$$

Differentiation of vector quantities, which is more complex than differentiation of scalars, is performed with the vector operator *Del*. In a three-dimensional, rectilinear, Cartesian coordinate system, the Del operation can be defined as

$$\nabla = \frac{\partial}{\partial x} i + \frac{\partial}{\partial y} j + \frac{\partial}{\partial z} k,$$ (2)

where i, j, and k are unit vectors in the 3 orthogonal spatial directions x, y, and z, respectively. This operator can be used with both scalar and vector functions to produce the field properties of gradient, divergence, and curl. If the operand is a scalar field such as P, the *gradient* of this field, which is itself a vector and therefore possesses both magnitude and direction, can be defined as

$$\nabla P = \frac{\partial P}{\partial x} i + \frac{\partial P}{\partial y} j + \frac{\partial P}{\partial z} k.$$ (3)

Thus defined, the direction of ∇P represents the direction of greatest spatial rate of change of P, while the magnitude of ∇P represents its rate of change in that direction.

To then find the rate of change of P in an arbitrary direction, say that of an infinitesimal displacement vector, dr, given as

$$dr = dxi + dyj + dzk$$ (4)

we need to perform the *dot product* vector manipulation. This results in a scalar quantity. For example, the total differential of P, which represents the change in P as one moves a distance dr is

$$dP = \nabla P \cdot dr = \frac{\partial P}{\partial x} dx + \frac{\partial P}{\partial y} dy + \frac{\partial P}{\partial z} dz.$$ (5)

A vector field changes in a more complicated manner from point-to-point than does a scalar field. The *divergence* of a vector field is one way of expressing such a rate of change. For the arbitrary vector field V with components v_x, v_y and v_z in some Cartesian coordinate system, we can write

$$V = v_x i + v_y j + v_z k.$$ (6)

The divergence is then defined as

$$\nabla \cdot V = \frac{\partial v_x}{\partial x} + \frac{\partial v_y}{\partial y} + \frac{\partial v_z}{\partial z}.$$ (7)

The divergence itself is a scalar associated with the vector field V. It is particularly meaningful when applied to the mass flow per unit volume, ρu, where ρ is the density of a moving fluid, which is a scalar, and u is its velocity, a vector with components $u = (u, v, w)$. The divergence of ρu provides the mass rate of change per unit control volume in the flow field. This must be equal to the accumulation of the mass in that unit control volume, so that

$$-\frac{\partial \rho}{\partial t} = \nabla \cdot \rho u = \frac{\partial \rho u}{\partial x} + \frac{\partial \rho v}{\partial y} + \frac{\partial \rho w}{\partial z},$$ (8)

where the right hand side is the divergence of the vector ρu and the left hand side is taken negative by convention to indicate mass *influx* into the control volume. If the fluid is incompressible, which is an excellent assumption for both liquids and gases moving at velocities that are small compared with the speed of sound, and if there are no sources or sinks of mass within the unit control volume, then ρ = constant and the net gain or loss of mass in the unit volume is zero. Equation 8 therefore becomes the *continuity equation*, an expression of conservation of mass for a flow field with constant density

$$\nabla \cdot u = 0 . \tag{9}$$

Equation 9 does not hold for compressible fluids, since the fluid density ρ is then not constant.

Another way of describing the rate of change of a vector field is through the *curl*, which is proportional to the rate of rotation in the vector field. The curl of the vector field V can be expressed as a determinant

$$\nabla \times V = \begin{vmatrix} i & j & k \\ \frac{\partial}{\partial x} & \frac{\partial}{\partial y} & \frac{\partial}{\partial z} \\ v_x & v_y & v_z \end{vmatrix} \tag{10}$$

which when expanded to its full form is

$$\nabla \times V = \left(\frac{\partial v_z}{\partial y} - \frac{\partial v_y}{\partial z} \right) i + \left(\frac{\partial v_x}{\partial z} - \frac{\partial v_z}{\partial x} \right) j + \left(\frac{\partial v_y}{\partial x} - \frac{\partial v_x}{\partial y} \right) k . \tag{11}$$

In particular, the curl of the fluid velocity field u can be interpreted as the tendency to impart a rotation upon a fluid element that enters this velocity field. Such rotation is given the formal name *vorticity*, where vorticity = $\nabla \times u$. Then the direction of the vorticity vector indicates the axis about which the flow field rotates, while its magnitude indicates the rate of rotation. It can be shown that the magnitude of the vorticity is exactly twice the local angular velocity of the fluid elements.

The divergence and curl are both spatial derivatives of vector fields; physically, the divergence represents the net change of the vector field while the curl is the rate of rotation of the field. If the divergence is zero the field is termed *solenoidal* and if the curl is zero it is *irrotational*. Additional properties that are useful in analysis of fluid flow problems are that for any arbitrary scalar field q and vector field V, the curl of the gradient $(\nabla \times \nabla q)$ and the divergence of the curl $(\nabla \cdot \nabla \times V)$ are always identically equal to zero. Of particular importance is the situation in which a vector field V is determined by the gradient of a scalar field q, $V = \nabla q$; such a field is said to be *conservative*. Mathematically, if V is irrotational, then there exists a scalar function q such that V can be expressed as a gradient of that scalar. Thus, we have

$$\nabla q = \frac{\partial q}{\partial x} i + \frac{\partial q}{\partial y} j + \frac{\partial q}{\partial z} k = V . \tag{12}$$

In fluid dynamics, a conservative flow field occurs when the flow is irrotational. Then $\nabla \times u = 0$, implying that u can be equated to the gradient of some scalar ∇q. If the flow field is also incompressible, continuity gives

$$\nabla \cdot u = \nabla \cdot \nabla q = \nabla^2 q = 0 . \tag{13}$$

Equation 13, $\nabla^2 q = 0$, is known as *Laplace's* equation. After determination of appropriate boundary conditions, solution of Eq. 13 identifies the function q; the velocity field may then be easily obtained since $u = \nabla q$. This approach leads to the study of so-called *potential flow*. Along with their importance in aerodynamics, potential flows have applications to extracorporeal devices.

A final mathematical concept of great importance, particularly in the study of pulsatile blood flow, is the *Euler–deMoivre decomposition* of complex numbers. That is, any complex number z can be represented as the sum of its real and complex parts, $z = x + iy$, where x and y are themselves real and $i = \sqrt{-1}$. The Euler–deMoivre formula shows that this Cartesian representation of z is entirely equivalent to representation of z in a complex exponential form, since

$$r e^{i\theta} = r \left(\cos \theta + i \sin \theta \right) , \tag{14}$$

where $r = |z| = \left(x^2 + y^2 \right)^{1/2}$ represents the magnitude of the complex variable. Use of the complex exponential form permits expansion of complex variables in McLaurin series, and often simplifies mathematical operations. However, the choice of form in which to express a complex number or function is arbitrary, and may be made based on convenience for a given calculation. In discussing the hydrodynamics of pulsatile flows below (Sect. 6), extensive use is made of complex exponential formulations.

2.2
Elements of Continuum Mechanics

The theory of fluid flow and the theory of elasticity both belong to the domain of continuum mechanics, the study of the mechanics of continuously distributed materials. Such materials may be either solid or fluid, or may have intermediate visco-elastic properties. Since the concept of a continuous medium, or continuum, does not take into consideration the molecular structure of matter, it is inherently an idealization. However, as long as the smallest length scale in any problem under consideration is very much larger than the size of the molecules making up the medium and/or the mean free path within the medium, for mechanical purposes we may safely assume all mass to be continuously distributed in space. As a result, the density of materials can be considered to be a continuous function of spatial position and time.

Continuum mechanics principles are stated through two types of equations; constitutive equations, which express the properties of particular materials, and field equations, which express general mechanical principles that apply to all materials. For example, general conservation principles of mass, momentum and energy are field equations valid for all types of continuous media. In contrast, stress–strain relationships, which predict the response of particular materials to applied loads and stresses, are constitutive equations and as such pertain only to the specific material in question.

Although the distinction between constitutive and field relations seems straightforward, it is often the case that both types of relations are needed to analyze physical events. For instance, in pulsatile blood flow, the motion of blood can deform the surrounding vessel, and the extent of such deformation depends on the wall stiffness. Consequently, the material properties of both the vessel wall and the flowing blood must be considered for proper analysis of this coupled fluid-wall system, along with the field equations governing flow in general. Although some conservation equations apply equally to both the wall and the fluid, the constitutive equations pertaining to each are quite different. For a perfectly elastic body, the stress is a linear function of its small strains. However, for viscous fluids the stress is related to the rate of strain, and that relation depends on the nature of the fluid. For example, for Newtonian fluids the stress-rate of strain relation is linear, while for non-Newtonian fluids a nonlinear relation is more appropriate.

Here, basic concepts of stress, strain, and rate of strain are developed so as to obtain generally valid expressions for these quantities. Focus is restricted to displacements and flows resulting from imposed forces. Both surface and body forces are considered, as both are subject to Newton's second law; the net force on a body is equal to the product of its mass and acceleration.

Forces such as gravitational attraction that act throughout the volume of a material body and therefore are proportional in magnitude to its volume are termed *body forces*, while forces that act on a body's surface are *surface forces*. The latter category includes applied mechanical loads, tension, pressure (which acts perpendicularly to the surface) and shear stress (which acts parallel to the surface), A unit cube in a 3-dimensional Cartesian coordinate system is therefore acted on by three net surface forces. Each of these is itself a vector, and can be resolved into three components. It is customary to describe the associated *stress* components, here denoted s_{ij}, rather than the forces themselves, where *stress = force/unit area*. If desired, forces can be recovered by multiplying the stresses by the area of the surface. We then have

$$F_x = (s_{11}i + s_{12}j + s_{13}k) \cdot \Delta A_x$$
$$F_y = (s_{21}i + s_{22}j + s_{23}k) \cdot \Delta A_y$$
$$F_z = (s_{31}i + s_{32}j + s_{33}k) \cdot \Delta A_z .$$

$$(15)$$

The first subscript of these stress components indicates the axis to which the surface subjected to the stress is perpendicular, while the second denotes the direction of the stress. The values 1, 2, and 3 refer to the x-, y-, and z-directions, respectively, as do i, j, and k. The nine stress components, each of which is a scalar, can be grouped into a 3×3 matrix representation called the *stress tensor*. Of the nine components, the three diagonal components, which show repeated indices, represent *normal stresses* and act *perpendicularly* to their associated surfaces. The other six off-diagonal components represent *shear stresses*, and act *along* their surfaces. The stress tensor is *symmetric*; that is $s_{ij} = s_{ji}$. This information allows one to write for the *resultant* surface force,

$$F_s = \left(\left(\frac{\partial s_{11}}{\partial x} + \frac{\partial s_{12}}{\partial y} + \frac{\partial s_{13}}{\partial z} \right) i + \left(\frac{\partial s_{12}}{\partial x} + \frac{\partial s_{22}}{\partial y} + \frac{\partial s_{23}}{\partial z} \right) j \right.$$
$$\left. + \left(\frac{\partial s_{13}}{\partial x} + \frac{\partial s_{23}}{\partial y} + \frac{\partial s_{33}}{\partial z} \right) k \right) \cdot \Delta A_s . \tag{16}$$

Equation 16 can be substituted into Newton's second law to obtain an equation governing the motion of the unit cube. If the mass of the cube is $\rho \, dV$, and this equation is divided by dV, the result is three-component scalar equations:

$$\rho \frac{d^2 x}{dt^2} = \rho G_x + \frac{\partial s_{11}}{\partial x} + \frac{\partial s_{12}}{\partial y} + \frac{\partial s_{13}}{\partial z}$$
$$\rho \frac{d^2 y}{dt^2} = \rho G_y + \frac{\partial s_{12}}{\partial x} + \frac{\partial s_{22}}{\partial y} + \frac{\partial s_{23}}{\partial z}$$
$$\rho \frac{d^2 z}{dt^2} = \rho G_z + \frac{\partial s_{13}}{\partial x} + \frac{\partial s_{23}}{\partial y} + \frac{\partial s_{33}}{\partial z} , \tag{17}$$

where the terms G_x, G_y, and G_z refer to the respective components of the net body force.

If the continuum consists of an ideal frictionless fluid, all shear stresses s_{12}, s_{13}, etc. vanish. At the same time, the normal stresses are equal and define the pressure at a given point. For a viscous fluid, however, the normal stresses are not necessarily equal. The mean pressure is then defined as the arithmetic mean normal stress, taken with a negative sign. In some instances it is helpful to break the stress tensor into two parts, a *deviatoric* stress tensor associated with changes in shape and the pressure tensor associated with volume changes.

Changes in the *relative* position of the elements of a continuum are termed *deformations* or *strains*. The strain system of a continuous body can be described by expressing the deformation of a given volume as three elongations and three angular displacements. Elongations, which are denoted as ε_{11}, ε_{22}, and ε_{33}, denote a relative change in length of an elementary volume in the direction of the respective coordinate axes, while angular displacements ε_{12}, ε_{13}, etc. are associated with shear motions. The relative change in volume of a body during a deformation, its *volume dilatation*, consists of the sum of the elongations;

$$\frac{\mathrm{d}\Delta V}{\Delta V} = e = \varepsilon_{11} + \varepsilon_{22} + \varepsilon_{33} \,. \tag{18}$$

Typically, strain is a concept used to describe the response of a solid to an applied stress field. The corresponding concept for a fluid is *rate of strain*, or velocity. The *rate of strain tensor* describes the velocities of the different elements of a fluid. If $u = ui + vj + wk$ is the velocity field of a continuous fluid, the elements of the rate of strain tensor ξ_{ij} can be expressed in terms of u, v, and w. For example,

$$\xi_{12} = \frac{1}{2} \left(\frac{\partial u}{\partial y} + \frac{\partial v}{\partial x} \right) \tag{19}$$

with similar definitions for the other eight components.

2.2.1
Constitutive Equations

The response of any material to applied disturbances including forces and temperature can be used to characterize the material. Functional relationships are needed between applied stresses and the resulting strain or rate of strain field of the material to make the field equations usable with respect to specific materials. For example, Eq. 17 can be used to determine the displacements of a solid in response to an applied stress loading if a relation between the stress components s_{ij} and the resulting strain components ε_{ij} is known. One of the simplest forms that relates stress components to strain components in an elastic solid is the linear relationship developed by Hooke. This constitutive expression relates each stress component to only one strain component

$$s_{ij} = k_{ij} \cdot \varepsilon_{ij} \,. \tag{20}$$

An ideal elastic material shows no time or history effects of any sort. That is, when the stress is removed, the deformations disappear. Unfortunately, most materials of interest to tissue engineers show a more complex, time-dependent behavior. For example, Eq. 20 is too elementary to describe the visco-elastic behavior of the vasculature walls, a topic discussed further below.

In the most general hypothetical case, each component of strain would be expected to be related to *every* stress component. In that case, fully characterizing the material would require 36 elastic constants, each relating one stress component to one strain component. The total strain in one direction would then be determined by summing all the components that are the result of each of the six stress components. Fortunately, due to considerations based on the internal energy of deformations, all 36 constants are not independent. In the most general case of an anisotropic medium only 21 constants are independent. For a completely isotropic body, the set of elastic constants depends on only two parameters known as the Lamé constants, λ and η. The fundamental elastic moduli of an isotropic body are given as,

$$k_{11} = k_{22} = k_{33} = \lambda + 2\eta$$
$$k_{12} = k_{23} = k_{13} = \lambda$$
$$k_{44} = k_{55} = k_{66} = \eta \tag{21}$$

where

k_{11} relates s_{11} to ε_{11}

k_{22} relates s_{22} to ε_{22}

k_{33} relates s_{33} to ε_{33}

k_{12} relates s_{11} to ε_{22} and s_{22} to ε_{11}

k_{23} relates s_{22} to ε_{33} and s_{33} to ε_{22}

k_{13} relates s_{11} to ε_{33} and s_{33} to ε_{11}

k_{44} relates s_{23} to $\left(\varepsilon_{23} + \varepsilon_{32}\right)$

k_{55} relates s_{13} to $\left(\varepsilon_{13} + \varepsilon_{31}\right)$

k_{66} relates s_{12} to $\left(\varepsilon_{12} + \varepsilon_{21}\right)$

and all other $k_{ij} = 0$.

Although these relationships do identify the physical significance of the Lamé constants, they are inconvenient for experimental purposes. Four more directly measurable quantities, *bulk modulus, Young's modulus, shear modulus*, and *Poisson's ratio* are typically used to obtain values for λ and η from experimental data. The mean normal stress is related to the volume dilation, e, via the bulk modulus k.

$$\bar{s} = k \cdot e, \tag{22}$$

where $\bar{s} = \frac{1}{3}\left(s_{11} + s_{22} + s_{33}\right)$.

Young's modulus, E, relates a normal stress to a normal strain, for example,

$$s_{ii} = E\varepsilon_{ii}. \tag{23}$$

Shear stress and shear strain are related through the shear modulus, G, as

$$s_{ij} = G\varepsilon_{ij} \quad (i \neq j). \tag{24}$$

Poisson's ratio, σ, is a measure of the compressive strain associated with a tensile strain at right angles, for example, compression in the y-direction due to elongation in the x-direction;

$$\varepsilon_{22} = -\sigma\varepsilon_{11} = -\sigma\frac{s_{11}}{E}. \tag{25}$$

With Eq. 25 in mind, Eq. 23 should properly be modified to give

$$
\begin{aligned}
s_{11} - \sigma \left(s_{22} + s_{33}\right) &= E\varepsilon_{11} \\
s_{22} - \sigma \left(s_{11} + s_{33}\right) &= E\varepsilon_{22} \\
s_{33} - \sigma \left(s_{22} + s_{11}\right) &= E\varepsilon_{33}
\end{aligned}
\tag{26}
$$

and the dilatation e should be expressed as

$$
e = \frac{1}{E}(1 - 2\sigma)\left(s_{11} + s_{22} + s_{33}\right) .
\tag{27}
$$

With Eqs. 24–27 established, the Lamé constants can be derived from G, σ, E, and k since

$$
k = \lambda + \frac{2}{3}\eta
\tag{28a}
$$

$$
G = \frac{E}{2(1 + \sigma)} = \eta\mu
\tag{28b}
$$

$$
E = \frac{\eta(3\lambda + 2\eta)}{\lambda + \eta} = \frac{9k\eta}{3k + \eta}
\tag{28c}
$$

and

$$
\sigma = \frac{\lambda}{2\left(\lambda + \eta\right)} = \frac{3k - 2\eta}{2\left(3k + 2\eta\right)} .
\tag{28d}
$$

Using Eqs. 28a–d, the fundamental equations relating stress and strain in a homogeneous isotropic elastic solid can be rewritten as a *generalized Hooke's law*

$$
s_{ij} = \lambda e\delta_{ij} + 2\eta\varepsilon_{ij} ,
\tag{29}
$$

where $\delta_{ij} = 1$ for $i = j$ and zero for $i \neq j$.

An analogous expression to Eq. 29 can be developed that relates the rate of strain tensor of a fluid flow to the applied stress, following the method just described. The shear modulus is replaced by the coefficient of viscosity, μ, and the normal stress by the negative of the fluid pressure, $-P$. The result for an incompressible fluid is

$$
s_{ij} = - P\delta_{ij} + 2\mu\xi_{ij} .
\tag{30}
$$

2.2.2
Conservation (Field) Equations

Conservation of mass for a continuous material is expressed through the well-known continuity condition (Eq. 9). As seen in our discussion of the divergence of a vector field (Sect. 2.1), the net flow of mass from a unit control volume per unit time must be equal to any change in density within that volume. In vector form this accumulation term must equal the divergence of the mass flow field so that

$$\frac{\partial \rho}{\partial t} + \nabla \cdot \rho \mathbf{u} = 0 . \tag{8}$$

When the fluid is incompressible, Eq. 8 reduces to

$$\nabla \cdot \mathbf{u} = 0 . \tag{9}$$

In terms of Cartesian components, the continuity condition can be equally well expressed as

$$\frac{\partial u}{\partial x} + \frac{\partial v}{\partial y} + \frac{\partial w}{\partial z} = 0 . \tag{31}$$

In cylindrical polar coordinates (r, θ, z), for which $\mathbf{u} = \left(u_r, u_\theta, u_z \right)$ it is written as

$$\frac{\partial v_r}{\partial r} + \frac{v_r}{r} + \frac{1}{r}\frac{\partial v_\theta}{\partial \theta} + \frac{\partial v_z}{\partial z} = 0 . \tag{32}$$

The basic equation of motion (Eq. 19) can be applied to fluid motion by substitution of the constitutive relationship for a Newtonian fluid (Eq. 30). This substitution results in the *Navier–Stokes equations*, the fundamental statement of Newton's second Law for a Newtonian fluid. It is important to note that in writing the second Law for a continuously distributed fluid, care must be taken to correctly express the acceleration of the fluid particle to which the forces are being applied. That is, the velocity of a fluid particle may change for either of two reasons, because the particle accelerates or decelerates with time (*temporal acceleration*) or because the particle moves to a new position, at which the velocity has different magnitude and/or direction (*convective acceleration*). The total time derivative of the velocity is a sum of both effects, which yields the *material*, or *substantial*, *derivative*, written $D()/Dt$. Thus,

$$\begin{aligned}
\frac{Du}{Dt} &= \frac{\partial u}{\partial t} + u\frac{\partial u}{\partial x} + v\frac{\partial u}{\partial y} + w\frac{\partial u}{\partial z} \\
\frac{Dv}{Dt} &= \frac{\partial v}{\partial t} + u\frac{\partial v}{\partial x} + v\frac{\partial v}{\partial y} + w\frac{\partial v}{\partial z} \\
\frac{Dw}{Dt} &= \frac{\partial w}{\partial t} + u\frac{\partial w}{\partial x} + v\frac{\partial w}{\partial y} + w\frac{\partial w}{\partial z} .
\end{aligned} \tag{33}$$

A flow field for which $\partial/\partial t = 0$ for all possible properties of the fluid and its flow is described as *steady*, to indicate that it is independent of time. From Eq. 33, it is clear that the statement $Du/Dt = 0$ does not imply that $\partial u/\partial t = 0$, and similarly $\partial \mathbf{u}/\partial t = 0$ does not imply $D\mathbf{u}/Dt = 0$.

Using the material derivative, the Navier–Stokes equations for an incompressible fluid can be written in vector form as

$$\frac{D\boldsymbol{u}}{Dt} = \boldsymbol{G} - \frac{1}{\rho}\nabla P + \upsilon\nabla^2\boldsymbol{u}, \tag{34}$$

where υ is the kinematic viscosity, $= \mu/\rho$.

Expanded in full, the Navier–Stokes equations are three simultaneous, nonlinear scalar equations, one for each component of the velocity field. In Cartesian coordinates, Eq. 32 takes the form

$$\frac{\partial u}{\partial t} + u\frac{\partial u}{\partial x} + v\frac{\partial u}{\partial y} + w\frac{\partial u}{\partial z} = G_x - \frac{1}{\rho}\frac{\partial P}{\partial x} + \upsilon\left(\frac{\partial^2 u}{\partial x^2} + \frac{\partial^2 u}{\partial y^2} + \frac{\partial^2 u}{\partial z^2}\right) \tag{35a}$$

$$\frac{\partial v}{\partial t} + u\frac{\partial v}{\partial x} + v\frac{\partial v}{\partial y} + w\frac{\partial v}{\partial z} = G_y - \frac{1}{\rho}\frac{\partial P}{\partial y} + \upsilon\left(\frac{\partial^2 v}{\partial x^2} + \frac{\partial^2 v}{\partial y^2} + \frac{\partial^2 v}{\partial z^2}\right) \tag{35b}$$

$$\frac{\partial w}{\partial t} + u\frac{\partial w}{\partial x} + v\frac{\partial w}{\partial y} + w\frac{\partial w}{\partial z} = G_z - \frac{1}{\rho}\frac{\partial P}{\partial z} + \upsilon\left(\frac{\partial^2 w}{\partial x^2} + \frac{\partial^2 w}{\partial y^2} + \frac{\partial^2 w}{\partial z^2}\right). \tag{35c}$$

In cylindrical coordinates, with velocity components u_r, u_θ, u_z, it is written as

$$\frac{\partial u_r}{\partial t} + u_r\frac{\partial u_r}{\partial r} + \frac{u_\theta}{r}\frac{\partial u_r}{\partial \theta} - \frac{u_\theta^2}{r} + u_z\frac{\partial u_r}{\partial z}$$
$$= G_r - \frac{1}{\rho}\frac{\partial P}{\partial r} + \upsilon\left(\frac{\partial^2 u_r}{\partial r^2} + \frac{1}{r}\frac{\partial u_r}{\partial r} - \frac{u_r}{r^2} + \frac{1}{r^2}\frac{\partial^2 u_r}{\partial \theta^2} + \frac{2}{r^2}\frac{\partial u_\theta}{\partial \theta} + \frac{\partial^2 u_r}{\partial z^2}\right) \tag{36a}$$

$$\frac{\partial u_\theta}{\partial t} + u_r\frac{\partial u_\theta}{\partial r} + \frac{u_\theta}{r}\frac{\partial u_\theta}{\partial \theta} - \frac{u_r u_\theta}{r} + u_z\frac{\partial u_\theta}{\partial z}$$
$$= G_\theta - \frac{1}{\rho}\frac{1}{r}\frac{\partial P}{\partial \theta} + \upsilon\left(\frac{\partial^2 u_\theta}{\partial r^2} + \frac{1}{r}\frac{\partial u_\theta}{\partial r} - \frac{u_\theta}{r^2} + \frac{1}{r^2}\frac{\partial^2 u_\theta}{\partial \theta^2} + \frac{2}{r^2}\frac{\partial u_r}{\partial \theta} + \frac{\partial^2 u_\theta}{\partial z^2}\right) \tag{36b}$$

$$\frac{\partial u_z}{\partial t} + u_r\frac{\partial u_z}{\partial r} + \frac{u_\theta}{r}\frac{\partial u_z}{\partial \theta} + u_z\frac{\partial u_z}{\partial z}$$
$$= G_z - \frac{1}{\rho}\frac{\partial P}{\partial z} + \upsilon\left(\frac{\partial^2 u_z}{\partial r^2} + \frac{1}{r}\frac{\partial u_z}{\partial r} + \frac{1}{r^2}\frac{\partial^2 u_z}{\partial \theta^2} + \frac{\partial^2 u_z}{\partial z^2}\right). \tag{36c}$$

Flow fields may be determined by solution of the Navier–Stokes equations, provided the body force $\boldsymbol{G}$ is known. This is generally not a difficulty, since the only body force normally significant in tissue engineering applications is gravity. For an incompressible flow, there are then four unknown dependent variables, the three components of velocity and the pressure P, and four governing equations, the three components of the Navier–Stokes equations and the continuity condition. It is important to emphasize that this set of equations is *not* sufficient to calculate the flow field when the flow is compressible or involves temperature changes, since pressure, density, and temperature are then interrelated, which introduces new dependent variables to the problem.

Solution of the Navier–Stokes equations also requires that boundary conditions, and sometimes initial conditions as well, be specified for the flow field of interest. By far the most commonly used boundary condition in physiologic flows and other tissue engineering settings is the so-called *no slip condition*, which requires that the layer of fluid elements in contact with a boundary have the same velocity as the boundary itself. For an unmoving, rigid wall, as in a pipe, this velocity is of course zero. However, in the vasculature, vessel walls expand and contract during the cardiac cycle.

Flow patterns and accompanying flow field characteristics depend largely on the values of governing dimensionless parameters. There are many such parameters, each relevant to specific types of flow settings, but the principle parameter of steady flows is the *Reynolds number*, *Re*, defined as

$$Re = \frac{\rho UL}{\mu} , \tag{37}$$

where U is a characteristic velocity of the flow field and L is a characteristic length. Both U and L must be selected for the specific problem under study, and in general both will have different values in different problems. For pipe flow, U is most commonly selected to be the mean velocity of the flow with L the pipe diameter.

It can be shown that the Reynolds number represents the ratio of inertial forces to viscous forces in the flow field. Flows at sufficiently low *Re* therefore behave as if highly viscous, with little to no fluid acceleration possible. At the opposite extreme, high *Re* flows behave as if lacking viscosity.

One consequence of this distinction is that very high Reynolds number flow fields may at first thought seem to contradict the no slip condition, in that they seem to "slip" along a solid boundary exerting no shear stress. This dilemma was first resolved in 1905 with Prandtl's introduction of the *boundary layer*, a thin region of the flow field adjacent to the boundary in which viscous effects are important and the no slip condition is obeyed [9–11].

In the boundary layer concept, flow at sufficiently high Reynolds number is thought of as having two regions, each with its own distinct flow regime. Velocity gradients in the inner region, the boundary layer itself, are very large, permitting a transition from wall velocity on one side to a free stream velocity on the opposite side. Within the boundary layer, the effects of viscosity are therefore always felt, regardless of the magnitude of *Re* outside. In the second or outer region, inertial forces dominate, velocity gradients are essentially zero and viscous effects are negligible. As a result, flow in the outer layer is normally irrotational, and the fluid element paths are determined by the pressure gradient. The transition region between the two zones can be described mathematically using the technique of matched asymptotic expansion.

In general, the assumptions on which boundary layer equations are based become increasingly more valid as the Reynolds number increases, since the boundary layer thickness is inversely proportional to *Re*. This allows simplification of the Navier–Stokes equation, so that within the boundary layer flow is approximately governed by

$$\rho\left(\frac{\partial u}{\partial t} + u\frac{\partial u}{\partial x} + v\frac{\partial u}{\partial y}\right) = -\frac{dP}{dx} + \mu\frac{\partial^2 u}{\partial y^2}\,.$$ (38)

Conversely, the outer region is primarily governed by potential flow conditions.

The position of the boundary layer–outer layer interface can be variable, since the thickness of the boundary layer may grow in the flow direction due to the diffusion of vortices. Flow in the boundary layer can even reverse direction, if the fluid is moving against a sufficiently large adverse pressure gradient and the energy and momentum of the fluid in the boundary layer are insufficient to overcome viscous forces. Decelerated fluid elements are then driven to the mainstream, with a concomitant separation of the layer from the wall and reversal of the flow direction distal to the separation point. Such separation is always associated with the formation of vortices and large energy dissipation in the region of strongly decelerated flow. This process can be particularly important when the flow field encounters an expansion, for the resulting decrease in velocity leads to an increase in pressure that can induce flow reversals and thickening of the boundary layer. This process is well illustrated in the aneurysmal flow discussed in the capstone illustration (Sect. 7).

2.2.3
Turbulence and Instabilities

Flow fields are broadly classified as either *laminar* or *turbulent* to distinguish between smooth and irregular motion, respectively. Fluid elements in laminar flow fields follow well-defined paths indicating smooth flow in discrete layers or "laminae". Furthermore, there is minimal exchange of material between layers due to the lack of macroscopic mixing. The transport of momentum between system boundaries is thus controlled by molecular action, and is dependent on the fluid viscosity.

In contrast, many flows in nature as well as engineered applications are found to fluctuate randomly and continuously, rather than streaming smoothly, and are classified as turbulent. These are characterized by a vigorous mixing action throughout the flow field, which is caused by *eddies* of varying size within the flow. Physically, the two flow states are linked, in the sense that any flow can be stable and laminar if the ratio of inertial to viscous forces is sufficiently small. Turbulence results when this ratio exceeds a *critical value*, above which the flow becomes unstable to perturbations and breaks down into fluctuations. Because of these fluctuations, the velocity field *u* in a turbulent flow field is not constant in time. Although turbulent flows therefore do not meet the above definition for steady, the velocity at any point presents a statistically distinct time-average value that is constant. Turbulent flows are therefore described as *stationary*, rather than truly unsteady.

Fully turbulent flow fields have four defining characteristics [11, 12]:

1. They fluctuate *randomly*. The velocity at any point is found to fluctuate continuously and unpredictably.
2. They are *three-dimensional*. Turbulent eddies appear with all possible random orientations, so that all components of the velocity field, u, v, and w, are nonzero and each component itself fluctuates randomly.
3. They are *dissipative*. Because of the large number of eddies and accompanying high shear regions in the flow field, turbulent flows dissipate energy at a much greater rate than do laminar flows.
4. They are *dispersive*. Transport of dissolved solutes, suspended particles, and fluid momentum throughout the flow field by turbulent eddies is much more efficient than molecular transfer in laminar flow.

The *turbulence intensity I* of any flow field is defined as the ratio of velocity fluctuations u' to time-average velocity $\bar{u}$, $I = u'/\bar{u}$.

Turbulence can be generated either by friction forces at boundaries, as in pipe flows, or by the flow of layers of fluids with differing velocities past or over one another. These are designated *wall turbulence* and *free stream turbulence*, respectively. Although both meet all the above criteria, they show different properties otherwise. In all flow fields, boundary layer properties are critical to the stability of outer flow fields. Inflection points in the velocity and/or pressure profiles tend to destabilize the flow, as does convective or temporal deceleration.

Steady flow in straight, rigid pipes is characterized by only one dimensionless parameter, the Reynolds number. It was shown by Osborne Reynolds that for $Re < 2000$, incidental disturbances in the flow field are damped out and the flow remains stable and laminar. For $Re > 2000$, brief bursts of fluctuations appear in the velocity separated by periods of laminar flow. As Re increases, the duration and intensity of these bursts increases until they merge together into full turbulence. Laminar flow may be achieved with Re as large as 20 000 or greater in extremely smooth pipes, but it is unstable to flow disturbances and rapidly becomes turbulent if perturbed.

It is important to note that many flow fields show stable intermediate states at Reynolds numbers greater than their critical value for laminar stability. These states should not be confused with true turbulence, which is randomly unstable. For example, flow between two concentric cylinders, of which the inner is in motion and outer stationery, exhibits stable stratifications. These are now known as Taylor vortices after G.I. Taylor, who explained their existence through linear stability analysis [13]. At a critical Reynolds number of 94.5, a series of flow vortices appears with axes oriented circumferentially around the cylinder, rotating in alternating opposite directions. These ring-like vortices remain laminar up to $Re = 3900$, at which point turbulent flow develops.

Since the Navier–Stokes equations govern all the behavior of any Newtonian fluid flow, it follows that turbulent flow patterns should be predictable through analysis based on those equations. However, although turbulent flows have been investigated for more than a century and the equations of motion analyzed in great detail, no general approach to the solution of problems in turbulent flow has been found. Statistical studies invariably lead to a situation in which there are more unknown variables than equations, which is called the *closure problem* of turbulence. Efforts to circumvent this difficulty have included phenomenologic concepts such as *eddy viscosity* and *mixing length*, as well as analytical methods including dimensional analysis and asymptotic invariance studies. Even formal statistical analyses, however, have been unable to produce a generally valid solution. The mathematical complexity of these methods is beyond the scope of this work.

Although a general method for predicting turbulent flow fields has not been achieved, a number of methods exist for investigating the conditions under which any specific flow field can be expected to become unstable. Flow stability can be analyzed by both linear and nonlinear techniques based on perturbation and averaging analyses, optimization by variational calculus, and energy and collocation methods [13–23]. Of these, linear theory is the simplest approach, and has been successfully used to predict the onset of numerous flow regime changes. Linear theory can predict whether a disturbance will dampen out or grow, but it cannot predict system behavior and dynamics beyond infinitesimal perturbations. If disturbance growth is predicted, linear theory cannot determine how the system will behave once far from its original stationary state. In fact, linear theory cannot determine whether a given disturbance will result in the system moving to a different stationary state or to some unbounded behavior leading to a possible system failure. It should also be emphasized that a system determined to be stable to infinitesimal perturbations may actually be unstable to finite sized disturbances. The inability to account for the influence of disturbance magnitude and direction on system stability is a major deficiency of linear theory.

Nonlinear stability analysis can address all of these limitations with some degree of reliability. Nonlinear theory also establishes methods to determine system response trajectories to various stimuli, permitting development of control strategies to either return the system to its original state or direct it to a new state as quickly as possible.

An integral component of any stability analysis is the development of realistic models that provide the basis for perturbation analysis used to predict the system dynamics and determine the location of the stationary states. The models must also include the trajectories to and from each stationary state, since those trajectories determine hysteresis and stability limits as well as potential control strategies. The most successful stability solution methods employ *modal analysis* [14–23], incorporating nonlinear terms in a perturbation analysis developed from the inherent eigenvalue problem associated with the linearized system to determine the finite amplitude stability of systems [23]. The approach taken is to first determine the mechanics of the steady-state flow, then examine the nonlinear dynamics of the most unstable spatial mode obtained from linear theory. The initial linear analysis requires that the system's conservation equations be linearized about the steady state, and a separation-of-variables solution sought in the form of an exponential time function multiplying an associated spatial function. Substitution of this assumed form into the linearized partial differential equations leads to an eigenproblem for the flow system. The eigenvalues obtained from solution of this problem are the reciprocal of the modal time constants, and the associated eigenfunctions are the orthogonal spatial functions of the separation solution. The real part of the dominate (first) eigenvalue will pass from negative to positive at a critical shear rate. Below this value, all infinitesimal disturbances decay in time; above this value, the first mode in the eigenfunction expansion grows and the system is unstable to arbitrarily small disturbances.

Viscous and non-Newtonian fluids are readily simulated by this approach. Such systems can exhibit persistent oscillations. A nonmonotonic relationship between steady shear stress and strain rate has been successfully utilized to show that flow system changes start in a thin layer near the wall, giving the appearance of a slip layer [14–16]. Extending these ideas to multiple layers of the flow that differ in characteristic relaxation times and rheological properties permits sub-region dynamics to be integrated into the system overall global response. At shear rates higher than some critical value, these systems are unstable to all disturbances. However, it is possible that the response to these perturbations can approach limit cycle behavior with stable finite-amplitude oscillations.

For nonlinear analysis, the full field equations are expressed in terms of perturbation variables, as in the linear analysis, to obtain the base orthogonal function set. A solution is sought in the form of a series expansion in these eigenfunctions multiplied by associated amplitude functions. When substituted into the original nonlinear field equations, a set of ordinary differential equations can be generated [22]. The essential assumptions here are that the eigenfunctions form a complete set and the nonlinearities are "lumped" into the amplitude functions that replace the simple exponential form assumed in the linear approach. These amplitude functions are complex conjugate functions of time. In the limit of infinitesimal perturbations they reduce within a normalizing constant to the complex exponential time dependence of the linear system. If the real parts of the eigenvalues from the linear analysis are widely spaced, indicating that the dynamics are dominated by the first, or most unstable mode, we may truncate after the lead term in the series. This leads to a form containing residuals, which reflects the fact that the deterministic equations are not satisfied at each point by this approximate solution. A coupled set of autonomous ordinary differential equations is obtained through the use of Galerkin's method [22, 23] in which the residuals are made orthogonal to each approximating function. These equations yield the requisite amplitude functions; each a complex conjugate pair. The trajectories obtained track the system dynamics with a relatively high degree of accuracy and thus determine the stability information sought. This includes how rapidly the system responds, not just simply if these finite disturbances decay or grow. For example, some possible scenarios are: (i) if stable; is the path direct or a somewhat oscillatory decay, (ii) if unstable; does it approach a limit cycle, even if the disturbance is larger than the amplitude of that steady oscillation (in other words, the disturbance could either grow or decay to this limit cycle but not approach the original state), or is the growth bounded taking the system to a new stationary point or unbounded leading to system failure. This technique has been successfully applied to reaction systems [20] as well as these hydrodynamic systems [17, 22, 23]. Thus, the behavior after a disturbance could account for a shift in the metabolic state of a tissue system, disassociation of cells from clusters, tissue disruption, detachment of cells and/or other components from the scaffolding in the extra-cellular matrix, and cell lysis.

2.3
Flow in Tubes

Flow in a tube is the most common fluid dynamic phenomenon in the physiology of living organisms, and is the basis for transport of nutrient molecules, respiratory gases, hormones, and a variety of other important solutes throughout the bodies of all complex living plants and animals. Only single-celled organisms, and multi-celled organisms with small numbers of cells, can survive without a mechanism for transporting such molecules, although even these organisms exchange materials with their external environment through fluid-filled spaces. Higher organisms, needing to transport molecules and materials over larger distances, rely on organized systems of directed flows through networks of tubes to carry fluids and solutes. Thus, fluids provide the critical medium in which convection and diffusion can accomplish both exchange of necessary chemicals and biochemicals with the external environment, as well as transport of those materials throughout the organism. Tubes provide the means for containing and moving those fluids.

In human physiology, the circulatory system, which consists of the heart, the blood vessels of the vascular tree and the fluid, blood, and which serves to transport blood throughout the body tissues, is perhaps the most obvious example of an organ system dedicated to creating and sustaining flow in a network of tubes. However, flow in tubes is also a central characteristic of the respiratory, digestive, and urinary systems. Furthermore, the immune system utilizes systemic circulatory mechanisms to facilitate transport of antibodies, white blood cells and lymph throughout the body, while the endocrine system is critically dependent on blood flow for delivery of its secreted hormones. In addition, reproductive functions are also based on fluid flow in tubes. Thus, seven of the ten major organ systems depend on flow in tubes to fulfill their functions. It has been estimated that there are more fluid-filled tubes in a single human body than in any other body or place on this planet [24].

In addition, an enormous variety of commercial and industrial processes involve delivery of fluid through tubes and pipes of circular cross-section. Included in this category are feeds and product removals from reactors and bioreactors of all types. Efficient use of all such devices requires selection of tube sizes for appropriate rates of delivery. Consequently, there is a strong incentive for tissue engineers and biotechnology investigators to understand flow in tubes in detail.

2.3.1
Steady Poiseuille Flow

The most basic state of motion for fluid in a pipe is one in which the motion occurs at a constant rate, independent of time. The pressure-flow relation for laminar, steady flow in round tubes is called *Poiseuille's Law*, after J.L.M. Poiseuille, the French physiologist who first derived the relation in 1840 [25]. Accordingly, steady flow through a pipe or channel that is driven by a pressure difference between the pipe ends of just sufficient magnitude to overcome the tendency of the fluid to dissipate energy through the action of viscosity is called *Poiseuille flow*.

Strictly speaking, Poiseuille's Law applies only to steady, laminar flow through pipes that are straight, rigid, and infinitely long with uniform diameter, so that effects at the pipe ends may be neglected without loss of accuracy. However, although neither physiologic vessels nor industrial tubes fulfill all those conditions exactly, Poiseuille relationships have proven to be of such widespread usefulness that they are often applied even when the underlying assumptions are not met. As such, Poiseuille flow can be taken as the starting point for analysis of cardiovascular, respiratory, and other physiologic flows of interest, which is the approach taken here. Modifications to the basic Poiseuille flow description will then be added in later sections, below, to indicate how difficulties associated with departures from ideal conditions that occur in real vessels can be described.

Consider a straight, rigid round pipe, as shown in Fig. 1. Let x denote the pipe axis, with y and z its transverse coordinates, and let a be the pipe radius. Flow in the pipe will be governed by the Navier–Stokes equations (Eq. 38a–c). We seek solutions to these equations in which the flow is parallel to the pipe axis, with no nonzero transverse components of velocity. Such flow fields take the form $u = \left(u(y, z), 0, 0\right)$, where u is the axial component of the velocity field, which may depend on transverse but not axial position. It is then clear that the continuity equation (Eq. 9), is satisfied by definition, since $\partial u/\partial x$ is au-

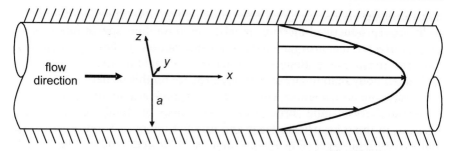

Fig. 1 Parabolic velocity profile characteristic of Poiseuille in a round pipe of radius a. x,y,z – Cartesian coordinate system with origin on the pipe centerline. x – axial coordinate, y,z – transverse coordinates

tomatically zero, as are $\partial v/\partial y$ and $\partial w/\partial z$. Moreover, since the flow is steady, Du/Dt is also automatically zero for all components of u, so that neglecting gravitational effects, the Navier–Stokes equations become

$$0 = -\frac{1}{\rho}\frac{\partial P}{\partial x} + \upsilon \left(\frac{\partial^2 u}{\partial y^2} + \frac{\partial^2 u}{\partial z^2}\right) \tag{39a}$$

$$0 = -\frac{1}{\rho}\frac{\partial P}{\partial y} \tag{39b}$$

$$0 = -\frac{1}{\rho}\frac{\partial P}{\partial z}. \tag{39c}$$

From (b) and (c), $P = P(x$ only). But from (a), $\partial P/\partial x =$ function of y and z. Hence, $\partial P/\partial x$ can be at most a constant, which we designate $-\kappa$.

It is important to be clear that the *only* way for pressure to vary along the tube, $P = P(x)$, but for the velocity field to not vary with x is for the tube to be *rigid*. In a nonrigid tube, any local variation in pressure will lead to a local change in the cross-sectional area of the tube, and thus to a change in velocity.

Under these conditions, the governing equation becomes

$$\nabla^2 u = -\frac{\kappa}{\mu}. \tag{40}$$

For a round pipe, it is easiest to solve Eq. 40 in cylindrical coordinates, for which $\nabla^2 u = \frac{\partial^2 u}{\partial r^2} + \frac{1}{r}\frac{\partial u}{\partial r}$. The flow is then governed by the ODE

$$\frac{d^2 u}{dr^2} + \frac{1}{r}\frac{du}{dr} = -\frac{\kappa}{\mu} \tag{41}$$

with the conditions that the flow field must be symmetric about the pipe center line, i.e. $du/dr|_{r=0} = 0$, and the no-slip boundary condition applies at the wall, $u = 0$ at $r = a$. Under these conditions, the solution of Eq. 41 is

$$u(r) = \frac{\kappa}{4\mu}\left(a^2 - r^2\right). \tag{42}$$

The velocity profile described in Eq. 42 has the familiar parabolic form known as Poiseuille flow (Fig. 1). The velocity at the wall ($r = a$) is clearly zero, as required by the no-slip condition, while as expected for physical reasons the maximum velocity occurs on the axis of the tube ($r = 0$) where $u_{max} = \kappa a^2/4\mu$. At any position between the wall and the tube axis, the velocity varies smoothly with r, with no step change at any point.

The above discussion represents a mathematical derivation of the velocity field in a round tube. Although correct, such a derivation does not give physical insight into the origin of the forces that give rise to the motion. To appreciate how those forces interact to produce the observed velocity profile, we can re-derive an expression for the velocity field using a physically based approach. Consider, then, an infinitely long, round pipe in which a steady flow has been established, and consider an imaginary cylindrical free body of length L and radius r centered on the pipe center line and moving with the flow, as shown in Fig. 2. Let P_1 denote the pressure at the upstream end of this free body, and P_2 denote the pressure at its downstream end. The forces acting on the free body are:

pushing it forward:
the pressure on the upstream end, $P_1 \cdot \pi r^2$
pushing it backward:
the pressure on the downstream end, $P_2 \cdot \pi r^2$, and
shear stress on the body's surface due to the viscosity of the fluid, $\mu \frac{\partial u}{\partial r} 2\pi rL$.

Since the flow is steady, the cylindrical volume must be in equilibrium and the *net* force acting on it must be zero. That is, the total force pushing the body forward must just balance the total force pushing the body backward. Consequently, for equilibrium,

$$(P_1 - P_2)\,\pi r^2 = \mu \frac{\partial u}{\partial r} 2\pi rL \tag{43}$$

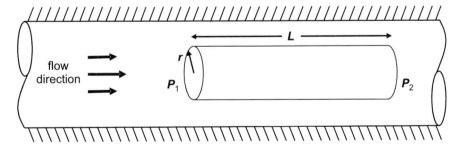

Fig. 2 Pipe flow, with cylindrical free body of length L and radius r centered on the pipe axis. P_1 – pressure on the upstream face of the free body, P_2 – pressure on its downstream face

or

$$\frac{\partial u}{\partial r} = \frac{1}{2\mu} \frac{(P_1 - P_2)}{L} r .$$ (44)

However, the quantity $(P_1 - P_2)/L$, which is called the *pressure gradient* of the flow field, is just $\partial P/\partial x = -\kappa$. Hence by integrating and applying the no-slip condition at the wall, we again have

$$u = \frac{\kappa}{4\mu} \left(a^2 - r^2 \right) .$$ (42)

From this physical analysis, it can be seen that the parabolic velocity profile results from a *balance* of the forces on the fluid in the pipe. The pressure gradient along the pipe, or alternatively the pressure difference between the pipe ends, *not* the pressure itself, accelerates fluid in the forward direction through the pipe. At the same time, viscous shear stress retards the fluid motion. A parabolic profile is created by the balance of these effects.

Although the velocity profile is important and informative, it is difficult to measure experimentally without highly expensive, noninvasive optical techniques, since any attempt to position a probe in the flow field and thereby measure the velocity inevitably disturbs the flow, creating a new and different flow field. In practice, one is therefore apt to be more concerned with measurement of the *discharge rate*, or total rate of flow in the pipe, Q, which can far more easily be accessed. The volume flow rate is given by area-integration of the velocity:

$$Q = \int_A u \cdot dA$$ (45)

$$= \int_0^a \int_0^{2\pi} u r \, d\theta \, dr$$ (46)

$$= \frac{\pi \kappa a^4}{8\mu} ,$$ (47)

which is Poiseuille's Law.

For convenience, the relation between pressure and flow rate is often re-expressed in an Ohm's Law form, *driving force = flow × resistance*, or

$$\kappa = Q \cdot \frac{8\mu}{\pi a^4} ,$$ (48)

from which the resistance to flow, $8\mu/\pi a^4$, is seen to be inversely proportional to the 4th power of the tube radius.

This result is the basis for the Poiseuille viscometer (Fig. 3), once a widely used device for measurement of fluid viscosity based on delivering flow through a long, narrow bore capillary tube of known radius and length. From

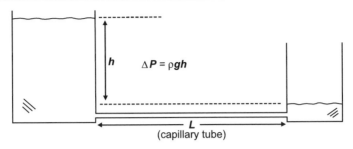

Fig. 3 Schematic diagram of a Poiseuille viscometer

measurement of the resulting flow rate, μ can be calculated without difficulty. For the Poiseuille viscometer to provide accurate results, it is important
to minimize the capillary tube diameter, in part to keep flow rates small so
that velocities will remain laminar. In addition, Poiseuille's Law fails at the
tube ends, where the velocity profile may depart significantly from its ideal
parabolic shape. To minimize the influence of end effects, the ratio a/L must
be maintained as small as possible. Typical diameters range from a few hundred micrometers to 1 mm.

A further point about Poiseuille flow concerns the area-average velocity, U.
Clearly,

$$U = \frac{Q}{\text{cross-sectional area}} = \frac{\pi \kappa a^4 / 8\mu}{\pi a^2} = \frac{\kappa a^2}{8\mu} . \tag{49}$$

But, as was pointed out, the maximum velocity in the tube is $u_{\max} = \kappa a^2 / 4\mu$.
Hence

$$U = \frac{u_{\max}}{2} = \frac{1}{2} u|_{r=0} = \frac{1}{2} u_{CL} . \tag{50}$$

Finally, the shear stress exerted by the flow on the wall can be a critical parameter, particularly when it is desired to control the wall's exposure to shear.
From Eq. 30, it can be shown that wall shear stress, τ_w, is given by

$$\tau_w = -\mu \frac{du}{dr}|_{r=a} . \tag{51}$$

Using Eq. 42, this becomes

$$\tau_w = \frac{\kappa a}{2} . \tag{52}$$

Alternatively, in terms of the flow rate,

$$\tau_w = \frac{4\mu Q}{\pi a^3} . \tag{53}$$

To summarize this section briefly, Poiseuille's Law, Eq. 47 provides a relation
between the pressure drop and net laminar flow in any tube, while Eq. 53 pro-

vides a relation between the flow rate and wall shear stress. Thus, physical forces on the wall may be calculated from knowledge of the flow fields.

2.3.2
Entrance Flow

It can be shown that a Poiseuille velocity profile is the velocity distribution that minimizes energy dissipation in steady laminar flow through a rigid tube. That is, in principle there are an infinite number of different profiles by which a given total flow rate through the tube can be achieved. However, the one that actually forms, the Poiseuille profile, is the one that minimizes the net dissipation of energy and therefore requires the least mechanical power to maintain. Consequently, it is not surprising that if the flow in a tube encounters a perturbation that alters its profile, such as a branch vessel or a region of stenosis, immediately downstream of the perturbation the velocity profile will be disturbed, perhaps highly so, away from a parabolic form. However, if the Reynolds number is low enough for the flow to remain stable as it convects downstream from the site of the original distribution, a parabolic form is gradually recovered. Consequently, at a sufficient distance downstream, a fully developed parabolic velocity profile again emerges.

For example, if fluid enters a tube with a nearly uniform velocity distribution, the velocity profile will consist of a straight line across most of the tube with only very narrow regions near the wall in which the velocity drops off to zero in conformance with the no-slip boundary condition. As the flow moves downstream, the boundary layer thickens, due to the loss of energy by viscous dissipation. Simultaneously, the velocity near the tube center must increase in compensation, to satisfy conservation of mass. Eventually, the effect of the wall is felt fully across the cross-section of the tube and a parabolic velocity distribution is established. At this point the flow is considered to be "fully developed", and no further evolution is possible. This process is depicted in Fig. 4. By extension, the upstream region, in which the flow field is not yet fully developed, is denoted the "entrance" region.

Both blood vessels and bronchial tubes of the lung possess enormous numbers of branchings, each of which produces its own flow disturbance. As a result, many physiologic flows may not be fully developed over a significant fraction of their length. It therefore becomes important to ask, what length of tube is required for a perturbed velocity profile to recover its parabolic form, i.e. how long is the entrance length in a given tube? In fact, the entrance length should be expected to depend in part on how rapidly the fluid is flowing, since a fast flow can propagate further before evolving to full development than can a slow flow. It should also be anticipated that the entrance length will depend on the fluid in question, since a very viscous fluid, such as molasses, will dissipate energy and thus evolve more rapidly than a less viscous fluid such as air or water.

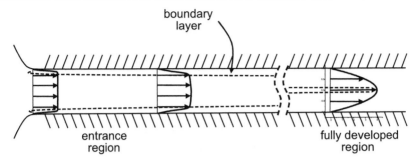

Fig. 4 Progression of the flow field from uniform to fully developed at the entrance of a round pipe

The question of entrance length can be formally posed as, if x is the coordinate along the tube axis, for what value of x does $u|_{r=0} = 2U$? Actually, no analytic solution to this question has been found, in part because the flow progresses only asymptotically towards a fully developed state, and strictly does not achieve full development in a finite length of the tube. However, for practical purposes, differences between the actual velocity field and the idealized state of a fully developed Poiseuille flow become negligibly small at finite distances downstream of a disturbance. Dimensional reasoning can be used to develop an order-of-magnitude estimate of the tube length over which this occurs.

For this purpose, the boundary layer may be defined to be the region of the flow for which $u(r)$ is less than some agreed upon cutoff, say 0.95, the magnitude of u on the tube axis (Fig. 4). Because of the no-slip requirement, in a tube flow there will at least always be a region near the wall in which the velocity becomes very small and a boundary layer exists. Near any disturbance, or at the entrance to a pipe as in Fig. 4, the boundary layer thickness, δ, may be very small. As the flow field develops, however, δ increases until eventually, when the flow is fully developed, the boundary layer occupies all but a thin region at the tube center.

Consider a small fluid element in the boundary layer, of width dy and area dA (Fig. 5). The shear stress on this element is

on the top face, pushing it forward:

$$\mu \left(\frac{\partial u}{\partial y}|_y + \frac{\partial}{\partial y} \left(\frac{\partial u}{\partial y} \right) dy|_{y+dy} \right) \tag{54}$$

on the bottom face, pulling it back:

$$\mu \frac{\partial u}{\partial y}|_y . \tag{55}$$

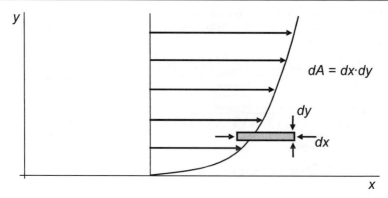

Fig. 5 Free body taken in the boundary layer of the entrance region in a round pipe

Hence the net viscous force acting on the element is

$$\mu \frac{\partial}{\partial y} \left(\frac{\partial u}{\partial y} \right) dy \, dA \, . \tag{56}$$

The scale over which u changes in the y-direction is δ, while the scale of u is U. Consequently, the net viscous force has magnitude

$$\mu \frac{U}{\delta^2} dy \, dA \, . \tag{57}$$

At the same time, the inertia of this fluid element is given by its *mass × acceleration*. The mass is just density × volume = $\rho \, dy \, dA$, while the order of magnitude of acceleration is given by

$$\frac{U}{\text{time for the fluid element to travel a distance } x} \tag{58}$$

$$= \frac{U}{x/U} = \frac{U^2}{x} \, . \tag{59}$$

Thus, the inertia has magnitude

$$\rho \, dy \, dA \frac{U^2}{x} \, . \tag{60}$$

Since these must balance, we have

$$\mu \frac{U}{\delta^2} dy \, dA = \rho \, dy \, dA \frac{U^2}{x} \tag{61}$$

or

$$\delta = C_1 \left(\frac{\mu x}{\rho U} \right)^{1/2} , \tag{62}$$

where C_1 is a constant yet to be determined.

When the flow is fully developed, $\delta \to a$. However, it is more common in practice to take $\delta = d$, the tube diameter, and absorb the resulting factor of 2 into the constant. Then

$$x = C_2 \frac{\rho d^2 U}{\mu} \qquad (63)$$

or

$$\frac{x}{d} = C_2 \frac{\rho d U}{\mu} = C_2 Re . \qquad (64)$$

Thus, the length of tube over which the flow develops is

$$\text{const} \times Re \times d . \qquad (65)$$

The constant must be determined by experiment, and is found to be in the range 0.03–0.04.

From this derivation, it can be seen that the entrance length, in units of tube diameters, is proportional to the Reynolds number. The mean Reynolds number for flow in large blood vessels such as the aorta, as well as in the trachea, is of the order of 500–1000. Thus, the entrance length in these vessels can be expected to be as much as 20–30 diameters. Since there are few if any segments of these vessels even close to that length without a branch or curve that perturbs their flow, flow in them can be expected to almost never be fully developed. In contrast, flow in the smallest bronchioles, arterioles, and capillaries may take place with $Re < 1$. As a result, their entrance length is $\ll 1$ diameter, and flow in them will virtually always be nearly or fully developed.

2.3.3
Mechanical Energy Equation

Flow fields in tubes with more complex shapes than simple straight pipes, such as those possessing bends, curves, orifices, and other intricacies, are often analyzed with an *energy balance* approach, since they are not well described by Poiseuille's Law. Understanding such flow fields is significant for in vitro studies and perfusion devices, to establish dynamic similitude parameters, as well as for in vivo studies of curved and/or branched vessel flows. For any system of total energy E, the *first law of thermodynamics* states that any change in the energy of the system ΔE must appear as either heat transferred to the system in unit time Q or as work done by the system W, so that

$$\Delta E = Q - W . \qquad (66)$$

Here a sign convention is taken such that Q, when positive, represents heat transferred *to* the system and W, when positive, is the work done *by* the sys-

tem on its surroundings. The first law can also be expressed in terms of the
rate of change of E

$$\frac{\mathrm{d}E}{\mathrm{d}t} = \dot{Q} - \dot{W} \,. \tag{67}$$

The total energy of a fluid system includes the fluid kinetic and potential en-
ergies, as well as the internal energy associated with motion of the individual
molecules. Accordingly, the rate of change of the total energy of an arbitrary
volume of fluid V, bounded by the surface S (Fig. 6), flowing through a pipe
is given by

$$\frac{\mathrm{d}E}{\mathrm{d}t} = \dot{Q} - \dot{W}$$

$$= \frac{\mathrm{d}}{\mathrm{d}t} \int_V \left(\frac{U^2}{2} + gz + e \right) \rho \, \mathrm{d}V + \int_S \left(\frac{U^2}{2} + gz + e \right) \rho u \cdot \mathrm{d}S \,, \tag{68}$$

where $U^2/2$ is the kinetic energy per unit mass of the fluid within V,
gz is its potential energy per unit mass, with z the vertical coordinate
and g gravitational acceleration, e is its internal energy per unit mass
and the density ρ is assumed to be constant. The first integral in Eq. 68
therefore represents the total energy contained within V at any time t,
while the second integral represents the net rate of transport of energy
across S.

Work done by a fluid system consists of two types, flow work W_f, i.e. work
done by the pressure field in moving fluid through space, and shaft work W_s,
work done on the fluid by pumps, turbines or other external devices through
which power is often transmitted by means of a shaft. For example, pumping
requirements are often needed by the tissue engineer to design extracorporeal devices, as well as to analyze the work done by the heart and lungs in

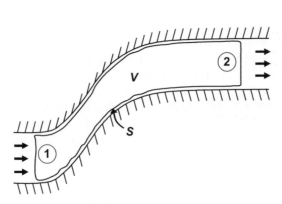

Fig. 6 Control volume V for steady flow through a rigid wall pipe of arbitrary shape.
S – boundary of V

vivo. It can be shown that the total rate of flow work done as the fluid contained in V moves in space is

$$\dot{W}_f = \int_S \left(\frac{p}{\rho}\right) \rho \boldsymbol{u} \cdot d\boldsymbol{S}. \tag{69}$$

Substitution of this expression into Eq. 68 gives the general form of the energy equation for a fluid system

$$\dot{Q} - \dot{W}_s = \frac{d}{dt} \int_V \left(\frac{U^2}{2} + gz + e\right) \rho \, dV$$

$$+ \int_S \left(\frac{p}{\rho} + \frac{U^2}{2} + gz + e\right) \rho \boldsymbol{u} \cdot d\boldsymbol{S}. \tag{70}$$

The general equation can be simplified greatly when the flow is steady, since the total energy contained within any prescribed volume is then constant, and $d/dt = 0$. To develop an energy balance equation that applies specifically to pipe flows, Eq. 70 is applied between sections 1 and 2 of Fig. 6 assuming the flow to be steady. The result is

$$\dot{Q} - \dot{W}_s + \int_{A1} \left(\frac{p_1}{\rho} + gz_1 + e_1\right) \rho U_1 \, dA_1 + \int_{A1} \frac{\rho U_1^3}{2} \, dA_1$$

$$= \int_{A2} \left(\frac{p_2}{\rho} + gz_2 + e_2\right) \rho U_2 \, dA_2 + \int_{A2} \frac{\rho U_2^3}{2} \, dA_2. \tag{71}$$

It is common to assume that the flow is uniform and the internal energy constant at sections 1 and 2. Under those assumptions, the quantity $(p/\rho + gz + e)$ can be considered to be constant and removed from the integrals. This leaves integrals of the form $\int_A \rho U \, dA$, which is simply $\dot{m}$, the rate of mass transport by the system. Using this result, and dividing by $\dot{m}$ then gives

$$\frac{1}{\dot{m}} \left(\dot{Q} - \dot{W}_s\right) + \frac{p_1}{\rho} + gz_1 + e_1 + \beta_1 \frac{U_1^2}{2} = \frac{p_2}{\rho} + gz_2 + e_2 + \beta_2 \frac{U_2^2}{2}, \tag{72}$$

where the coefficients β_1 and β_2 are kinetic energy correction factors introduced to simplify the integration of $\rho U^3/2$. Calculations show that $\beta = 1$ when the velocity is uniform across the section, and $\beta = 2$ for laminar Poiseuille flow with a parabolic velocity profile. For other velocity distributions, $\beta > 1$. For most cases of turbulent flow, $\beta \approx 1.05$, and it is common to approximate this with $\beta = 1$ for turbulent flow calculations.

Although in principle, the shaft work term in Eq. 72 can include the effects of both turbines and pumps in the flow system, in tissue engineering applications only the latter are normally significant. Since work performed by a pump is work done *on* the fluid, we have $\dot{W}_s = -\dot{W}_p$. Substituting this expression for shaft work into Eq. 72 and dividing by g gives

$$\frac{\dot{W}_p}{\dot{m}g} + \frac{p_1}{\gamma} + z_1 + \beta_1\frac{U_1^2}{2g} = \frac{p_2}{\gamma} + z_2 + \beta_2\frac{U_2^2}{2g} + \frac{e_2 - e_1}{g} - \frac{\dot{Q}}{\dot{m}g},\qquad(73)$$

where $\gamma = \rho g$. This is a particularly convenient means of expressing the energy balance, since in this form all of the terms have dimensions of *length*. Accordingly, the term involving pump work is often designated as h_p, head supplied by a pump. Equation 73 can then be written as

$$\frac{p_1}{\gamma} + \beta_1\frac{U_1^2}{2g} + z_1 + h_p = \frac{p_2}{\gamma} + \beta_2\frac{U_2^2}{2g} + z_2 + \left[\frac{(e_2 - e_1)}{g} - \frac{\dot{Q}}{\dot{m}g}\right].\qquad(74)$$

In this form, Eq. 74 presents a separation of the energy equation into a mechanical part and a thermal part, the latter of which is represented by the last term, in brackets. The thermal term itself represents the conversion of mechanical energy to thermal energy through the action of viscosity, as well as gain or loss of thermal energy through the system boundaries. The bracketed term in Eq. 74 therefore represents a loss of mechanical energy from the system. This lost energy is usually lumped together as a single term called *head loss*, and symbolized by h_L. Thus, Eq. 74 becomes

$$\frac{p_1}{\gamma} + \beta_1\frac{U_1^2}{2g} + z_1 + h_p = \frac{p_2}{\gamma} + \beta_2\frac{U_2^2}{2g} + z_2 + h_L.\qquad(75)$$

Equation 75 is the form of energy balance expression most commonly utilized in practice. A variety of methods have been developed to estimate h_L in different flow situations; these are discussed in most fluid mechanics texts. For flow in a rigid pipe of length L and diameter d, h_L is well represented by

$$h_L = f\frac{L}{d}\frac{U^2}{2g},\qquad(76)$$

where f is called the *friction factor* of the pipe, and depends on both the pipe roughness and the flow Reynolds number. It can be shown that for laminar flow, $f = 64/Re$. Then

$$h_L = \frac{32\mu LU}{\gamma d^2}.\qquad(77)$$

It is worth repeating that Eq. 75 is only correct when the fluid density is constant, as is normally the case in tissue engineering applications and even for air flow in the lung. Compressibility effects require separate energy considerations.

3
Models and Computational Techniques

3.1
Approximations to the Navier–Stokes Equations

The Navier–Stokes equations (Eqs. 34, 35a and 36), together with the continuity condition (Eq. 9), provide a complete set of governing equations for the motion of an incompressible Newtonian fluid. If appropriate boundary and initial conditions can be specified for the motion of such a fluid in a given flow system, in principle a full set of governing equations and conditions for the system will be known. It may then be expected that the fluid motion can be deduced simply by solution of the resulting boundary value problem. Unfortunately, however, the mathematical difficulties resulting from the nonlinear character of the acceleration terms Du/Dt in Eq. 34 are so great that only a very limited number of exact solutions have ever been found. The simplest of these pertain to cases in which the velocity has the same direction at every point in the flow field, as in the steady pipe flows discussed above (Sect. 2.3) and the pulsatile flows described below (Sect. 6).

Accordingly, there is a strong incentive to seek conditions under which one or more of the terms in Eq. 34 are negligible or nearly so, and therefore an approximate and much simpler governing equation can be generated by neglecting them altogether. For example, we have seen that the Reynolds number represents the ratio of inertial to viscous forces in the flow field. Accordingly, in flows for which $Re \gg 1$, it can be shown that the viscous term $\upsilon\nabla^2 u$ is very much smaller than the acceleration Du/Dt. This leads to a new governing equation of the form

$$\frac{Du}{Dt} = -\frac{1}{\rho}\nabla P + \frac{1}{Re}\nabla^2 u . \tag{78}$$

As a first approximation, $1/Re$ may be set to zero everywhere in the flow, reducing Eq. 78 to the inviscid *Euler equation*. However, as we have seen, omission of the viscous terms cannot be valid near a solid boundary, since inviscid flow cannot satisfy the no slip boundary condition. Hence, viscous forces must remain important in the boundary layer near the solid surface.

Conversely, when $Re \ll 1$, the viscous term $\upsilon\nabla^2 u$ is much larger than the acceleration Du/Dt, which gives the form

$$Re \frac{Du}{Dt} = -\frac{1}{\rho}\nabla P + \nabla^2 u . \tag{79}$$

In the limit $Re \to 0$, Eq. 79 becomes the linear equation

$$\nabla P = \mu\nabla^2 u . \tag{80}$$

Flows at $Re \ll 1$ are called *creeping motions*. They can be due to small velocity, large viscosity, or often small size of the body. Examples of such flows are blood flow in capillaries or the motion of air in the smallest bronchioles and alveoli.

In summary, these approximations show that viscosity is important in three situations:

1. When the overall Reynolds number is *low*, since then viscous effects act over the full flow field;
2. When the overall Reynolds number is *high*, viscosity is important in thin boundary layers, and;
3. When the flow is *enclosed*, as in a pipe flow, since then the available diffusion time is very large, and viscous effects can become important in the whole flow after some initial region or time.

An alternative approach to seeking simplifications to the Navier–Stokes equations is to accept the full set of equations, but approximate each term in the equation with a simpler form that permits solutions to be developed. Although the resulting equations are only *approximately* correct, the advent of modern digital computers has allowed them to be written with great fineness, so that highly accurate solutions are achieved. These techniques are called *computational fluid dynamics* (CFD).

3.2
Computational Fluid Dynamics (CFD)

In the last few decades CFD has become a very powerful and versatile tool for the analysis of complex problems of interest in the engineering biosciences. Many of these problems combine thermodynamics, reaction kinetics and, transport phenomena to perform fluid flow analyses, potentially in highly complex geometries. Consequently, they may be far too difficult to study accurately and in detail without computationally complicated models.

A representative example of widespread interest to tissue engineers is the analysis of perfusion devises. Before employing newly developed perfusion devices such as chemostats and other bioreactor systems, and/or transport devices, characterization studies need to be conducted to substantiate applicability. Cellular metabolic rates in encapsulated and free states, as well as pertinent transport phenomena, can be evaluated in selected membrane bioreactor configurations. These data, coupled with computational fluid dynamics modeling, provide the basis for redesign/reconfigurations as apropos. CFD is a very powerful and versatile tool for an analysis of this type.

Furthermore, computational fluid dynamics methods are finding many new and diverse applications in tissue engineering and biomimetics. For example, CFD techniques can be used to predict (1) velocity and stress distribution maps in the complex reactor performance studies mentioned above as well as in vascular and bronchial models; (2) strength of adhesion and dynamics of detachment for mammalian cells; (3) transport properties for nonhomogeneous materials and nonideal interfaces; (4) multi-component diffusion rates using the Maxwell–Stefan transport model, as opposed to the limited traditional Fickian approach, incorporating interactive molecular immobilizing sites; and (5) materials processing capabilities useful in encapsulation technology and designing functional surfaces.

3.2.1
Theory and Software Packages

The numerical aspects of CFD are discussed in detail in a number of excellent references [7, 8]. A wide variety of computational techniques for solution of the Navier–Stokes equations have been developed, including *finite difference*, *finite element*, *finite volume*, and *spectral* methods, in both *velocity-pressure* and *vorticity-stream function* formulations, as well as many other approaches. All such techniques share a general framework in which the Navier–Stokes equations (Eq. 34), or an equivalent governing equation together with initial and boundary conditions are first restated in nondimensional form, then are replaced by systems of algebraic equations. In *local methods*, including finite difference and finite volume techniques, replacement proceeds by *discretization* of the governing equation, that is, replacement of differentiation operations with algebraic operations at a large series of time and spatial *node* points forming a *discrete grid* in the flow field. The resulting system of algebraic equations is then solved simultaneously at each point. For a *global method*, such as the spectral method, the dependent variables are replaced with a series of amplitudes associated with different frequencies and a continuous representation of the solution is maintained.

Since an algebraic expression of the form $\Delta u / \Delta x$ is only an approximation of the differential term du / dx, the algebraic equations can only approximate the actual governing differential equation. However, if the number of node points is made very large and the solution grid very fine, subsequent errors can become very small, so that highly accurate solutions are achieved. Furthermore, because of this approach, boundary positions may be specified as a list of points rather than a closed-form mathematical expression. This permits flow field analysis within boundaries far too complex to specify in closed form.

To illustrate this process very briefly with a simple, one-dimensional finite difference example, consider a time-dependent simple shear flow in which the velocity u depends on time t and on the cross-stream position y, so that $u = u(y, t)$. The governing equation for this flow field, in dimensionless form, can be shown to be

$$\frac{\partial u}{\partial t} = \frac{\partial^2 u}{\partial y^2} . \tag{81}$$

Let y be divided into a large number of intervals of width h each assigned an index $i = 1, 2, 3, \ldots$ etc. and let t be divided into intervals of width τ and given the index $j = 1, 2, 3, \ldots$. The time derivative of the velocity u at any position i and time j may be represented by a *backwards difference*

$$\left. \frac{\partial u}{\partial t} \right|_{i,j} \approx \frac{u_{i,j} - u_{i,j-1}}{\tau} , \tag{82}$$

where higher order terms, of order τ^2 and higher, are neglected since τ is presumed to be small, as in a Taylor series expansion. Spatial derivatives can be represented by a *central difference*

$$\frac{\partial^2 u}{\partial y^2} \approx \frac{u_{i-1,j} - 2u_{i,j} + u_{i+1,j}}{h^2} , \tag{83}$$

again neglecting higher order terms. The new governing equation is then

$$\frac{u_{i,j} - u_{i,j-1}}{\tau} \approx \frac{u_{i-1,j} - 2u_{i,j} + u_{i+1,j}}{h^2} \tag{84}$$

or, letting $R = \tau/h^2$,

$$u_{i,j-1} \approx \left(1 + 2R/ \, u_{i,j} - R \left(u_{i-1,j} + u_{i+1,j}\right) \right) . \tag{85}$$

Equation 85 represents a set of simultaneous algebraic equations in the three unknown variables $u_{i,j}$, $u_{i-1,j}$, and $u_{i+1,j}$. As i varies from 1 to its maximum value, it thus defines the full velocity field at time j in terms of the velocity field at time $j - 1$. These equations may be written in a matrix form

$$\begin{pmatrix} 0 & (1+2R) & -R & 0 & \cdots & 0 \\ 0 & -R & (1+2R) & -R & \cdots & 0 \\ 0 & 0 & -R & (1+2R) & \cdots & 0 \\ \cdots & \cdots & \cdots & \cdots & \cdots & \cdots \\ 0 & 0 & 0 & \cdots & -R & (1+2R) \end{pmatrix} \cdot \begin{pmatrix} u_{0,j} \\ u_{1,j} \\ u_{2,j} \\ \cdots \\ \cdots \end{pmatrix} = \begin{pmatrix} u_{1,j-1} \\ u_{2,j-1} \\ u_{3,j-1} \\ \cdots \\ \cdots \end{pmatrix} . \tag{86}$$

The unknown variables $u_{i,j}$ may be then be found by standard matrix inversion techniques to give the velocity field $u(y)$ at any time, a process that is well suited to digital computing. In practice, the velocity field at time zero ($j = 0$) is normally prescribed as an initial condition. One therefore solves initially for the new velocity field at the first time interval ($j = 1$), then repeats for the next time ($j = 2$), and continues marching forward in time until the desired solution is reached. The number of spatial and time intervals may be chosen by the user based on the desired accuracy of the solution.

Modern software packages available from numerous industrial and academic sources are far more sophisticated than this simple example suggests, and permit investigation of a great variety of fluid flow problems, including turbulent and time-dependent flow fields, Newtonian and non-Newtonian fluids, compressible and hypersonic flows, particulate transport, and many other flow phenomena. Other physical effects, including heat transfer, chemical reaction kinetics, multi-scale diffusion, and the types of problems discussed above may be studied as well.

In the last two decades, CFD and finite element analysis, once restricted to specialized users, have become standard techniques for many bioengineering research teams. Two representative problems are discussed below to illustrate the power and flexibility of CFD modeling. One is the hemodynamics of abdominal aortic aneurysms (AAAs). This problem is deferred to the Capstone section (Sect. 7). The other focuses on the analysis of transport and biomimicry with nonhomogeneous surfaces. The emphasis is on the development of membrane technologies [26–28] for use in bioreaction engineering applications such as the design, development, and delivery of pharmaceutics [29], and for extra-corporeal medical devices [30] useful for biofluid detoxification and to bridge transplantation.

3.2.2
Predicting Surface and Interfacial Phenomena

The primary concern in the biomimetic membrane transport problem described here is ion transport through the micro-pores in supported liquid membrane systems (SLMs) [27]. The configuration studied, however, is applicable to any heterogeneous interface/surface where reaction or transport occurs [31], as in cell receptor–ligand interactions, immobilized fixed carrier systems, etc. Experimentally, a flat plate configuration, consisting of a microporous membrane as the support and an organic solution with a mobile carrier as the liquid membrane, was used to obtain permeation rate data for metal ions to be compared to CFD predictions. The inclusion of convective mass transfer to and from this heterogeneous surface leads to a complex analysis procedure. The local Peclet number Pe was used to characterize the hydrodynamics and mass transfer occurring at the fluid-membrane interfaces, and to assess the magnitude of each resistance. This dimensionless parameter reflects the ratio of transport by bulk motion to that by molecular diffusion and is discussed in greater detail in another part of this volume and in any textbook on transport phenomena [e.g., 1, 2, 32]. Finite element methods were employed for the complex, high Pe situations studied, while a straightforward eigenvalue problem can be solved to obtain an asymptotic solution for the stagnant film approximation, $Pe = 0$. The appropriate field conservation equation and required boundary conditions for the transported species can be found in detail elsewhere [27, 31–33].

The symmetry of the concentration field, inherent in the stagnant film scenario, is distorted by flow parallel to the surface and thus mass transfer is enhanced. Depending on the Peclet number and the fraction of active area, the surface can behave as if it is homogeneous with negligible mass transfer resistance in the boundary layer. A series of scenarios are given in which various assumptions are made to simplify the Navier–Stokes and mass flux equations, which are coupled in these formulations. These assumptions lead to useful asymptotes, as mentioned above, to compare with the detailed CFD analysis. A brief qualitative analysis follows that stresses the important role of hydrodynamics and the power of CFD modeling.

Mass transfer and/or reaction can only occur at "active sites" along the heterogeneous surface, in this case the transverse pores of a micro-porous matrix filled with the liquid membrane barrier. Flow parallel to the surface will tend to destroy the concentration distribution symmetry above a pore that would be present in a truly stagnant fluid film. For the highly idealized situation of circular pores uniformly distributed in hexagonal domains, even a simple one-dimensional flow field parallel to the surface will give rise to a three-dimensional concentration field. To avoid analyzing the formidably complex transport problem associated with such a concentration field, we need to consider a simpler problem that will retain some of the key features of the real situation and yield exact solutions for some limiting cases. A two-dimensional model problem is therefore selected in which the concentration variations along the surface are periodic. The surface of the membrane is assumed to have uniform channels (pores) of width a, evenly distributed with spacing b. Distance from the surface into the fluid's dynamic boundary layer is the second spatial dimension of interest. The dimensions in the third spatial coordinate are assumed to be irrelevant in this first approximation, assuming the flow and concentration fields to be uniform in that direction.

The flow field in general consists of two nonzero velocity components, each of which may be a function of the two spatial coordinates. The dynamic boundary layer has thickness δ, and the concentration gradient is assumed to be confined to this region, i.e., the solute concentration at the outer edge of the boundary layer is equal to the bulk fluid concentration. For simplicity it is assumed that the flow field is fully developed before contact with the membrane (upstream is a completely inactive, nonporous surface) and the focus is only on the "fully" developed mass transfer region far from the boundary between this nonporous plate and the SLM. For a homogeneously active surface, the concentration profile in this moving film eventually becomes independent of flow direction and linear in distance from the surface, i.e., in the boundary layer, when either a constant concentration or a constant flux boundary condition is imposed. The local mass-transfer coefficient in this region will be determined by the molecular diffusivity divided by the diffusion length, in this case δ. For our spatially periodic heterogeneous surface, the local mass-transfer coefficient is once again defined in terms of the diffusivity and diffusion length, however, this is no longer δ. Since surface heterogeneity increases the diffusion path through the boundary layer, we expect the mass transfer coefficient to be lower. It is useful to develop a correlation between them since significant data are available for the homogeneous case, over a wide variety of configurations. The ratio of these mass transfer coefficients (i.e., a correction factor) R, will be a function of a Reynolds number to characterize the flow and the Peclet number to characterize mass transfer rates, and other dimensionless groupings descriptive of system geometry [32].

Two limiting cases are worthy of discussion. If the pores are widely spaced compared to the boundary layer thickness ($b \gg \delta$), then the pores will not be interactive. For this situation, the result is that the correction factor R approaches the ratio of pore area to total surface area ε, $R = \varepsilon$. When the spacing is small compared to the boundary layer thickness ($b \ll \delta$), the pores are highly interactive and the diffusional path approaches δ since the surface characteristics approach homogeneous behavior. Thus, $R = 1$. The goal is to use the two-dimensional model system to predict R values in the intermediate region. The effects of system geometry and hydro-dynamics on the design and implementation of transport mimics can then be evaluated effectively.

The important point to be made here is that by using the Peclet number as the key parameter to describe a two-dimensional flow over a heterogeneous interface, profiles of the concentration field are obtainable using CFD techniques. The mass transfer in this case is best described as an average Sherwood number $\langle Sh \rangle$ which is identical to our ratio R. It is therefore bounded by the same limits; i.e., $\varepsilon \leq \langle Sh \rangle \leq 1$.

The ratio of pore spacing relative to the boundary layer thickness (β), fraction of pore area to total surface area (ε), and Peclet number, with mass transport rates characterized by the average Schmidt number ($\langle Sh \rangle$), was obtained for a variety of scenarios. The CFD-generated results show that even for a very nonhomogeneous surface (e.g., $\varepsilon = 0.25$) and very large values of β, axial dispersion, represented by Pe, forces the surface to act as if it were homogeneous. The major point to be stressed here is that $\langle Sh \rangle$ approaches 1.0 for nonhomogeneous surfaces when the hydrodynamics cause significant pore interaction. A stagnant film assumption could lead to serious misinterpretation of the controlling resistance in various geometric configurations. For example, the data obtained in "well-mixed" batch experiments, where the membrane surface may approach homogeneous behavior because of the local hydrodynamic influence (i.e., no stagnant film present), are often used to design continuous contactors such as flat-sheet dialyzer units or hollow-fiber cartridges. The mass transfer characteristics are assumed to be better in these "flow" systems when, in reality, they may be significantly poorer due to local Pe differences.

3.2.3
Predicting Biomimetic Reactor Performance

The analysis and design of reactors is highly dependent upon knowing the nonideal flow patterns that exist within the vessel. In principle, if we have a complete velocity distribution map for the fluid, then we are able to predict the behavior of any given vessel as a reactor. Once considered an impractical approach, this is now obtainable by computing the velocity distribution using the CFD-based procedures where the full conservation equations are solved. Both the nonlinear nature of these equations themselves and the nonlinear constitutive relationships between a flux and a gradient of an appropriate state property (e.g. a non-Newtonian fluid) are taken into consideration [7, 8, 28–30]. Discussions of actual performance in biomimetic systems, compared to computational analysis, are presented in "Controlling Tissue Microenvironments" (in this volume) along with appropriate references. They therefore will not be repeated here. Furthermore, a few comments on the design criteria are given in the next section. The important points to consider are that these systems must account for the role of hydrodynamics in both the transport characteristics and the effects on cellular performance, which is the emphasis of this review.

3.3
Neural Networks

Thusfar, only deterministic techniques have been considered. That is, meaningful representations of physically relevant problems have been obtained through the use of well-posed mathematical formulations. Deterministic techniques have the advantage of producing repeatable outcomes from a definite set of input variables. Unfortunately, most realistic problems do not readily conform to this format, necessitating an excessive number of simulations to identify system responses to the multitude of input variances expected due to the inherent randomness of the system stimuli. One other approach that has been used by many workers is that of Monte Carlo Simulation. Monte Carlo analysis is a hybrid technique, utilizing both stochastic and deterministic approaches. A set of inputs are selected randomly, within expected ranges specified for each variable, using a random number generator program. This set of inputs then yields a corresponding set of output responses determined from the deterministic model equations. This procedure is repeated many times to obtain an averaged response and associated confidence interval. In effect, this is similar to the approach taken to analyze experimental results with their inherent stochastic nature. Unfortunately, to obtain statistically reasonable system behavior, an extremely large number of simulations must be performed. This can become quite cumbersome when highly nonlinear, multivariable systems are being studied.

An alternative approach is to use purely stochastic modeling techniques, such as Neural Network Analysis, that have been successful in Artificial Intelligence programs. The basic concept of such methods is that given a set of probable inputs, we can obtain a probable system response that is based on prior behavioral characteristics. Clearly this technique has similarities to a regression analysis, however its predictive capabilities are much stronger. The major effort is in the amount of training that must take place with the neural network (NN) to assure some level of confidence in the output responses, since the models are relatively simple and computationally straightforward. A brief overview of the technique is given below in the context of a system exhibiting segregated sub-regions due to material property and/or concentration gradients.

The concepts of molecular and/or cellular segregation into sub-regions within flow fields and the formation of thin "wall-like" layers can be particularly beneficial with respect to demonstrating the power of a "mechanistic" approach to understanding and mimicking phenomenological events. These regions can be uniquely defined by molecular weight subgroups identified by our modeling techniques. The coupling of changes in fluid properties with geometry variations and blood vessel wall characteristics due to deposition of components from the fluid is the main focus.

A key feature in our analysis is that as the molecular nature of the fluid changes, the characteristic constitutive properties change. If the molecular scale phenomena are averaged to determine a single lumped parameter then we are dealing with a nonsegregated model which can be unrealistic in many situations. When we consider the global effect as a coupling of all the individual responses, then the observance of nonmonotonic rheological behavior is possible in these various fluid regions. The size and character of these regions are not known a priori, but can be obtained from the NN. Changes from region to region need not be continuous since these discontinuities are easily handled.

A molecular scale response of interest could be the coiled/uncoiled nature of blood proteins and changes in positioning of functional groups that are accessible for interactions. These can be with other proteins (entanglement) and/or surfaces of cells, whether in suspension or as wall components. Representation of flow characteristics by a single parameter (for example, a relaxation time) is most inappropriate since clearly there is not a unique one to one stimulus/response correlation, as in any stochastic process. Furthermore, there is not an easily defined inversion process for the coiling/uncoiling, entanglement, or other interactive processes and nonmonotonic behavior should be expected in these various sub-regions and on the wall.

Of course, finite element computations could be used to handle the rheological discontinuities between these distinct fluid regions. However, incorporating this changing spectrum of rheological properties, both spatial and dynamic in nature, can be more readily accomplished using neural network modeling and parallel processing techniques [34, 35]. Furthermore, the ability to de-convolute the contributions from individual relaxation times permits incorporation of nonmonotonic constitutive relationships [14, 15] easily into these distinct fluid regions. When cells and/or proteins are present in a fluid one should expect functional group/receptor interactions with each other (entanglement) and/or surfaces. Thus, changes in the population density (distribution) of relaxation modes can be observed even though no change in global composition has occurred. When dealing with a wide distribution of relaxation times, the probability of occurrence for an event associated with a single parameter will diminish, masked by all the other characteristic responses observed globally. Neural network (NN) modeling is particularly attractive to uncouple these masking effects and has been successful in predicting behavior by learning the conditions necessary for a response to physical property changes by analysis of existing data. This is particularly important here with respect to surface characteristics and thus interactions forming the "thin layer" as discussed previously. A NN accounts for drift, i.e., changes in system response due to "character" changes in the media/process. One can go back to see how these changes occurred and what phenomena were responsible, and thus to develop deterministic models to explain NN model outputs and gain further insights into the mechanisms involved. Deterministic models are superior to NN models when used for this type of analysis. However, NN's have the advantage that in the predictive mode they are computationally faster and more robust to uncertainties in system parameter values.

A typical feedforward, multilayer NN consists of S layers of neural elements, an input layer, S-2 hidden layers, and one output layer. In the jth layer there are M_j processing elements which are interconnected with elements in the $j - 1$th and $j + 1$th layers. Associated with the interconnection between the k_{j-1}th element of layer $j - 1$ and the k_jth element of layer j is the weighting factor $W(j - 1, j)$. The sigmoidal transfer function maps the cumulative input, X, to the output, Y, of any given processing element. The feedforward network is inherently parallelizable and may be implemented on a network of work stations. Training of the NN is done by systematically adjusting the interconnection weights such that the model output is as close as possible to the desired output.

The NN technique can also be coupled with the finite element packages that simulate the flow field within each sub-region, to evaluate system performance and "learn" how to divide the region into the most efficient sub-regions. The size and characteristics can then be determined using the approach developed for compartmental analysis as presented in another part of this volume; "Controlling Tissue Microenvironments".

4
Perfusion

The importance of maintaining the appropriate microenvironment in tissue culture systems, whether for in vivo, ex vivo, or in vitro applications, has been stressed throughout this book. Dynamic similitude between the cultured tissue and its native in vivo environment, both with respect to chemical and physical phenomena, is essential for optimum culture performance and is of particular concern in extra-corporeal devices. Perfusion, either with actual body fluids or with engineered surrogates, is crucial for establishment of the culture microenvironment, though the need for rapid delivery must be balanced with possible detrimental efforts from high shear forces and inappropriate signal generation. A properly designed and constructed perfusion system can eliminate many of the current problems associated with tissue culture systems. For example, although the production and longevity of hematopoietic cultures can be improved through optimal manual feeding protocols, such protocols are too labor intensive for large-scale use. Furthermore, subjecting the cultures to the frequent physical disruptions associated with manual feeding leads to large changes in culture conditions and possible contamination at each feeding. These complications restrict the optimization of the culture environment, and provided incentives to develop continuously perfused bioreactors. By delivery of appropriate substrates, perfusion bioreactors support the development and maintenance of accessory cell populations, resulting in significant endogenous growth factor production and enhanced culture success. In addition, perfusion difficulties encountered in vivo can lead to tissue necrosis and transplant failure, whether of a whole organ or an ex vivo system [36–38].

4.1
Design Specifications

The ability to design an effective perfusion system for a given application is highly dependent upon knowledge of the specific requirements of that tissue system, referred to by many authors as "tissue specs". These requirements have been discussed throughout this text and in many other sources [39]. Cellular function in vivo must be identified through these specifications, to characterize the tissue microenvironment and identify communication requirements. Key elements to consider in determining how tissue can best be built, reconstructed, and/or modified are based on the following axioms [40]: (1) in organogenesis and wound healing, proper cellular communications, with respect to each others activities, are of paramount concern since a systematic and regulated response is required from all participating cells; (2) the function of fully formed organs is strongly dependent on the coordinated function of multiple cell types, with tissue function based on multi-cellular aggregates; (3) the functionality of an individual cell is strongly affected by its microenvironment (typically within $100\,\mu m$ of the cell); (4) this microenvironment is further characterized by (i) neighboring cells via cell–cell contact and/or the presence of molecular signals such as soluble growth factors, (ii) transport processes and physical interactions with the extracellular matrix (ECM), and (iii) the local geometry, in particular its effects on microcirculation.

The importance of the microcirculation is that it connects all the microenvironments in every tissue to their larger whole body environment. Most metabolically active cells in the body are located within a few hundred micrometers from a capillary. This high degree of vascularity is necessary to provide the perfusion environment that connects every cell to a source and sink for respiratory gases, a source of nutrients from the small intestine, the hormones from the pancreas, liver, and other endocrine tissues, clearance of waste products, delivery of immune system respondents and so forth [41]. Culture devices must appropriately simulate and/or provide these functions while supporting the formation of appropriate microenvironments. Consequently, they must possess perfusion characteristics designed down to the 100-micrometer length scale. These are stringent design requirements that must be addressed with a high priority to properly account for the role of neighboring cells, the ECM, cyto/chemokine and hormone trafficking, geometry and the transport of respiratory gases, nutrients, and metabolic byproducts for each tissue system considered. Dynamic, chemical, and geometric variables must be duplicated as accurately as possible to achieve proper reconstitution even though the microenvironment can be quite complex [39, 42].

4.2
Devices and Performance

Dynamic similitude between the cultured tissue and its in vivo setting is the main design objective for a perfusion device. Close simulation of the native tissue microenvironment can foster optimized culture growth, protein expression and functionality and normoperative tissue performance in extracorporeal or transplantation systems. A common characteristic in most of these systems is that they are designed to achieve maximum movement of fluids with minimum shear. The purpose of this section is to identify the types of perfusion bioreactor systems available (or needing to be developed) that meet desired performance requirements for various tissue engineering applications. For example, when growth of individual cells in vitro is sought to increase cell density, a chemostat, possessing good back mixing, would be selected. When emulation of pancreatic behavior using encapsulated islets is desired, a microporous hollow fiber system, with plug flow characteristics could be considered apropos. Reactor inserts that improve contacting patterns and/or control transport, such as permeable supports, filter wells, gels and sponges, and capillary fibers, are also discussed in what follows. More in depth analyses of bioreactor types, for both research and production scale, are given in many alternate sources, such as in other parts of this volume and elsewhere [39, 40, 42–44].

There are many reactor types available that can be used as perfusion devices under restricted conditions. The most common types include chemostats, airlift fermentors, rotating chambers, perfused suspension culture (hollow fiber and folded membrane) systems, fluidized bed reactors, fixed bed reactors, and the NASA bioreactor. Each reactor was designed for specific application areas to address particular requirements. However, as often is the case, a compromise amongst multiple criteria and constraints imposed upon the systems was needed. Thus, each device is not without its limitations. Proper selection for a given application, that is, to stay within the design parameters of the device, should ensure good performance. The following brief discussion of the applicability of many of these bioreactor types is intended to aid in this selection process.

To maintain cells at a set concentration in a uniform environment a chemostat is often used. This bioreactor type is basically a continuous flow stirred tank reactor system that is characterized by good back mixing. The flow conditions are such that the cells may be purged from the vessel at a constant discharge rate matched by addition of medium at the same rate. To prevent the system from complete runoff of cells these rates must also match the growth rate of the cells. These systems are best used for production of cells in bulk, with vessels in the range 50–1000 liters. The required high degree of mixing is generally accomplished with impellers, typically slowly rotating large bladed paddles with a relatively high surface area to minimize the harmful effects of shear. Surfactants, and/or adjuvants such as pluronic F68, generally used as anti-foaming agents, can also be helpful in shear reduction. Minor changes in the base design of the chemostat to maintain the tissue in a constant metabolic state, through controlled changes to the chemical environment, produces a system often referred to as a biostat.

Fixed bed reactors utilizing glass beads for surface immobilization or hydrogel spheres encapsulating cells can be used if low flow rates and nonhomogeneous conditions in the reactor are tolerable. Medium can be perfused upward through the bed or percolated downward by gravity to minimize channeling versus that in a horizontal flow. The concentration environment within this reactor type varies in the flow direction. Consequently, the cells at different locations may be in varying metabolic states. This could be quite advantageous in certain situations, such as in the detoxification of media-based components, as will be seen in a later section.

Generally product is recovered from the spent medium downstream from both of these reactor types, although many alternative designs permitting in situ extraction of the products from the reactor have also been shown to be successful. One example is a hollow fiber technique in which cells are grown on the outer surface of bundles of polymer microcapillary tubes. The fiber is permeable, allowing diffusion of nutrients and products to and from the medium perfused through the capillary tubes. Culture growth may be up to several cells deep and thus an analogy with whole tissue is suggested.

A major deficiency in tissue architecture and organ culture is the absence of a vascular system, which can limit the size (by diffusion) and potentially the polarity of the cells within the organ culture. When cells are cultured as aggregates that grow beyond 250 micrometers in diameter (5000 cells), diffusional limitations begin. At or above 1000 micrometers in diameter (250000 cells), central necrosis usually occurs. To overcome this problem, cultures can be placed between the liquid and gas phases using reactor inserts similar to filter wells. Various attempts have been made to recreate tissue-like architecture from dispersed monolayer cultures. The formation of capillary tubules in cultures of vascular endothelial cells in the presence of growth factors such as vascular endothelial growth factor (VEGF) and medium conditioned by tumor cells demonstrates the applicability of this approach for some systems. Folded permeable membrane systems have been used to create appropriate compartments in the hope of providing a natural substrate for cell attachment. For example, collagen may exert some biological control of phenotypic expression due to the interaction of integrin receptors on the cell surface with specific sites in the ECM. Additional techniques include use of gel and sponge forms to support the cells and emulate tissue geometry. However, the most significant development has been the use of filter well inserts to create organotypic cultures utilizing their excellent features for cell interaction, stratification, and polarization. The permeability of the surface may induce this polarity in the cell through its ability to simulate the basement membrane. These filter well inserts are available in many different sizes, materials, and membrane porosities. Furthermore, they can be precoated with collagen, laminin, or other matrix materials such as synthetic hydrogels.

Airlift fermentors typically are used in the biotechnology industry, up to capacities of 20000 liters. This bioreactor type consists of two concentric cylinders with the inner being shorter at both bands to create two chambers. The bottom of the inner chamber has a sintered steel ring to disperse the gas as bubbles that rise up the center tube lifting the cell suspension with them. Spent gas is vented at the top and displacement insures return of the cell suspension down the outer chamber. In this manner, both aeration and mixing are accomplished simultaneously. Furthermore, as a result of this recirculation scheme, mixing is good and relatively gentle to the cellular system.

Rotating chamber systems achieve mixing and aeration by culturing the biosystem on the surface of a chamber that can be rotated in a larger vessel containing a well-mixed fluid. This system can be operated in batch or continuous fashion. A design that is somewhat successful consists of a series of disks on a rotating shaft. The cells can be immobilized on the disk [43, 44] coated with a thin layer of collagen, for example to maintain integrity, or they could be contained within "hollow" circular disks mounted on the shaft with semi-permeable membranes used to enclose each end of these low aspect ratio cylinders.

Perfused membrane systems have been discussed earlier with respect to folded membranes and compartmentalization, as were hollow fiber cartridge systems. The design of a more efficient system for use as an oxygenator and blood detoxification unit will be discussed in a later section on extracorporeal devices. These systems tend to be laboratory scale when used as a culture device, however they are extremely useful as recycle conditioning systems both for aeration and reaction vessels in the recycle loop [42, 45] of both laboratory and commercial scale continuous reactor configurations.

Fluidized bed reactors, utilizing microcarrier bead technology, have great promise for suspension cultures. By using high-density porous microspheres, in which suspended cells can become entrapped in the interstices of the matrix, either a well-mixed or fixed bed reactor system can be achieved dependent upon fluidization intensity. A balance of buoyant, shear, and gravitational forces can result in a stationary suspension or fluidized state when these high-density particles are perfused from the bottom of a vertical cylindrical vessel. The upward flow of liquid medium is constantly replenishing nutrients and collecting products for a downstream separation system. Gas exchange is generally external to the reactor and this could be with a hollow fiber exchange membrane system. Although no mechanical mixing is required, the fluid must be delivered at a pressure sufficient to overcome the bed resistance. If the fluidizing media has a velocity greater than that needed for stationary suspension pneumatic transport of the micro-carriers will occur and plug flow reactor characteristics will be obtained. A particle separation system will then be required; most often it is needed for recycle purposes. Advantages of this approach are that cellular activity within the reactor can be maintained through a recycle conditioning protocol and hydraulic residence time is no longer the only variable used to control reaction rates.

The NASA bioreactor [39] was developed to investigate cell growth in zero gravity. It consists of a rotating chamber in which cells growing in suspension culture achieve simulated zero gravity by continuously alternating the sedimentation vector. The cells remain stationary, subject to zero shear force, and tend to form three-dimensional aggregates, which reportedly enhances product formation. The limitations are similar to those encountered with roller bottle use; (a) the medium is replaced only after rotation is terminated and (b) product collection, whether within the aggregates or medium, is constrained by this protocol.

Since the use of microcarriers in cell culture systems and the general techniques of cell immobilization and encapsulation have become more common in recent years, the tools of reaction engineering can be applied more successfully to the biosciences. Many novel, innovative reactor systems have been developed by chemical reaction engineers to cope with the intricacies of heterogeneous catalysis research, development, and implementation [43, 44, 46, 47]. Consideration of living systems as biocatalytic processes and their multi-phase nature logically leads to the concept that retrofitting heterogeneous catalytic systems for use by tissue engineers is viable. Although many similarities are clear, there are sufficient differences, especially with aseptic processing, that present difficult challenges requiring innovation.

4.3
Stress Effects on Cellular Viability and Function

In recent years, there has been growing awareness that mechanical stimuli in general, and shear stress in particular, are important determinants of a great variety of cell activities. A large literature has developed concerning the effects of stress on cell viability, proliferation, cytoskeletal alteration, adhesion, receptor and other protein production and gene expression responses. In fact, the breadth of information now available on this subject is well beyond the scope of what can be described in this section. Here we restrict our attention to the important specific case of shear stress effects on *endothelial cells* of the circulatory system; even this is too large a topic for more than a very brief synopsis of the current state of knowledge. However, the summary that follows should suggest the importance of this subject to tissue engineers and give an idea of the wealth of issues involved. A number of recent reviews can provide much more detailed information to interested readers [48–51].

Vascular endothelial cells line the inner surface of blood vessels, providing a biological barrier between blood and other tissues and organs. The endothelium is a metabolically active layer, and is constantly exposed to both biochemical and biomechanical stimuli from blood flow over its surface. A significant complication is the fact that hemodynamic forces vary in magnitude, frequency, and direction over the vascular tree. Most of the studies published to date have therefore investigated endothelial cell responses to a representative time-average value for the wall shear, which for large arteries is in the range of 5–40 dynes/cm^2, depending on activity level and flow rate. Although in vivo studies have been carried out, particularly with regard to lipid uptake and leukocyte adhesion in response to disturbed or altered blood flow, most authors have elected to investigate in vitro systems, in which systematic variation of shear stress can be more easily controlled. The most widely used such system is the microscale parallel plate flow chamber, which allows a flow of desired shear rate to be established between two parallel plates, over the surface of a monolayer of cells grown to confluence on one of the plates.

Abundant evidence has now accumulated that the endothelial cell cytoskeleton is actively remodeled in response to shear stress changes. Adaptation is accompanied by changes in the cell morphology, alignment of cells in the flow direction and by rearrangement of the endothelial cytoskeleton. Cortical actin stress fibers are reshaped, as are microtubules and intermediate filaments. Furthermore, many studies indicate that actin fibers in the cytoskeleton appear to have specific functions related to sensing shear stress signals and transmitting that signaling to other cellular components, including the nucleus and the cell-ECM junction. Disruption of intracellular actin fibers has been shown to inhibit shear stress-mediated signaling and to obstruct the resulting changes in gene expression. The intermediate filament vimentin has also been suggested to play a role in shear stress transduction.

Some structures may function as both biomechanical and biochemical sensors. The VEGF receptor VEGFR2 is found to be activated by shear stress shortly after initiation of flow. Shear stress also stimulates VEGFR2 clustering and the formation of VEGFR2 complexes with the adherens cell–cell junction. This is particularly interesting in that VEGFR2 activation and complex formation also occur in response to VEGF itself. However, the outcomes of VEGF stimulation differ markedly from those of shear stress. VEGF induces endothelial cell proliferation and increases the permeability of the endothelial layer, whereas exposure to steady shear stress, although having an anti-apoptotic effect, does not produce a similar increase in cell division.

Shear stress appears to have significant influence on cell–ECM and cell–cell interactions. *Integrins* are a class of proteins involved in cell–ECM interactions in general; shear stress appears to activate specific integrins through the activation of specific kinases as well as through binding with VEGF receptors. In addition, cell adhesion molecules of the *PECAM* family are known to be rapidly phosphorylated after the application of shear stress, although not by elevation of intracellular calcium. The observation that the PECAM family of molecules undergoes rapid changes following the onset of flow suggests its possible importance as a mechanosensor. In addition, *VE-cadherin* is the major adhesive protein of the adherens junction, and is specific to endothelial cells. It has recently been shown that exposure of endothelial cells to physiologic levels of laminar shear stress produces a cadherin-containing complex within minutes that binds to the endothelial cytoskeleton.

It has been known for over a decade that many types of cell membrane ion channels are sensitive to shear stress. Those present in endothelial cells include a shear stress-responsive K^+ channel and a stretch-activated channel. Treatment of endothelial cells with barium chloride or tetraethylammonium, blocking agents for the K^+ channel, suppressed the induction of nitric oxide (NO) by shear stress. It also suppressed expression of NO synthase and the cytokine TGFβ, which supports the central role of this channel in mediating shear stress responses. Recently, a shear stress-sensitive Na^+ channel has been identified in endothelial cells and shown to affect activation of the ERK metabolic pathway. Other membrane features are responsive to shear stress as well. *Caveolae* are membrane domains rich in cholesterol, NO and tyrosine kinases. Cholesterol has been shown to modulate shear stress-induced transfer of NO from caveolae to the cytoplasm. Treatment of endothelial cells with antibodies against caveolin inhibited this transfer and suppressed activation of ERK 1/2. Shear stress has also been shown to increase activity of G proteins.

In the last few years, microarray technology has provided a powerful new method for investigating differential expression of endothelial cell genes after exposure to static vs. shear stress conditions, or under different shear stress levels. Initial studies from several years earlier had reported more than 40 genes to be regulated by shear stress, and their expression increased or decreased in response to flow. Promoter sequences that mediate shear stress transcriptional responses had been identified and termed *shear stress responsive elements* (SSREs). More recent, high-throughput gene array techniques have updated that figure and suggested several tentative conclusions: (1) hundreds of endothelial genes are regulated by physiologic levels of shear stress, (2) under chronic shear stress conditions (≥ 24 hours), more genes are suppressed than induced, (3) acute shear stress regulates the expression of genes related to activation of the endothelium [48]. In general, chronic exposure of cultured cells to flow seems to lead to activation of anti-proliferative genes. Simultaneously, it also seems to result in suppression of inflammatory or atherogenic genes such as VCAMs, ICAMs, and PECAM-1, apoptotic genes such as thrombospondin and signaling and matrix interaction genes such as integrins and VE-cadherin.

Along with research into identification of shear stress responsive genes, another line of investigation is concerned with the mechanisms by which biomechanical stresses lead to gene expression events. In spite of some success identifying promoter sequences and the specific transcription factors that bind to them, elucidation of mechanistic paths has not yet reached the point at which a comprehensive perspective has been achieved.

5
Flow Patterns, Mixing and Transport Phenomena

The importance of designing systems with appropriate hydrodynamic character with respect to transport phenomena and effects on cellular response has been stressed in previous sections. The purpose of this section is to extend these discussions to process innovations that may be considered to have a secondary impact on tissue engineering. These include methods to develop systems for (i) nano-scale particle formation, e.g., devices that utilize flow instabilities for mixing, (ii) improved contacting patterns to enhance interactions that promote better kinetics performance and detoxification rates in extra-corporeal devices, (iii) improved transport via mechanical turbulence promoters, and (iv) biotechnological processes for protein and pharmaceutical production.

5.1
Nano-Encapsulation

The ability to form nano-scale particles and/or emulsions that encapsulate active ingredients is finding applications in many facets of the engineering biosciences. Nano-technologies are having a major impact on drug delivery, molecular targeting, medical imaging and biosensor development, as well as on cosmetic and personal care products and the nutraceutics industry, to mention only a few of the uses to which these technologies have been applied [53, 54]. New techniques utilize high shear fields to obtain particle sizes in the range 50 to 100 nm, about the size of the turbulent eddies found in some of these processing units. Stable emulsions can be formed with conventional mixing equipment in which high shear, elongation flow fields are generated near the tip of high-speed blades, but only in the range 500 nm and larger. High shear stresses can also be generated by forcing the components of a microemulsion to flow through a microporous material. The resultant solution contains average particle sizes as small as 50 nm. Units that incorporate jet impingement on a solid surface or with another jet also perform in this size range. A unit combining jet impingement and high turbulence shear stress field is currently under development. A prototype of this mixer produces average particle sizes in the range of 50 to 100 nm, but with less energy dissipation than existing systems. It consists of concentric cylinders, with the outer cylinder fixed and the inner one rotating, as in the Taylor flow field configuration discussed above, but for this purpose operating in the highly turbulent regime and with a unique internal recirculation system.

Molecular self-assembly systems using novel, biocompatible synthetic polymer compounds are also under development. These have been used successfully to encapsulate chemotherapeutic drugs and perflourocarbons (PFC). The high oxygen solubility of PFC's makes them attractive as blood surrogates and also as additives in immuno-isolation tissue encapsulation systems to enhance gas transport. Using nano-scale encapsulation techniques, these compounds can be dispersed throughout the microencapsulating matrix and/or in the tissue ECM scaffold system.

A key objective for all these encapsulation systems is to provide effective and efficient contacting patterns with sufficiently rapid dynamics behavior. Appropriate shear stress fields must be developed and maintained for relevant time scales for the desired processes to occur. As discussed previously, CFD calculations can help in the design of systems to accomplish this. When coupled with experimental validation in flow emulators and/or prototype units, scale up for clinical relevance can be accomplished more effectively.

5.2
Extra-Corporeal Systems

Exploitation of membrane technology has been highly useful in artificial organ design. A prime example is the use of hydrophobic, micro-porous membranes for exchange of gases such as O_2 and CO_2 between two segregated fluids, which has proven advantageous in artificial lung applications. Current systems are expensive, however, due to the special design considerations required for patient comfort, health and contamination factors for overall safety, and reusability constraints.

5.2.1
Blood Oxygenation

This discussion of blood oxygenation units will be restricted to systems that utilize hydrophobic hollow fiber membranes, as an example of a system whose performance is highly dependent on hydrodynamic effects. It has been well documented that these effects play a major role in both the transport processes by which respiratory gases are exchanged with blood and the protection of the fragile blood cells. Furthermore, knowledge of the location and control of the gas/liquid interface within the pores of the hollow fiber is important not only for transport considerations, but also to maintain bubble free operation, which is essential for patient safety.

Sensitivity of blood to high shear forces is a significant consideration in oxygenation units, as are the needs for small hold-up volumes, short contact times, and high transfer rates. Since these parameters generally conflict with one another, design is usually a compromise [55–57]. However, it is estimated that these new designs can outperform existing systems by a factor of 10 [30, 45, 55–62]. The physical and chemical characteristics of hollow fiber membranes, including hydrophobility, thin walls, small internal diameter and large porosity, lead to excellent membrane transport. Consequently, the major mass transfer limitation is typically in the adjacent fluid boundary layers, not within the membrane. Considering the relatively poor transport performance of commercial blood oxygenators compared to a module carefully built by hand, one fiber at a time, it becomes apparent that module geometry is also crucial. Uneven spacing of the hollow fibers, resulting in channeling and the development of stagnant boundary layers adjacent to the membranes, is a major contribution to compromised performance. By using fibers woven into a hollow fiber fabric, improvements are obtained. However, severe flux decline is possible due to plugging from particle accumulation in the extra-capillary space [58].

A current practice used in many laboratories to obtain optimal pitch and spacing between fibers is to mount fibers in perforated plates. Although effective, this is not only very tedious and labor intensive, especially when high efficiency, super fine novel fibers are used, but also there is no mechanical support between the flexible fibers, rendering the designs impractical. The baffled membrane, hollow fiber fabric modules overcame these problems and maintained uniform flow patterns.

Novel systems are continuously being proposed that possess desirable features. For example, a system currently being evaluated [55, 56] is a spiral wound, hollow fiber module that has the following advantageous features: (1) high transfer efficiencies and low prime volume; (2) optimal pitch layout with fiber spacing readily adjustable in all three spatial dimensions; (3) baffles can be located wherever desired; (4) turbulence promoters and mechanical support features are added to disrupt the boundary layers and prevent fiber damage; (5) cross flow, parallel flow, and counter-current contacting patterns are obtainable; (6) antifouling characteristics with easy clean-up and disinfective procedures for reuse; and (7) ease of assembly; only slight modifications to existing fabrication methods are needed.

Application of this membrane technology to other areas is also quite promising. Current research and development activities utilizing these techniques include in situ extraction of pharmaceutics with simultaneous removal of the fermentation heat, utilization as a reactive/selective separator, biomimetic studies for ion transport, immuno-magnetic capture of pathogens and biochemical reactor design useful in biotransformation processes and cell culture analog systems. The use of a biomimetic hollow fiber configuration as a recycle conditioning reactor-separator to improve continuous fermentation systems for pharmaceutics has been validated [63].

5.2.2
Blood Detoxification

The ability to detoxify blood in a rapid, safe fashion, as in an emergency room or intensive care unit situation, can be essential for successful stabilization of a patient. In this application, the objective is to remove potentially pathogenic compounds present in the blood due to poisoning, intentional or accidental drug overdoses or unanticipated drug interactions. Treatment of sepsis is also a possibility. An extra-corporeal device could be capable of accomplishing these tasks at low cost and with high reliability. Furthermore, such a device could also be useful for in vitro detection and remediation of blood supplies that may have viral contamination. One possible approach to designing a functional extra-corporeal blood detoxifier might be to couple immuno-magnetic capture (IMC) techniques with chemical engineering process technologies. Such technologies include the coupling of mass transport and hydrodynamics to enhance contact, and the use of specific receptor–ligand binding sites to improve efficacy of sequestering toxic compounds or agents.

There are inherent problems with existing extra-corporeal devices used to remove toxic substances from blood that could result in serious consequences for the patient. There is always the danger of cellular damage, inappropriate substance removal, elicitation of immune response, and other complications due to bio-incompatibilities. Novel contacting configurations and advanced materials selection can mitigate the majority of these difficulties. A conceptual process to remove bacteria from blood, for example, would be the use of polyclonal antibodies attached to paramagnetic microspheres to capture the specific entity and to sequester the bead/microbe complex on nickel spheres in a magnetically stabilized fluid bed (MSFB) contactor. The bead/microbe complex is ultimately collected by releasing the magnetic field and identifying the microbes by ELISA assays.

A similar system has been successful for the detection and removal of water-borne pathogens, both small entities, such as the polio virus, and larger ones such as *Giardia* cysts and *Cryptosporidium* oocysts [64]. However, water systems are easier to deal with than biologically complex media such as blood. Comfort and safety issues for the patient and to the health care provider must be considered in addition to the scientific and technological challenges involved. These include prevention against cell damage (i.e., low shear contacting), elimination of platelet deposition and immune response, establishment of effective contacting and binding in the low shear and low concentration fields, utilization of selective removal strategies to ensure that essential substances/cells are not removed with the toxins, and minimization of residual effects from the magnetic collection scheme. The essential features that this process concept must possess are as follows:

- Specific receptors, such as antibodies for sepsis and MDR gene expressed proteins for drugs/endotoxins, bound to neutrally buoyant polystyrene-coated paramagnetic spheres (5 μm diameter) to remove the toxin from the blood stream;
- Use of transport reactor configurations to accomplish efficient contacting;
- Collection of toxins in an expanded fluidized bed system with a magnetic field imposed to enhance stability and immobilization.

This approach should meet the low shear requirement and accomplish physical separation/collection while providing concentration enhancement that improves kinetics. Additional advantages include: (1) the use of specific receptors for capture implies that capture also provides identification, (2) complete recovery is possible when excess immunobeads are present, and (3) magnetic susceptibility facilitates recovery of captured species simply by application of a magnetic field. Thus, with IMC, purification and concentration could potentially be accomplished in minutes with minimal effort.

Initial attempts to use fixed bed and conventional fluidized bed systems met with varying degrees of success due to the inherent disadvantages of each system. The fixed bed system works well for Polio virus collection since the bead (5 μm diameter) is larger than the virus. However, with 15 μm diameter *Giardia* cysts (or even larger *Cryptosporidium* oocysts), plugging becomes problematic. Furthermore, in all cases there exist solids removal problems. The fluidized bed systems experience bubbling and jetting effects that result in inefficient and nonuniform contact between the phases. Large residence time distributions exist in both phases and consequently early withdrawal of beads occurred.

The magnetically stabilized fluidized bed (MSFB) system eliminates many of these difficulties. Capture efficiencies for *Giardia* cysts of 100 percent were obtained at flow rates nearly an order of magnitude higher than previously reported [64]. The collection of the bead/cyst complex is through attachment to larger spheres that act as a moving filter in a low shear field. The flow field was independently verified through residence time studies, magnetic field mapping and pressure drop measurements to identify both fluidization and stability regions. Nearly ideal fluid-solid countercurrent contact was measured and bubbling and jetting in the fluid phase was eliminated. Thus, the MSFB combines the major advantages of the fixed bed (good countercurrent contact) with those of the fluidized bed (low pressure drop, flow of solids to prevent clogging, and ease of solids replacement). Clinical use of an appropriately modified system for the recovery of pathogens and/or other toxins from flowing blood seems reasonable. These modifications must focus on the role of hydrodynamics in transport performance and compatibility with the requirements for handling the blood in an extra-corporeal environment.

6
Pulsatile Flow

Flow in a straight, round tube driven by an axial pressure gradient that varies in time is the basis for blood transport in the arterial tree as well as respiratory gas transport in the trachea and bronchi. The nature of the flow fields developed in this situation depends on whether the tube in question has rigid or elastic walls. We focus first on fluid confined within a tube of perfectly *rigid*, undeformable walls. The direction of fluid motion will then always be parallel to the tube axis, so that there will only be an axial component of velocity $u = (u(r, t), 0, 0)$ (Fig. 7, below). In that case, all the fluid elements in the tube, regardless of axial position, will respond to any change in the pressure magnitude instantaneously and in unison. Consequently, the velocity profile will be the same at all positions along the tube. It is as if all the fluid in the tube moves as a single rigid body.

The energy balance for pulsatile flow in a rigid tube is such that energy input into the fluid from the pressure gradient is balanced by convective energy transport by the flow field, diffusion of energy from the action of viscosity and energy dissipation, also due to viscosity:

$$\text{energy input} \atop \text{(from the pressure gradient)} = \text{convective transport of energy} + \text{diffusion of energy due to viscosity} + \text{viscous dissipation} \,.$$

$$(87)$$

As a result of the flow field acceleration and decelerations in pulsatile flows, a special type of boundary layer known as the *Stokes layer* develops in pulsatile flows. This is best explained by considering the case that the pressure gradient varies sinusoidally in time, and the flow field responds accordingly. That is, as the pressure increases to its maximum, the flow increases, and as the pressure decreases, the flow also decreases. If the oscillations are of very low frequency, the pressure varies only gradually, the velocity field will essentially be in phase with the pressure gradient and the boundary layers will have adequate time to grow into the tube core region. In the limit of very low frequency, the velocity field must therefore approach that of a steady Poiseuille flow. As frequency increases, however, the pressure gradient changes more rapidly and the flow begins to lag behind it due to the inertia of the fluid. Because of this lag, the velocity in the pipe center is less than it would be in a steady flow of equivalent pressure gradient. At the same time, the Stokes boundary layers become confined to a region near the wall, lacking the time required for further growth. The loss in flow magnitude increases with increasing oscillation frequency, as pressure gradient reversals occur more and more rapidly. In the limit of very high frequency, fluid in the tube center hardly moves at all and the Stokes layers are confined to a very thin region along the wall.

Because of the inertia of the fluid, the Stokes layer thickness, δ, is expected to be inversely related to the frequency of the flow. In fact, it can be shown that

$$\delta \propto \left(\frac{\upsilon}{\omega}\right)^{1/2}, \tag{88}$$

where υ is the fluid kinematic viscosity, as above, and ω is the flow angular frequency (in radian/sec).

Clearly, this is a somewhat artificial situation, at least in the sense that no physiologic blood vessel or bronchial tube is ever perfectly rigid. However, it is a mathematically tractable flow field that can be analyzed using the Navier–Stokes equations (Eq. 34). As a result, much of what is known about pulsatile flows in general has been developed from understanding of flow in rigid tubes.

In contrast, flow within an *elastic-walled* tube has a very different character. A pressure surge within a nonrigid tube results in local bulging of the tube wall (Fig. 9, below), and radial fluid flow into the bulge. Therefore, the flow field is no longer purely axial. Subsequently, the stretched wall section recoils, pushing fluid and accompanying elevated pressure downstream in the tube. The axial velocity component u becomes a function of x, as well as of r and t, and the radial velocity component v is no longer zero. The result is a traveling pressure wave in the tube, and a flow field in which all fluid elements do not accelerate and decelerate in unison. Furthermore, the energy balance of this flow situation must now include a term due to deformation of the wall

$$
\begin{array}{l}
\text{energy input} \\
\text{(from the pressure gradient)}
\end{array}
=
\begin{array}{l}
\text{convective} \\
\text{energy} \\
\text{transport}
\end{array}
+
\begin{array}{l}
\text{diffusion of} \\
\text{energy due} \\
\text{to viscosity}
\end{array}
$$
$$
+
\begin{array}{l}
\text{wall} \\
\text{deformation}
\end{array}
+
\begin{array}{l}
\text{viscous} \\
\text{dissipation}
\end{array}
. \tag{89}
$$

6.1
Hemodynamics in Rigid Tubes: Womersley's Theory

The rhythmic contractions of the heart produce a pressure distribution in the arterial tree that includes both a steady component, P_s, and a purely oscillatory component, P_{osc}, as does the velocity field. In contrast, flow in the trachea and bronchi has no steady component, and thus is purely oscillatory. It is common practice to refer to these components of pressure and flow as *steady* and *oscillatory*, respectively, and to use the term *pulsatile* to refer to the superposition of the two. A very useful feature of these flows, when modeled as if they occurred in rigid tubes, is that the governing equation (Eq. 91) is linear, since the flow field is unidirectional and independent of axial position. The steady and oscillatory components can therefore be decoupled from each other, and analyzed separately. This gives

$$
P(x, t) = P_s(x) + P_{osc}(x, t)
$$
$$
u(r, t) = u_s(r) + u_{osc}(r, t) . \tag{90}
$$

Substituting (Eq. 90) into the Navier–Stokes equations (Eq. 34) gives

$$
\left[\frac{1}{\rho} \frac{\partial P_s}{\partial x} - \upsilon \left(\frac{\partial^2 u_s}{\partial r^2} + \frac{1}{r} \frac{\partial u_s}{\partial r} \right) \right]
$$
$$
+ \frac{\partial u_{osc}}{\partial t} + \frac{1}{\rho} \frac{\partial P_{osc}}{\partial x} - \upsilon \left(\frac{\partial^2 u_{osc}}{\partial r^2} + \frac{1}{r} \frac{\partial u_{osc}}{\partial r} \right) = 0 , \tag{91}
$$

where the pair of brackets separate terms that are independent of time while all other terms are time-dependent. This is a particularly helpful grouping of terms in that, since the terms in the square brackets are time-independent, they must sum to zero separately from the remaining time-dependent terms, which therefore must themselves equal zero. The bracketed term is therefore identical to Eq. 39a from Sect. 2.3.1, the governing equation for steady flow.

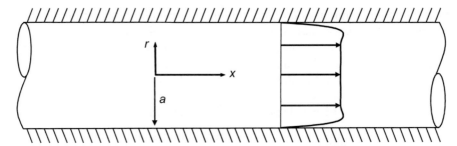

Fig. 7 Unidirectional flow field parallel to the tube axis for pulsatile flow in a straight, round rigid tube of radius *a*. *x* – axial coordinate, *r* – radial coordinate

Thus, all the results obtained in that section apply to the steady component of pulsatile flow as well.

Dropping the subscript, on the understanding that only the oscillatory component of velocity is referred to, then gives the following governing equation for flow in a straight, rigid tube, when the pressure gradient varies in time

$$\frac{\partial u}{\partial t} = -\frac{1}{\rho}\frac{\partial P}{\partial x} + v\left(\frac{\partial^2 u}{\partial r^2} + \frac{1}{r}\frac{\partial u}{\partial r}\right) \tag{92}$$

subject to the no-slip boundary condition at the tube wall, which for a round tube takes the form $u = 0$ for $r = a$. Before a solution can be developed for Eq. 92, two additional assumptions must be made. First, we assume that the flow is fully developed, so that entrance effects may be neglected. Second, we assume the flow to be driven by a purely oscillatory pressure gradient, which is conveniently expressed in the form

$$-\frac{1}{\rho}\frac{\partial P}{\partial x} = K\cos(\omega t) = Re\left(Ke^{i\omega t}\right), \tag{93}$$

where here *Re* indicates the *real part* of the expression $Ke^{i\omega t}$, with $i = \sqrt{-1}$. In fact, the pressure waveform in large blood vessels is far more complex than a sinusoidal wave of a single frequency [24, 65]. However, this assumption is actually not restrictive. It can be shown that any *periodic* function can be expressed as a sum of sine and cosine functions, known as a *Fourier series*. Briefly put, a function $f(t)$ can be said to be periodic if for any arbitrary time *t*,

$$f(t) = f(t + T), \tag{94}$$

where T is called the period of the function. A periodic function can be represented by the Fourier series

$$f(t) = \sum_{n=0}^{\infty} A_n \cos\left(\frac{2n\pi t}{T}\right) + \sum_{n=0}^{\infty} B_n \sin\left(\frac{2n\pi t}{T}\right)$$

$$= A_0 + A_1 \cos\left(\frac{2\pi t}{T}\right) + A_2 \cos\left(\frac{4\pi t}{T}\right)$$

$$+ \dots + B_1 \sin\left(\frac{2\pi t}{T}\right) + B_2 \sin\left(\frac{4\pi t}{T}\right) + \dots, \tag{95}$$

where the coefficients A_n and B_n are constants that are determined by the properties of the specific function $f(t)$, and are given by

$$A_0 = \frac{1}{2\pi} \int_0^{2\pi} f(t)\, dt \tag{96a}$$

$$A_n = \frac{1}{\pi} \int_0^{2\pi} f(t) \cos\left(\frac{2n\pi t}{T}\right) dt \tag{96b}$$

$$B_n = \frac{1}{\pi} \int_0^{2\pi} f(t) \sin\left(\frac{2n\pi t}{T}\right) dt . \tag{96c}$$

Since Eq. 93 is linear, a solution for any arbitrary periodic pressure waveform can be developed by *Fourier decomposition* of the waveform into a series of sine and cosine terms and superposition of the solutions to each individual Fourier component.

It is also convenient to introduce a new dimensionless parameter, the Womersley number, α [65], defined as

$$\alpha = \frac{\text{tube radius}}{\text{Stokes layer thickness}} = a\sqrt{\frac{\omega}{\upsilon}} . \tag{97}$$

Thus defined, the Womersley number can be used to calculate the fraction of the tube cross-section occupied by the boundary layer. In view of the comments above, it should not be surprising that α turns out to have a major governing influence on the nature of the flow patterns in any oscillatory flow.

Solution of Eq. 92 is achieved most simply through a separation of variables approach, assuming on physical grounds that the solution will have the form

$$u(r, t) = f(r) \cdot e^{i\omega t} , \tag{98}$$

where $f(r)$ is an unknown function expressing the dependence of the velocity on radial position, and is yet to be determined. To find it, the trial form Eq. 98

may be substituted into Eq. 92. The result is a new governing equation for f, which after multiplying through by $e^{-i\omega t}$, is

$$\frac{d^2 f(r)}{dr^2} + \frac{1}{r}\frac{df(r)}{dr} - \frac{i\omega}{\upsilon}f(r) = -\frac{K}{\upsilon} . \tag{99}$$

Equation 99 turns out to be a special case of *Bessel's equation*, which itself can be written as

$$r^2\frac{d^2 f}{dr^2} + r\frac{df}{dr} + \left(m^2 r^2 - p^2\right)f(r) = 0 , \tag{100}$$

where m is a nonzero constant and p is a parameter of the problem. Clearly, Eq. 99 is obtained when $p = 0$, $m = \left(-\frac{i\omega}{\upsilon}\right)^{1/2}$ and a nonzero forcing term is added to the right hand side. Solutions of Bessel's equations, called *Bessel functions* and often denoted as $J_p(r)$ and $Y_p(r)$, are well known, and are tabulated in many references [for example, 66]. In terms of Bessel functions, the solution to Eq. 99 that satisfies the no-slip condition at the wall is

$$u(r, t) = Re\left(-i\frac{K}{\omega}\left\{1 - \frac{J_0\left(r\sqrt{-\frac{i\omega}{\upsilon}}\right)}{J_0\left(a\sqrt{-\frac{i\omega}{\upsilon}}\right)}\right\} \cdot e^{i\omega t}\right) . \tag{101}$$

To this of course must be added a steady component if the flow field is pulsatile rather than purely oscillatory.

Although correct, in practice Eq. 101 is more cumbersome than is needed for evaluating the velocity distribution. Furthermore, although it is not necessarily obvious from Eq. 101, the velocity does not simply vary as $\cos \omega t$, since both the amplitude and the phase of u ($e^{i\omega t}$) are complex. Since the pressure gradient *does* vary as $\cos \omega t$ (Eq. 93), the fact that the amplitude of u is complex generates a phase difference between the oscillating pressure gradient and the resulting velocity field. Hence, finding the real part of u is itself a laborious task. What is needed is a simpler form for u, in which the amplitude of the real part is as self-evident as possible. Such an expression can be derived by making use of so called *ber* and *bei* functions [67], which themselves are defined through

$$ber(r) + i \cdot bei(r) = J_0\left(r \cdot i\sqrt{i}\right) . \tag{102}$$

Using these functions, the solution becomes

$$u(r, t) = \frac{K}{\omega}\left(B \cos \omega t + (1 - A) \sin \omega t\right) \tag{103}$$

where

$$A = \frac{ber\alpha \cdot ber\alpha\frac{r}{a} + bei\alpha \cdot bei\alpha\frac{r}{a}}{ber^2\alpha + bei^2\alpha} \tag{104a}$$

and

$$B = \frac{bei\alpha \cdot ber\alpha\frac{r}{a} - ber\alpha \cdot bei\alpha\frac{r}{a}}{ber^2\alpha + bei^2\alpha}.$$ (104b)

Representative velocity profiles derived from these expressions are shown in Fig. 8 for two values of α, at four phases of the flow cycle. In these figures the radial position, r, has been normalized by the tube radius, a. At $\alpha = 3$ (Fig. 8a), a value that under resting conditions can occur in the smallest arteries and larger arterioles as well as the middle airways, Stokes layers can occupy a significant fraction of the tube radius. The velocity at the wall ($r/a = \pm1$ in Fig. 8) is zero, as required by the no-slip condition, and as in steady flow the velocity varies smoothly with r, with no step change at any point. However, even at this low α, the velocity profile resembles a parabola only during peak flow rates. At other flow phases, a more uniform profile forms across the tube core.

In contrast, at $\alpha = 13$ (Fig. 8b), which characterizes rest state flow in the aorta and trachea, the velocity profile of the pipe core is nearly uniform at all flow phases, while flow in the boundary layer is out of phase with that in the core. Flow reversals are possible in the Stokes layer. The reason for these changes is clearly the inertia of the fluid, since as the flow frequency increases, less time is available to accelerate the fluid in each flow cycle.

As with steady flows, it is important to be able to use these expressions for the velocity field to determine the instantaneous total volume flow rate, Q_{inst}.

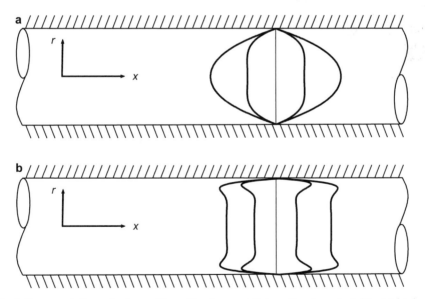

Fig. 8 Representative velocity profiles of laminar, oscillatory flow in a straight, rigid tube, at four phases of the flow cycle. **a** $\alpha = 3$, **b** $\alpha = 13$

At first thought this may seem odd, in that a purely oscillatory flow field simply moves back and forth, with no net fluid translation. It is clearly true that over a full flow cycle the *net* volume flow of Q_{osc} will be zero. However, oscillatory motion substantially alters the *instantaneous* shear stress to which the wall is exposed, compared to steady flow at an equivalent Reynolds number. It therefore has the potential to significantly alter the wall response.

For any pipe flow, knowledge of the volume flow rate is equivalent to knowledge of the instantaneous mean velocity, U_{inst}, since $Q_{inst} = U_{inst} \times$ pipe area. Following [67], it can be shown that the mean velocity is

$$U(t) = \frac{K}{\omega} \left(\frac{2D}{\alpha} \cos \omega t + \left(\frac{1 - 2C}{\alpha} \right) \sin \omega t \right)$$

$$= \frac{K}{\omega} \sigma \cos (\omega t - \delta) \tag{105}$$

where

$$C = \frac{ber\alpha \cdot bei'\alpha - bei\alpha \cdot ber'\alpha}{ber^2\alpha + bei^2\alpha} \tag{106a}$$

$$D = \frac{ber\alpha \cdot ber'\alpha + bei\alpha \cdot bei'\alpha}{ber^2\alpha + bei^2\alpha} \tag{106b}$$

$$\sigma^2 = \left(\frac{1 - 2C}{\alpha} \right)^2 + \left(\frac{2D}{\alpha} \right)^2 \tag{106c}$$

and

$$\tan \delta = \frac{(1 - 2C/\alpha)}{(2D/\alpha)} . \tag{106d}$$

The oscillatory shear stress at the wall, $\tau_{w,osc}$, is given by

$$\tau_{w,osc} = -\mu \frac{\partial u_{osc}}{\partial r}\Big|_{r=a} . \tag{107}$$

From Eq. 101, it can be shown that this is given by

$$\tau_{w,osc} = Re \left(\frac{\rho K a \sqrt{i}}{\alpha} \frac{J_1 \left(a\sqrt{-\frac{i\omega}{\upsilon}} \right)}{J_0 \left(a\sqrt{-\frac{i\omega}{\upsilon}} \right)} \cdot e^{i\omega t} \right) . \tag{108}$$

As with the oscillatory flow rate, the oscillatory wall shear stress lags the pressure gradient, reaching a maximum during peak flow.

6.2
Hemodynamics in Elastic Tubes

Because of the mathematical complexity of analysis of pulsatile flows in elastic tubes, and the variety of physical phenomena associated with them, space does not permit a full description of this topic in the present setting. The reader is instead referred to a number of excellent references for a more complete treatment [24, 68, 69]. Here we only briefly summarize the most important features of these flows, to give the reader a sense of the richness of the physics underlying them.

In a tube with a nonrigid wall, any pressure change within the tube will lead to bulging of the tube wall in the high pressure region (Fig. 9). It is then possible for fluid to flow in the radial direction into the bulge. Hence, not only is the radial velocity v no longer zero, but both u and v can no longer be independent of x even far from the tube ends. Thus, the flow field is governed by the continuity condition along with the full Navier–Stokes equations. Assuming axial symmetry of the tube, these become

$$\frac{\partial u}{\partial x} + \frac{\partial v}{\partial r} + \frac{v}{r} = 0$$

$$\frac{\partial u}{\partial t} + u\frac{\partial u}{\partial x} + v\frac{\partial u}{\partial r} = -\frac{1}{\rho}\frac{\partial P}{\partial x} + \upsilon\left(\frac{\partial^2 u}{\partial x^2} + \frac{\partial^2 u}{\partial r^2} + \frac{1}{r}\frac{\partial u}{\partial r}\right)$$

$$\frac{\partial v}{\partial t} + u\frac{\partial v}{\partial x} + v\frac{\partial v}{\partial r} = -\frac{1}{\rho}\frac{\partial P}{\partial r} + \upsilon\left(\frac{\partial^2 v}{\partial x^2} + \frac{\partial^2 v}{\partial r^2} + \frac{1}{r}\frac{\partial v}{\partial r} - \frac{v}{r^2}\right). \tag{109}$$

The most important consequence of this is that even if the inlet pressure gradient depends only on t, within the tube the pressure gradient depends on x as well as t. An oscillatory pressure gradient applied at the tube entrance therefore propagates down the tube in a wave motion. Both the pressure and the velocity fields therefore take on wave characteristics.

The speed with which these waves travel down the tube can be expected to depend on the fluid inertia, i.e. on its density, and on the wall stiffness. If the wall thickness is small compared to the tube radius and the effect of viscosity

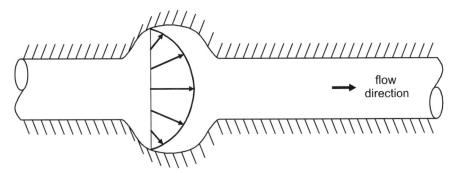

Fig. 9 Local bulging of the tube wall at regions of high pressure in pulsatile flow in an elastic tube

is neglected, the wave speed c_0 is given by the Moen–Korteweg formula

$$c_0 = \sqrt{\frac{Eh}{\rho d}}\,, \tag{110}$$

where E is the stiffness, or Young's modulus, of the tube wall and h is its thickness. As can be expected on physical grounds, the wave speed increases as the wall stiffness rises. When E becomes infinite, the wall is rigid. Thus, oscillatory motion in a *rigid* tube, in which all the fluid moves together in bulk, may be thought of as resulting from a wave traveling with infinite speed, so that any change in the pressure gradient is felt throughout the whole tube instantaneously. In an *elastic* tube, by contrast, pressure changes are felt locally at first and then propagate downstream at finite speed.

Because of the action of the pressure and shear stress on the wall position and displacement, oscillatory flow in an elastic tube is inherently a *coupled* problem, in the sense that it is not possible in general to determine the fluid motion without also determining the resulting wall motion; the two are intrinsically linked. It can be shown [24] that the motion of the *wall* is governed by

$$\frac{\partial^2 \varsigma}{\partial t^2} = \frac{E}{\left(1 - \sigma^2\right)\rho_w}\left(\frac{\partial^2 \varsigma}{\partial x^2} + \frac{\sigma}{a}\frac{\partial^2 \eta}{\partial x}\right) - \frac{\tau_w}{\rho_w h} \tag{111a}$$

$$\frac{\partial^2 \eta}{\partial t^2} = \frac{P_w}{\rho_w h} - \frac{E}{\left(1 - \sigma^2\right)\rho_w a}\left(\frac{\eta}{a} + \sigma\frac{\partial \varsigma}{\partial x}\right), \tag{111b}$$

where ς and η are the axial and radial displacement of the wall, respectively, (both of which may vary with axial position x), P_w and τ_w are the fluid pressure and shear stress at the wall, ρ_w is the wall density, and σ is Poisson's ratio, a wall material property defined above (Sect. 2.2.1). Equations 111aa and b indicate the coupling of the wall and fluid motions, since they explicitly describe ς and η, which are properties of the *wall*, in terms of P_w and τ_w, which are themselves properties of the *flow*. In addition, coupling is imposed by the no-slip boundary condition, since the layer of fluid in contact with the wall must have the same velocity as the wall. Hence

$$\frac{\partial \varsigma}{\partial t} = u(x, a, t)\,, \quad \text{the axial component of velocity at the wall} \tag{112a}$$

$$\frac{\partial \eta}{\partial t} = v(x, a, t)\,, \quad \text{the radial component of velocity at the wall.} \tag{112b}$$

With these governing equations and boundary conditions in place, and if the input pressure distribution that drives the flow field is known, it is possible to develop a formal solution for the axial velocity. For an oscillatory flow, the input pressure would normally be expected to be of a sinusoidal form

$$P(x, r, t) = \text{const} \cdot e^{i\omega t}\,. \tag{113}$$

Following [24], the method of characteristics shows the pressure distribution throughout the tube to be

$$P(x, r, t) = A(x, r)e^{i\omega(t-x/c)} \tag{114}$$

where c is the wave speed in the fluid and A is the pressure amplitude. Since the fluid must be taken to be viscous, c is not equal to c_0, the inviscid fluid wave speed given by Eq. 110. Instead,

$$c = c_0 \sqrt{\frac{2}{(1-\sigma^2)z}} \tag{115}$$

with z a parameter of the problem that depends on a, ω, υ, σ, ρ, ρ_w, and h. It can also be shown that the pressure amplitude A depends on x, but not on r, and therefore the pressure is uniform across any axial position in the tube [24]. Under these conditions, the solution for u, the principal velocity component of interest, can be stated as

$$u(x, r, t) = Re\left(\frac{A}{\rho c}\left\{1 - G\frac{J_0\left(r\sqrt{-\frac{i\omega}{\upsilon}}\right)}{J_0\left(a\sqrt{-\frac{i\omega}{\upsilon}}\right)}\right\} \cdot e^{i\omega(t-x/c)}\right) \tag{116}$$

with G a factor that modifies the velocity profile shape compared to that in a rigid tube due to the wall elasticity. G is given by

$$G = \frac{2 + z(2\upsilon - 1)}{z(2\upsilon - g)} \tag{117}$$

with

$$g = \frac{2J_1\left(a\sqrt{-\frac{i\omega}{\upsilon}}\right)}{\left(a\sqrt{-\frac{i\omega}{\upsilon}}\right)J_0\left(a\sqrt{-\frac{i\omega}{\upsilon}}\right)}. \tag{118}$$

It is apparent from inspection of Eq. 116 that the difference between the velocity field in a rigid tube and that in an elastic tube is contained in the factor G. However, since G is complex, and both its real and imaginary parts depend on the flow frequency ω, the difference is by no means readily evident. The reader is referred to [24] for detailed depiction of representative velocity profiles. However, it is important to note here that because the pressure distribution in an elastic tube takes the form of a traveling wave, *two separate* periodic oscillations can be derived from Eq. 116. The first is that at any given axial position in the tube, the velocity profile varies sinusoidally with time, just as it does in a rigid tube. The second, however, is that at any instant of time during the flow cycle, the velocity field also varies sinusoidally in space. Fluid flows *away* from regions in which the pressure is greatest and *towards* regions in which it is least. In a rigid tube, there is only one region of maximum pressure, the upstream tube end, and only one region of minimum pressure, the downstream end. Between them, the pressure varies linearly with axial position *x*. In contrast, in an elastic tube the pressure varies *sinusoidally* with *x*, so that many high pressure regions can exist along the tube, and these lead to a series of flow reversals at any specific time. This is *not* a violation of conservation of mass, as it would be if such reversals were to happen in a rigid tube, since in an elastic tube radial velocity *v* can replace or remove fluid as needed to satisfy continuity.

A final word about oscillatory flow in an elastic tube concerns the possibility of *wave reflections*. In a rigid tube, there is no wave motion as such, and flow arriving at an obstruction or branch is disturbed in some way, but otherwise progresses through the obstruction. In contrast, the wave nature of flow in an elastic tube leads to entirely different behavior at an obstacle. At an obstruction such as a bifurcation or a branch, some of the energy associated with pressure and flow is transmitted through the obstruction, while the remainder is *reflected*. This leads to a highly complex pattern of superposing primary and reflected pressure and flow waves, particularly in the arterial tree since blood vessels are elastic and vessel branchings are ubiquitous throughout the vascular system. Such wave reflections may be analyzed in terms of transmission line theory [24, 70].

6.3
Turbulence in Pulsatile Flow

Transition to turbulence in oscillatory pipe flows occurs through fundamentally different mechanisms than transition in steady flows, for two separate reasons. The first is that the oscillatory nature of the flow leads to a unique base state, the most important feature of which is the formation of an oscillatory Stokes layer on the tube wall. This layer has its own stability characteristics, which are not comparable to the stability characteristics of the boundary layer of steady flow. The second reason is that temporal deceleration destabilizes the whole flow field, so that perturbations of the Stokes layer can cause the flow to break down into unstable, random fluctuations. In steady flow, perturbations either grow or decay. In contrast, even when unstable, oscillatory flow is characterized by a fully laminar motion during its acceleration and peak flow phases. Instability occurs during the deceleration phase of the flow cycle, and is immediately followed by relaminarization as the net flow decays to zero prior to reversal. Because of these characteristics, during deceleration phases of the flow cycle instabilities are observable in the Stokes layer even at much lower Reynolds numbers than those for which they would be found in steady flow [71].

Since the Stokes layer thickness δ itself depends on the flow frequency, it is not surprising that transition to turbulence should depend on the Womersley number α, as well as on the Reynolds number. In fact, for sufficiently large values of α, laminar theory predicts that viscous effects are confined to the

Stokes layer, since the tube central core approaches an irrotational plug flow condition in the limit of large α. It would be expected, then, that flow instabilities would initially appear in the region of the velocity field where velocity gradients are largest, i.e. in the Stokes layer.

Experimental measurements of the velocity made by noninvasive optical techniques have verified that expectation [71]. Over a range of values of $\alpha \geq 8$, the flow was found to be fully laminar for $Re_\delta \leq 500$, where Re_δ is the Reynolds number based on the Stokes layer thickness rather than tube diameter. That is,

$$Re_\delta = \frac{U\delta}{\upsilon}. \tag{119}$$

For $500 < Re_\delta < 1300$, the core flow remained laminar while the Stokes layer became unstable during the deceleration phase of fluid motion. This turbulence was most intense in an annular region near the tube wall. These results are in accord with theoretical predictions of instabilities in Stokes layers. For higher values of Re_δ, instability can be expected to spread across the tube core.

The spatial distribution of instability in oscillatory flows is significantly different from its distribution in steady flow. In addition, instability occurs during the deceleration phase of each half-cycle of oscillatory flow, and is immediately followed by relaminarization of the flow field. Taken together, these facts suggest that models of mass transport and/or heat exchange in unstable, oscillatory pipe flow should incorporate the concepts that the flow becomes unstable only during part of the oscillatory cycle and it is only the boundary layer that exhibits the instability, at least for $Re_\delta < 1300$. Only at sufficiently large Re_δ is turbulent mixing not confined to a bounded annular region near the wall.

7
Capstone Illustration:
Understanding Arterial Diseases; Diagnosis and Therapy

To conclude this review, we discuss a capstone example intended to show how tissue engineering combined with an understanding of vascular hemodynamics can provide significant insight into blood flows in health and disease, as well as make substantial contributions to clinical practice.

The major diseases of the arterial tree include *atherosclerosis*, *aneurysm*, and *dissection*. Of these conditions, the most prevalent is *atherosclerosis*, which is defined by the characteristic accumulation of lipid and lipoprotein plaques in the arterial wall. Rather than being distributed continuously throughout the arterial tree, plaque formation occurs predominantly at specific isolated sites, including the femoral bifurcation, the femoral arteries, and at branchings of the tibial artery. However, the most common site for atherosclerotic deposits is in the aorta, and it is thought that the prevalence of atherosclerotic deposits in the infrarenal aorta, between the renal arteries and the femoral bifurcation, in patients over age 50 is 100%. In contrast, occlusive deposits are not normally found in vessels such as the internal iliac or peroneal arteries. In view of the large size of the vessels at risk, it is not surprising that accumulation in them often does not progress to the point at which it becomes symptomatic or compromises health. However, patients with measurable atheroma of the large vessels have a high incidence of lipid deposition in the cerebrovascular and coronary arteries as well. This is potentially of much higher clinical risk, since the small diameter of these vessels permits only limited deposition before the vessels become fully occluded and unable to sustain flow.

One common characteristic of the sites prone to atherosclerotic disease is that they tend to be regions of arterial curvature and branching, where shear stress is altered from its levels in straight vessels. Consequently, the possible role of fluid mechanics in plaque localization has been debated for many years [36, 72]. It is now believed that the limiting resistance to lipid transport is derived from the endothelium wall rather than the blood. This in turn has given rise to the suggestion that the observed fluid mechanical effects on lipid deposition are the result of direct mechanical influences on the transport characteristics of the endothelium. As a result, in the last decade there has been a large number of investigations of the relations between hemodynamic flow patterns, shear stress distribution, and endothelial disruption. Lipid uptake has been found to occur primarily in regions of low or disturbed shear stress, and to be accompanied by monocyte adhesion and expression of inflammatory and atherogenic mediators. Over time, the wall becomes damaged and its inner layer, the *tunica intima*, thickens. Since these effects retard diffusion of nutrients through the endothelium, smooth muscle cells begin to die, leading to formation of recognizable plaque.

Aortic dissection, the most common acute aortic event requiring admission to a hospital, is characterized by a tear of the tunica intima which allows blood to perfuse into the wall, creating a false lumen embedded within the wall. Perhaps not surprisingly, nearly all patients presenting such dissections are hypertensive. Over 95% of dissections occur in the aortic arch just distal to the aortic valve, a position at which the aorta is well fixed, exposing it to risk of injury due to shear stress. The false lumen can rupture in the aortic arch, or the dissection can propagate distally into the nearby branch vessels, exposing those vessels to rupture risk as well. In either case, aortic dissection is normally a fatal disease if left untreated, and is associated with 90% mortality within one month.

Upon diagnosis, treatment of aortic dissection is based on the dual approaches of immediate control of blood pressure through drugs along with immediate surgical repair in almost all cases. The damaged ascending aorta and associated damaged vessels are replaced with a synthetic graft, restoring normal blood flow. Five-year survival rates following surgical intervention are 70–80%.

Since rupture of a dissection is a biomechanical event, it might be thought that biomechanical modeling of the dissection could become an important component of diagnosis and management decisions. However, in fact the risk of rupture is so large and the mortality of unoperated dissection so high that in effect there is rarely a management decision to be made. Virtually all diagnosed dissections are subject to immediate surgical repair.

7.1
Selection of AAA as a Representative Case Study

Of the diseases of the aorta, *aortic aneurysm* is second only to atherosclerosis in its prevalence and health implications, since AAAs have been estimated to occur in as much as 2–3% of the population [73, 74], with a mortality rate on rupture between 78–94%. Rupture of these aneurysms currently ranks as the 13th leading cause of death in western societies, producing 15 000 deaths annually in the U.S. alone. In spite of improvements in surgical procedures, this mortality rate has not declined in recent years. However, the rate of post-operative mortality following open-abdomen elective resection has declined to a level normally reported to be less than 10%. Nevertheless, a very large number of patients are potentially faced with this problem. In addition, most of the patients are elderly, and surgical resection, even of unruptured aneurysms, is not a risk-free procedure. Thus, there exists a major incentive for (a) early detection of AAAs and (b) careful assessment of risk of rupture, in order to minimize the need for unnecessary or high-risk surgery.

Although it is widely anticipated that endovascular surgical stent graft placement will significantly improve these statistics when it receives approval for widespread use, the long-term success of stent placement is far from known. Moreover, even if successful, stent placement as a treatment will merely change the bulge size at which surgical management becomes a viable option, not eliminate the need for understanding rupture probability. Hence for the foreseeable future, there will remain an important clinical need for an accurate method for evaluating rupture risk of specific aneurysms.

Although the causes of AAA are not fully understood, it is thought that biochemical degradation of the aorta wall leads to local weakness that causes the wall to bulge under the influence of hemodynamic pressure. It has been shown that elastin fibers provide the aorta wall with compliance and maintain its normal size, while collagen fibers provide it with tensile strength. The walls of aneurysmal aortas show marked decreases in elastin and collagen content compared to the walls of normal aortas, and it has also been reported that aneurysmal patients show a type III collagen associated with decreased elastic modulus compared with controls. Autopsies of aneurysmal patients have demonstrated decreased hepatic copper, a known cross-linking agent for collagen and elastin fibers. Also, there are known irregularities of copper metabolism in so-called blotchy mice, a strain that is prone to spontaneous aortic aneurysms. Finally, it has been reported that aortic aneurysmal wall tissue is characterized by destructive remodeling of the ECM and increased local expression of matrix metalloproteinase enzymes.

Current clinical practice is to base treatment and intervention decisions of AAAs entirely on the assumption that risk of rupture outweighs surgical risk when bulge diameter exceeds 5 cm. Although this approach is unambiguous and clinically simple to apply, it has the drawback that as many as 25% of AAAs less than 5 cm do rupture. More importantly, from a biomechanical standpoint, rupture is caused by wall stress, not size. Thus, aneurysmal rupture may be thought of as a material failure, in which the wall separates when exposed to stress exceeding its strength to resist. Failure of the wall therefore may be expected when wall stress, not diameter, surpasses critical values. Though AAAs in vivo differ in their tensile strength, a mean failure strength of $65 \, \text{N}/\text{cm}^2$ has been reported; i.e. rupture may be expected to be increasingly probable as wall stress magnitude exceeds that level.

Physiologically, stress develops in any vessel wall as flow-induced forces act on the wall. As discussed above (Sect. 2.2), these forces include both a static component due to blood pressure and the deviatoric forces generated by blood flow. Consequently, from a biomechanical perspective, knowledge of AAA flow patterns and their relation to wall stress evolution is the central issue necessary to understand AAA wall stress magnitude and thereby predict rupture risk. Factors expected to affect wall stress development for any specific patient lesion include the size, shape, and tortuosity of the aneurysmal bulge, the diameter of the proximal and distal nondilated aorta, the volume flow rate, the nature of the flow (laminar or turbulent, steady or unsteady), the elasticity of the aneurysm wall, and the presence or absence of deposited thrombus on the wall. For mechanical purposes, variations in the biochemical or histological makeup of individual aneurysms may be grouped together as differences in the wall elasticity.

The appearance of turbulence in the aneurysmal flow field is particularly important for several reasons. Turbulence can increase the time-average wall shear stress by a factor of 10 over what it would be for laminar flow at a similar flow rate, and can instantaneously produce brief wall stress events another order of magnitude larger than that. In addition, turbulent flow can generate wall vibrations that are thought to be especially dangerous to the diseased aneurysmal vessel.

At present, development of more accurate rupture risk evaluative criteria is limited by poor understanding of (i) the relation of AAA size and shape (which vary widely between patients) to lumenal flow patterns, (ii) the relations between bulge size, shape, flow and wall stress distribution, and (iii) the mechanical and biologic response of the diseased wall to flow-induced stress.

For example, it has been reported that even under resting conditions, AAAs of equal bulge diameter can present widely differing flow fields in vivo. Accordingly, one objective of this illustrative example is to show that fluid motion can be quantitatively investigated in anatomically accurate, patient-based AAA models, and the flow fields, which are physiologically realistic, can be determined. Stress levels on the model walls produced by the motion can then be accurately deduced from the flow measurements. Once evaluated, measured wall stress can be related to risk of rupture.

The ultimate goal for this approach is to replace the simplistic, patient-average evaluation of AAA prognosis with a risk stratification analysis on an individualized, patient-by-patient basis that has equal ease of application for the clinician, but is quantitatively accurate. One can imagine an eventual system in which a patient presenting clinically with AAA will be imaged, for example by CT or MR, and an exact depiction of the patient aorta delivered along with relevant vital statistics into a computer model. That model will predict the flow field in the lesion and calculate the resulting wall stress distribution, thereby making possible a quantitatively based determination of rupture probability.

Creating such a biomechanical model with reliable predictive capabilities will require several major steps, each a substantial project in itself. Numerous research groups over the past decade have made significant process in both qualitative and quantitative investigation of aneurysmal flow and wall stresses. In some instances, mechanisms have been identified, and attempts have been made to develop the needed predictive models. Quantitative studies of flow patterns and pressure distribution have provided a widely accepted picture of the basic laminar flow through rigid, axisymmetric aneurysm models [75–82]. It should be stressed that such in vitro models are dynamically similar to real aneurysms only if (i) their geometry is a scale model of the geometry of in vivo aneurysms, (ii) their wall has similar material properties to real aneurysm walls and (iii) the flow through them generates values of governing dimensionless parameters equal to those of the in vivo flow. For steady flows, the only governing flow parameter is the Reynolds number. In spite of not fulfilling either of conditions (i) or (ii) exactly, experimental results furnished a qualitative match to an actual flow [81, 82]. The flow was laminar up to a critical Re of 1600–2000, after which it became turbulent. Under laminar conditions, the flow consisted of a jet of fluid passing through the core of the model aneurysm, surrounded by an outer annulus of fluid rotating as a recirculating vortex.

Unsteady, pulsatile flows in rigid wall models have been examined both experimentally [83] and computationally [84–87]. When the flow has an oscillatory component, as is true of in vivo aortic flows, an additional dimensionless parameter that influences the flow is the Womersley number (Sect. 6.1). Using flow waveforms matching those of the in vivo abdominal aorta, the appearance of recirculating vortices outside the core of the model aneurysm has been demonstrated and migration patterns of the vortex center along the aneurysm over the cardiac cycle shown.

A separate line of investigation into aneurysm mechanics has been based on evaluating stress development within the aneurysm wall. In an elegant series of papers over the last 5 years [88–95], a finite element analysis of patient-based wall internal stress distributions was developed. Maximum stresses were found to vary between 25–50 N/cm^2 [88], and to not necessarily occur in the largest bulges. Although powerful and quantitatively appealing, this approach is limited at present by the need to make mathematically simplifying assumptions neglecting both instantaneous spatial pressure variations and wall pressure variation over the cardiac cycle.

7.2
Coupling Tissue Engineering and Hydrodynamics

The effectiveness of using tissue culture systems to understand the role of flow fields on cellular behavior was discussed earlier, as were the fundamentals of hydrodynamic analysis. Exact, closed-form mathematical solutions are obtainable for idealized models with respect to fluid properties, dominate kinematics, and system geometries. These solutions provide the rational basis for extending analysis to realistic geometries and wall properties. The complex flow fields to be expected from these analyses can be simulated accurately using CFD techniques and emulated experimentally in representative geometric devices with dynamic similitude. The most realistic behavior will be obtained when hydrodynamics and tissue culture techniques are coupled in these devices. It will then be possible to assess their influence upon each other and possible synergism in response to system perturbations. In the discussion below, successes to date are described and a proposed computational and experimental program advancing some of these key concepts to clarify the importance of hydrodynamic analyses in tissue engineering is presented.

7.2.1
Clinical Evaluation of Patient Perfusion

Color-flow ultrasonography has been used to evaluate in vivo AAA flow in a series of patients at rest [96]. Two distinctly different patterns were observed, one smooth, unidirectional flow and apparently laminar while the other was irregular, rapidly fluctuating and apparently turbulent with both anterograde and retrograde flow regions. Significantly, in this study the presence of turbulent flow was uncorrelated with AAA size. As was described above, cases were found in which patients of equal AAA diameter presented dissimilar, even contradictory, flow regimes. However, there was a strong correlation between disturbed, irregular flow and the presence of mural thrombus.

7.2.2
Biomimetic Flow Emulation

To explain the above observation, flow fields were evaluated in phantoms that exactly replicate the in vivo geometry of specific patient aneurysms. A method for fabricating such phantoms was developed that produces a clear, hollow, flow-through cast, suitable for optical measurements, whose interior exactly replicates the patient aneurysm lumen [97, 98]. Flow fields within these models were interrogated via particle image velocimetry (PIV). Representative steady flow patterns from one such phantom, with $D/d = 1.7$, are given in Fig. 10a. In this particular phantom, a small narrowing of the lumen immediately proximal to the bulge caused the flow to accelerate into the bulge, forming a jet along the anterior wall. Such free jets are highly unstable [99], even at low flow rates. The corresponding instability intensity I is depicted in Fig. 10b, through a color scale superposed on the velocity profiles. Strong instabilities clearly following the main jet are apparent. In this phantom, I_{max} was 0.24 even at Reynolds numbers less than 1000 [98], at least an order of magnitude larger than would be expected to be found in a nondilated tube at a similar flow rate.

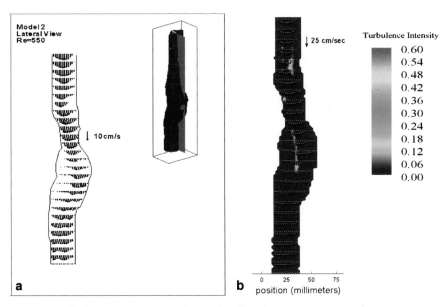

Fig. 10 Steady flow in a representative patient-based aortic aneurysm phantom, at $Re = 550$. The plane in which these profiles were measured is shown in *red* in the *inset*. **a** Mid-bulge velocity field, **b** turbulence intensity distribution

7.2.3
Physics of Flow in Axisymmetric Bulges

Earlier experiments with idealized, axially symmetric AAA replicas had also revealed strong instabilities [100–103], albeit not at such low flow rates. Under steady conditions in those models, the flow was always laminar for $Re \leq 1800$ and fully unstable for $Re > 2300$. Between those limits, as Re increased, the frequency and duration of unstable bursts increased until eventually a fully unstable flow emerged. In the axisymmetric phantoms, values of the centerline turbulence intensity increased rapidly in the distal bulge half, indicating that expansion of the core flow resulted in very strong amplification of velocity instabilities [81–83]. This is apparently a characteristic feature of aneurysmal bulges; as fluid travels through the bulge, turbulence intensity grows at a significantly more rapid rate than it would in a straight tube. In that sense, the bulge acts as a turbulence amplifier. The largest phantoms produced velocity fluctuations as great as 40% of the mean centerline velocity. Moreover, under unstable conditions the peak instantaneous wall shear stress at the bulge exit was 3–20 times the already elevated mean shear, suggesting peak instantaneous wall shear stresses of 100–150 dynes/cm^2 can occur in large AAAs. It has been known for many years that erosion of aortic endothelium can be acutely produced by exposure to continuous wall shear stresses of 300–450 dynes/cm^2 [100–103]. Consideration of all these data suggest the hypothesis that over long times, accumulated bursts of shear stress can eventually produce lesions of the aneurysmal endothelium, denuding the wall and initiating a thrombotic cascade [82]. Thrombus deposition would then limit further endothelial erosion, thereby protecting the underlying tissue.

Additional studies have investigated pulsatile flow simulating rest-state aortic conditions in the same axisymmetric set of phantoms. Strong instabilities were found in these flow fields as well, even though a rest-state flow was simulated.

In summary, preliminary experiments demonstrate that AAA replicas can successfully be fabricated and their flow properties characterized. Wide shape variation and a complex dependence of the resulting flow patterns on lumenal geometry are to be expected between patient-based phantoms, which underscores the difficulty of establishing one single parameter that discriminates rupture risk. Nevertheless, accurate evaluation of bulge flow fields can significantly improve understanding of in vivo wall stress magnitude, as well as ability to predict its evolution on a patient-by-patient basis, and therefore can form the basis of a biomechanical, quantitatively accurate approach to rupture risk analysis.

7.2.4
Pulsatile Flow in Compliant Blood Vessels: Computation and Experiments

A key future step in aortic aneurysm investigation, as well as in studies of hemodynamics in other vessels and vascular pathologies, will be to develop a wall internal stress model that is coupled with flow properties and subsequent fluid stresses at the fluid–wall interface, to evaluate their effect upon each other in both rigid and nonrigid wall models. To analyze this coupled problem will require calculation of the wall deformation simultaneously with the flow. Furthermore, the use of tissue culture techniques and advanced biomimetic materials in patient-based experimental phantoms can be exploited to evaluate the in vitro response to steady and pulsatile flows, in an emulated exercise-state, to establish a credibility link to in vivo behavior and use of these models in a clinically meaningful manner.

An important underlying hypothesis for this analysis is the concept that the elasticity of vascular tissue, i.e. the compliance of the vessel walls, changes the nature of the flow and creates a dynamic coupling of the fluid motion and wall strain. Better modeling of this coupling will assist in understanding the dynamics of arterial hemodynamics in general and aortic aneurysms in particular. Current models that treat the vessel wall as a fixed boundary suppress the governing dynamics of this coupled system. Consequently, there is a need for a computer model of pulsatile flow surrounded by compliant walls that can be validated with biomimetic flow-through systems. These flow emulators will need to use realistic geometries extracted from CT scans of actual patients, constructed of materials that mimic the mechanical properties of healthy and/or diseased vascular tissue as appropriate. Pulsatile flows must be considered in these computations and experiments, not only because of the deformation coupling but also since the elasticity of the vessel wall can create slow-moving wave propagation in long vessels (Sect. 6.2). Furthermore, wall deformation could either damp or stimulate flow instabilities of various frequencies. Although all of these phenomena are beyond the predictive capability of current CFD models, the evolving model variations discussed here are needed to accurately represent fluid/structure interactions and produce greater understanding of coupled hemodynamic systems.

The major output from such a computational model will be the predicted time-varying blood flow field in three dimensions through the deforming shape of the vessel. Particularly interesting features of this flow with respect to tissue engineering are the concepts of fluid-wall property coupling, segregated flow regions, identification of wall slip phenomena, and retrograde flow driven by wall stress variations. These efforts will require implementation of the requisite fluid equations, including Womersley theory and non-Newtonian constitutive equations, to calculate all the pertinent details of the flow. Shear stress on the walls will then be determined along with tensile stresses in the tissue and the time rate-of-change of these quantities. Such a model will also provide all the requisite data for experimental studies in a biomimetic flow emulator, including its design parameters, materials selection criteria, and predictive performance for protocol development together with the endothelial response studies. A high-quality biomimetic material will be needed for representing the vessel wall in these experiments. To design or select such a material, candidate biocompatible materials such as collagen and poly(lactic acid) will need to be evaluated both for their biomechanical suitability as a model for human vascular tissue and as scaffolding for tissue development on the wall surface. These substances must also offer the ability to create areas of local stiffness which can represent the behavior of diseased tissue.

One possible approach for solution of the three-dimensional, time-varying Navier–Stokes equations in complex geometries is through the embedded-boundary method. This technique has proven to be successful in a number of other complicated flow problems, including two-phase flow in a polymer processing stability study [16, 21, 33, 104]. In this method, vessel walls are represented to the fluid as a set of forces. Rather than impose the walls as boundary conditions, the walls become sources and sinks of momentum, sufficient to represent the no-slip boundary. One may then use a fixed, regular, Cartesian mesh, which avoids meshing problems that hinder finite volume methods in complex shapes. As the walls move, so do the location and magnitude of the forces in the fixed mesh. Thus, the difficulties of re-meshing the fluid flow domain are avoided. Thus, the embedded-boundary method can provide accurate solutions of the fluid flow without the obstacles that arise in finite-volume methods.

Simultaneously with the equations of motion for the fluid, a force balance can be solved for the deformation of the vessel walls, determining the stress/strain relationship by experimental investigation. This stress/strain constitutive relationship is necessary and sufficient to close the coupled flow-wall deformation problem mathematically. The results will provide a complete description of the fluid-wall system, including time-varying resolution of shear stresses, normal stresses, and material strain.

Candidate wall materials that mimic aspects of the mechanical properties of healthy vascular tissue for fabrication of the experimental validation include collagen gels, poly(L-lactic acid)-poly(glycolide) systems, poly(glycolic acid)-poly(4-hydroxybutyrate) blends, and poly(acrylic acid)-based networks. Many of these materials have already been used for tissue engineering of vascular grafts. Many also offer the possibility of rigidification or stiffening via glycation reactions or chemical crosslinkers. Since the resulting mechanical properties are directly related to the degree of crosslinking, it may be possible to create tubes with tunable, well-defined "pockets" of rigidity by controlling crosslinking in those regions. This method could be used to simulate tissue that has been damaged by plaque formation in certain regions. It could also be used to create walls with varying degrees of thickness to mimic stenotic vessels.

Experimentally measured properties of selected wall materials, including tensile strength, elastic modulus, loss modulus, recoverable compliance and Poisson's ratio will be needed as initial inputs into the CFD calculations. The experimental system will be subjected to pulsatile flow, and wall profiles will be monitored and compared to predictions from CFD calculations. These results will provide the requisite information for the studies of cellular-level response, particle deposition, and cellular attachment.

Ultimately, the evaluation of time-varying fluid/solid interface conditions as influenced by particulate and/or cellular deposition and removal is needed. A cellular detachment technique can be used to measure variations of the stress field with respect to time and position in these patient-based phantoms, as in the radial flow detachment assay technique currently used to measure bonding strengths [5]. However, biomimetic devices possess more flexibility in generating physiologically meaningful flow fields and understanding the effect of stress on tissue systems in general and endothelial cell behavior in particular.

References

1. Bird RB, Stewart WE, Lightfoot EN (2002) Transport Phenomena, 2nd edn. Wiley, New York
2. Lightfoot EN (1974) Transport Phenomena and Living Systems. Wiley-Interscience, New York
3. Cooney DO (1976) Biomedical Engineering Principles. Dekker, New York
4. Lightfoot EN, Duca KA (2000) The Roles of Mass Transfer in Tissue Function. In: Bronzino JD (ed) The Biomedical Engineering Handbook, 2nd edn. CRC Press, Boca Raton, Fl, Ch. 115
5. Goldstein AS, DiMilla PA (1997) Biotech Bioeng 55:616
6. Lauffenburger DA, Linderman JJ (1993) Receptors: Models for Binding, Trafficking, and Signaling. Oxford University Press, New York
7. Mueller TJ (1978) Applications of Numerical Methods in Fluid Dynamics. In: Wirz, Smolderen (eds) A von Karmon Institute Book, McGraw-Hill, New York
8. Kulpers JAM, Van Swaaij WPM (1998) Advances in Chem Eng 24:227
9. Lamb H (1945) Hydrodynamics. Dover Publishing Inc., New York
10. Schlichting H (1979) Boundary Layer Theory, 7. edn. McGraw-Hill, New York
11. Hinze JO (1986) Turbulence, (Reissued). McGraw-Hill, New York
12. Taylor GI, von Karman Th (1937) J Roy Aeronaut Soc 41:1109
13. Taylor GI (1923) Phil Trans A 223:289–343
14. Malkas DS, Nohel JA, Plohr BJ (1991) SIAM J Appl Math 51(4):899
15. Malkas DS, Nohel JA, Plohr BJ (1990) J Comp Phys 87(2):464
16. Malkas DS, Nohel JA, Plohr BJ (1993) CMS Tech Sum Reprt #93-9
17. Fisher RJ, Denn MM (1976) AIChE J 22:236
18. Fisher RJ, Denn MM (1975) Chem Engg Sci 26:44
19. Fisher RJ, Denn MM (1977) AIChE J 23:197
20. Fisher RJ, Denn MM (1978) AIChE J 24:519
21. Fisher RJ, Denn MM, Tanner RI (1980) Ind Engg Chem Funda 19(2):195

22. Denn MM (1975) Stability of Reaction and Transport Process. Prentice-Hall, Englewood Cliffs, NJ
23. Eckhaus W (1965) Studies in Non-linear Stability Theory. Springer, Berlin Heidelberg New York
24. Zamir M (2000) The Physics of Pulsatile Flow. AIP Press, Springer, Berlin Heidelberg New York
25. Poiseuille JLM (1840) Comptes Rendus 11:961
26. Kalachev AA, Kardivarenko LM, Plate NA, Bargreev VV (1992) J Memb Sci 75:1
27. Thoresen K, Fisher RJ (1995) Biomimetics 3:31
28. Chen X, Fenton JM, Fisher RJ, Peattie RA (2004) J Electrochem Soc 151:E56
29. Converti A, Perego P, Lodi A, Fiorito G, Del Borghi M, Ferraiolo G (1991) Bioproc Eng 7:3
30. Shuler ML, Ghanem A, Quick D, Wang MC, Miller P (1996) Biotech Bioeng 52:45
31. Juhasz NM, Deen WM (1991) Ind Eng Chem Res 30:556
32. Deen WM (1996) Analysis of Transport Phenomena. Oxford Press, New York
33. Ovaici H, Mackley MR, McKinley GH, Crook SJ (1998) J Rheol 42:125
34. Achenie LEK, Fisher RJ (1994) SIAM Conf; CPFD
35. Koufas D (1996) MS Thesis, University of Connecticut, Storrs, CT
36. Tarbell JM, Qui Y (2000) Arterial Wall Mass Transport: The Possible Role of Blood Phase Resistance in the Localization of Arterial Disease. In: Bronzino JD (ed) The Biomedical Engineering Handbook, 2nd edn. CRC Press, Boca Raton, Fl, Ch. 100
37. Galletti PM, Colton CK, Jaffrin M, Reach G (2000) Artifical Pancreas. In: Bronzino JD (ed) The Biomedical Engineering Handbook, 2nd edn. CRC Press, Boca Raton, Fl, Ch. 134
38. Lewis AS, Colton CK (2004) Tissue Engineering for Insulin Replacement in Diabetes. In: Ma PX, Elisseeff J (eds) Scaffolding in Tissue Engineering. Marcel Dekker, New York (in press)
39. Freshney RI (2000) Culture of Animal Cells: A Manual of Basic Technique, 4th edn. Wiley-Liss, New York
40. Palsson B (2000) Tissue Engineering. In: Enderle J, Blanchard S, Bronzino JD (eds) Introduction to Biomedical Engineering. Academic Press, Orlando, Fl, Ch. 12
41. Jain RK (1994) Transport Phenomena in Tumors. In: Advances in Chemical Engineering. Academic Press, Orlando, Fl, 19:129–194
42. Shuler MJ (2000) Animal Surrogate Systems. In: Bronzino JD (ed) The Biomedical Engineering Handbook, 2nd edn. CRC Press, Boca Raton, NJ, Ch. 97
43. Fisher RJ (2000) Transport Phenomena and Biomimetic Systems. In: Bronzino JD (ed) The Biomedical Engineering Handbook, 2nd edn. CRC Press, Boca Raton, NJ, Section XII
44. Fisher RJ (2000) Biomimetic Systems. In: Bronzino JD (ed) The Biomedical Engineering Handbook, 2nd edn. CRC Press, Boca Raton, NJ, Ch. 95
45. Grasso D, Strevett K, Fisher RJ (1995) Chem Eng J 59:195
46. Levenspiel O (1989) The Chemical Reactor Omnibook. OSU Book Stores Inc., Corvallis, OR
47. Carberry JJ (1976) Chemical and Catalytic Reaction Engineering. McGraw-Hill, New York
48. Resnick N, Yahav H, Shay-Salit A, Shushy M, Schubert S, Zilberman LC, Wofovitz E (2003) Prog Biophys Mol Biol 81(3):177
49. Griffith TM (2002) Biorheology 39(3–4):307
50. Barbee KA (2002) Ann Biomed Eng 30(4):472
51. Ali MH, Schumacker PT (2002) Crit Care Med 30(5 Suppl):S198

52. McIntire L, Stamatas G (2001) Biotech Prog 17:383
53. Moraru CI, Panchapakesan CP, Huang Q, Takhistov P, Liu S, Kokini JL (2003) Food Tech 57(12):24
54. D'Aquino R (2003) Chem Eng Prog Supplement: 42S
55. Zhang Y (1998) MS Thesis, University of Connecticut, Storrs
56. Cussler EL, Aris R, Brown A (1989) J Membrane Sci 43:149
57. McGregor WC (1986) Membrane Separations in Biotechnology. Marcel Dekker Inc., New York
58. Zhang Q, Cussler EL (1985) J Membrane Sci 23:321
59. Zhang Q, Cussler EL (1985) J Membrane Sci 23:333
60. Gaylor JDS (1988) J Biomed Eng 10:541
61. Wickramasinghe SR, Semmens MJ, Cussler EL (1992) J Membrane Sci 69:235
62. Yang MC, Cussler EL (1986) AIChE J 32:1910
63. Sweeney LM, Shuler MJ, Babish JG, Ghanem A (1995) Toxicol In Vitro 9:307
64. Koufas D, Fisher RJ (1998) Proc 24th NEBC/IEEE Trans 9:12
65. Womersley JR (1955) J Physiol 127:553
66. Dwight HB (1961) Tables of Integrals and Other Mathematical Data. McMillan Publishing Co., New York
67. Gerrard JH (1971) J Fluid Mech 46(1):43
68. Womersley JR (1955) Phil Mag 46:199
69. Atabek SC, Lew HS (1966) Biophys J 6:481
70. Fung YC (1997) Biomechanics: Circulation. Springer, Berlin Heidelberg New York
71. Eckmann DM, Grotberg JB (1991) J Fluid Mech 222:329
72. McIntyre L, Stamatas G (2001) Biotech Prog 17:383
73. Darling R, Messina CR, Brewster DC et al. (1977) Circulation 56(suppl 2):II-161
74. Galland RB, Whitely MS, Magee TR (1998) Eur J Endovasc Surg 16:104
75. Scherer PW (1973) J Biomech 6:695
76. Tam MK, Melbin J, Knight D (1976) IEEE Trans on Biomed Eng 23:453
77. Fukushima T, Matsuzawa T, Homma T (1989) Biorheology 26:109
78. Yu SCM (2000) Int J Heat Fluid Flow 21:74
79. Budwig R, Elger D, Hooper H, Slippy J (1993) J Biomech Engr 116:418
80. Asbury CL, Ruberti JW, Bluth EI, Peattie RA (1995) Ann Biomed Engr 23:29
81. Peattie RA, Asbury CL, Bluth EI, Ruberti JW (1996) J Ultrasound Med 15:679
82. Peattie RA, Asbury CL, Bluth EI, Riehle TJ (1996) J Ultrasound Med 15:689
83. Peattie RA, Riehle TJ, Bluth EI (2004) J Biomech Engr 126(4):438
84. Finol EA, Amon CH (2001) J Biomech Engr 123:474
85. Perktold K (1987) J Biomech 20:311
86. Taylor TW, Yamaguchi T (1993) J Biomech Engr 116:89
87. Yu SC, Chan WK, Ng BT, Chua LP (1999) J Med Eng Technol 23:228
88. Raghavan ML, Webster MW, Vorp DA (1996) Ann Biomed Eng 24:573
89. Vorp DA, Mandarino WA, Webster MW, Gorcsan J (1996) Cardiovasc Surg 4:732
90. Vorp DA, Raghavan ML, Webster MW (1998) J Vasc Surg 27:632
91. Vorp DA, Trachtenberg JD, Webster MW (1998) Semin Vasc Surg 11:169
92. Sacks MS, Vorp DA, Raghavan ML, Federie MP, Webster MW (1999) Ann Biomed Eng 27:469
93. Raghavan ML, Vorp DA (2000) J Biomech 33:475
94. Smith DB, Sacks MS, Vorp DA, Thornton M (2000) Ann Biomed Eng 28:598
95. Raghavan ML, Vorp DA, Federle MP, Makaroun MS, Webster MW (2000) J Vasc Surg 31:760
96. Bluth EI, Murphey SM, Hollier LH et al. (1990) Int Angiol 9:8

97. Atkinson SJ, Feller KJ, Peattie RA (2001) ASME-BED 50:729
 98. Feller KJ, Atkinson SJ, Peattie RA (2001) ASME-BED 50:753
 99. Batchelor GK (1967) An Introduction to Fluid Dynamics. Cambridge University Press, Cambridge, UK
100. Fry DL (1968) Circ Res 22:165
101. Ling SC, Atabek HB, Fry DL, Patel DJ, Janicki BS (1968) Circ Res 23:789
102. Patel DJ, Fry DL (1969) Circ Res 24:1
103. Fry DL (1969) Circ Res 24:93
104. Inn YW, Fisher RJ, Shaw MT (1998) Rheol Acta 37:573

Adv Biochem Engin/Biotechnol (2006) 103: 157–187
DOI 10.1007/b137204
© Springer-Verlag Berlin Heidelberg 2005
Published online: 25 October 2005

Biopreservation of Cells and Engineered Tissues

Jason P. Acker[1,2]

[1]Department of Laboratory Medicine and Pathology, University of Alberta,
8249-114 Street, Edmonton, AB T6G 2R8, Canada
jason.acker@bloodservices.ca

[2]Canadian Blood Services, Research and Development, 8249-114 Street,
Edmonton, AB T6G 2R8, Canada
jason.acker@bloodservices.ca

Abstract The development of effective preservation and long-term storage techniques is a critical requirement for the successful clinical and commercial application of emerging cell-based technologies. Biopreservation is the process of preserving the integrity and functionality of cells, tissues and organs held outside the native environment for extended storage times. Biopreservation can be categorized into four different areas on the basis of the techniques used to achieve biological stability and to ensure a viable state following long-term storage. These include in vitro culture, hypothermic storage, cryopreservation and desiccation. In this chapter, an overview of these four techniques is presented with an emphasis on the recent developments that have been made using these technologies for the biopreservation of cells and engineered tissues.

Keywords Cryopreservation · In vitro culture · Hypothermic storage · Desiccation · Dry storage

1
Introduction

The development of effective preservation and long-term storage techniques is a critical requirement for the successful clinical and commercial application of emerging cell-based technologies [1–3]. As cell-based therapeutics approach clinical utility, many fundamental and practical issues involving the isolation and manipulation of cells are being addressed to allow translation of these technologies from bench to bedside [4, 5]. With the efficacy of tissue engineering, cell and tissue transplantation, and genetic technologies dependent on the native and induced characteristics of living cells, preserving the functional viability of engineered cells and tissues remains one of the most important challenges facing reparative medicine.

In the body natural processes preserve the physiological function of cells, tissues and organs. As cells are damaged, or age and die, biological events ensure that the cells are repaired or replaced. Unfortunately, when cells are removed from the body, changes in the external environment not only result in cell damage, but also an inhibition or elimination of the natural repair and replacement processes. As isolated cells become damaged and die the absence of replacement cells results in a gradual reduction in the biological activity of the overall system. Therefore, the biopreservation sciences aim to (1) develop techniques that preserve the integrity and functionality of cells, tissues and organs held outside the native environment and (2) extend the storage time of the preserved biological material.

Biopreservation is an important tool for clinical cell and tissue banking and the biotechnology industry in that it provides the necessary time required to produce and distribute engineered cells and tissues. Maintaining intact, functional cells through the isolation and screening process, product manufacturing, inventory control, distribution and end use is essential for successful development of an engineered product [2–4]. Delivery of cell-based therapeutic products in a regulated environment further requires component archival for quality control testing and validation of the engineering process. Testing for transmissible diseases and bacterial contamination and, if necessary, donor–recipient compatibility all require that the cells or engineered tissues be stored for a finite time prior to release. To offset differences in production capacity and end user demand, optimized inventory management requires the capability to stockpile and store the product at the manufacturing site or at the end-user location. All of these elements, when combined, require a significant amount of time for which cell and tissue function must be preserved ex vivo prior to transplantation or transfusion.

Biopreservation can be categorized into four different areas on the basis of the techniques used to achieve biological stability and to ensure a viable state following long-term storage. These include in vitro culture, hypothermic storage, cryopreservation and desiccation. In this chapter, an overview of these four techniques is presented with an emphasis on the recent developments that have been made using these technologies for the biopreservation of cells and engineered tissues.

2
In Vitro Culture

In vitro culture is the process of preserving the normal phenotypic properties of a cell population or tissue for extended times at physiological temperatures by replicating ex vivo the native environment. As cell proliferation and differentiation is dependent on the physical environment [6] and is regulated by signals from soluble factors [7, 8], extracellular matrix proteins [9–11] and cell interactions [9, 12–15], tight control of these variables is essential for successful in vitro culture. Over the past century, efforts to identify key cell and tissue-specific physiological and physicochemical determinants have resulted in this technique being widely adopted by the basic sciences and the biotechnology industry.

The ability to preserve cell viability and function ex vivo is an essential technology used in basic and applied research. In vitro culture allows researchers to develop well-characterized, homogenous cell lines that can be perpetuated over several generations (primary cultures) or indefinitely (transformed or continuous cell lines). With standardized cell culture conditions, in vitro expansion of a uniform cell population can quickly produce the necessary biological material to perform multiparameter studies and/or perform sensitive biochemical or genetic manipulation and analysis. In addition, culture of cells, native tissue or tissue explants allows for precise environmental control and manipulation, thereby minimizing experimental variability. For these reasons, in vitro culture has been instrumental in advancing virology [16], immunology [17], hematology [18], molecular genetics [19], pharmacology [20] and other basic and applied disciplines [21].

Large-scale in vitro culture and expansion of cells and tissues has been extensively used by the biotechnology industry for the production of commercial products. Vaccines, monoclonal antibodies, recombinant proteins, cytokines and other therapeutic agents are routinely produced from transfected prokaryotic and eukaryotic cells. Industrial microbiology [22] and mycology [23, 24] are used to manufacture a wide variety of products, including antibiotics, enzymes, amino acids, oligosaccharides, alcohols, insecticides and herbicides. In addition, commercial plant tissue culture produces a number of secondary products which are used in the food (i.e. flavouring agents) and pharmaceutical (i.e. terpenoids, quinines, lignans, flavonoids, alkaloids) industries [25, 26].

2.1
Trends in in Vitro Culture

The importance of in vitro culture to cell-based bioengineering and to the basic and applied sciences has been the motivation for active research in this area. Efforts to improve the productivity of large-scale cell culture have focused on engineering novel methods for the addition of nutrients, elimination of waste, component mixing and aeration of the cultured cells. Continuous culture of suspension cells in fed-batch bioreactors [27–29], hollow-fibre perfusion systems [30, 31], fluidized bed reactors [32] or microgravity culture systems [33, 34] and the development of microcarriers [35] and large-surface-area culture devices for use with anchorage-dependent cells has allowed for significant scaling of production to be achieved. As it is becoming increasingly clear that the cellular microenvironment has an important role in cell function, efforts have been taken to better understand the role of soluble factors and cell–cell and cell–matrix interactions on cell proliferation and differentiation.

The addition of animal serum to culture media has traditionally been a requirement to maintain cells in vitro. Serum contains systemic components (nutrients, hormones, growth factors, protease inhibitors) involved in the homeostatic regulation of cell-cycle progression. However, the high cost and the fluctuating quality and composition of serum and the potential introduction of adventitious agents into the culture process has motivated the development of serum-free, animal protein-free media [36–38]. Identifying specific nonproteinaceous substitutes for the proteins in conventional media has been a challenge [37]. As a result, current chemically defined media are not protein-free, but rely on recombinant growth factors and hormones to eliminate components of animal or human origin [36].

Cell adhesion to extracellular matrices [39–41] and homotypic and heterotypic cell–cell interactions [9, 15] are critical elements that modulate the genetic regulation of cell proliferation and differentiation. Studying the phenotypic changes that accompany alterations to culture conditions has been an effective means to better understand the regulatory mechanisms responsible for "normal" function and to improve techniques for preserving the in vivo phenotype of cultured cells. Over the past few years, microfabrication technologies [42–44] have emerged as extremely useful tools for constructing patterned extracellular matrices and for controlling cell–cell interactions. This emerging technology has already significantly enhanced the in vitro preservation of hepatocytes [15] and neurons [45], and will continue to impact the preservation sciences [46].

2.2
In Vitro Culture of Engineered Cells and Tissues

Efforts have been made recently to extend the in vitro culture approach to clinically important cells and engineered tissues. In vitro culture is being developed to preserve the cellular components used in the engineering of tissue constructs and for the ex vivo expansion of native and metabolically engineered and genetically engineered cells used in cellular therapies. A number of excellent reference texts have recently been published which discuss current methods used for the in vitro culture of a variety of different cells and engineered tissues [47, 48]. The development of dermal replacements and the ex vivo expansion of hematopoietic progenitor cells provide two examples of how this technology is being developed, and the impact it will have in clinical medicine.

Artificial skin substitutes were the first engineered tissue to be successfully constructed and preserved using in vitro culture [49, 50]. While the development of dermal models has traditionally been motivated by the clinical need for skin substitutes to treat traumatic skin defects (i.e. burns), there is considerable interest in using these engineered tissues to accelerate or manipulate the wound healing process [51], or as platforms for gene therapy [52]. The preservation and ex vivo expansion of human keratinocytes [50, 53, 54], fibroblasts [55], melanocytes [56, 57] and Langerhans cells [58] as purified monolayers in chemically defined media has allowed for the development of a number of artificial skin constructs [50, 59–62]. Typically, fibroblasts are cultured in a three-dimensional extracellular matrix, resulting in a simplified dermis that can be used as a foundation for the growth of a multilayered epidermis using keratinocytes. As a complex, interacting system, these dermal models have been used to study the relationship between the extracellular matrix and fibroblast differentiation [63], and the role that fibroblasts have in remodelling the extracellular matrix [64] and promoting keratinocyte growth and differentiation [65, 66]. Understanding the ability of cell–cell and cell–matrix interactions to regulate cell proliferation and differentiation will advance wound healing research and have a dramatic effect on improving existing methods for the in vitro culture of skin cells and engineered dermal replacements.

The capacity for hematopoietic progenitor cells to proliferate and differentiate into all of the blood cell lineages provides an attractive means to produce the cellular components needed for the treatment of a variety of malignant and nonmalignant disorders. As the absolute number of hematopoietic cells found in mobilized peripheral blood, bone marrow or umbilical cord blood is low, there has been an active interest in developing in vitro culture methods that would selectively increase specific hematopoietic progenitors [18, 67]. With the identification and development of recombinant cytokines that can induce both proliferation and differentiation of hematopoietic cells and the ability to selectively separate populations of progenitor and mature cells of interest, controlling the experimental conditions required to expand hematopoietic progenitor cells has been achieved [68, 69]. For example, the addition of the cytokines Flt-3 ligand, stem cell factor, thrombopoietin and specific interleukins has been used to increase the number of long-term culture initiating cells from umbilical cord blood that are used in the repopulation of the bone marrow following myeloablative therapy [70–72]. The ex vivo expansion of megakaryocytic cells has been actively pursued as a means to decrease the demand on donor-derived platelets used in the treatment of thrombocytopenia [73, 74]. Similarly, the ex vivo expansion of antigen-presenting dendritic cells [75] and cytotoxic lymphocytes [76] is being explored owing to the potential use of these cells for immunotherapy. The ability to preserve the phenotypic properties of a hematopoietic cell population and to manipulate the differentiation of the cells into specific mature lineages demonstrates the significant progress that has been made in advancing in vitro culture technology.

2.3
Limitations of in Vitro Culture

While in vitro culture has been used effectively for the long-term preservation of a wide variety of cells and tissues used in science and industry, it is not an ideal strategy for large-scale and/or long-term storage of cells and engineered tissues. Extended in vitro culture is an extremely expensive process owing to the high cost of the components used in culture media and the requirement for continued media replenishment to maintain cell proliferation or differentiation. As cells and tissues in culture are susceptible to contamination [77] and prone to phenotypic and genetic drift [6, 78], reproducibility of the culture and/or manufacturing processes requires expensive quality control measures that need to be performed regularly over the storage term. These costs quickly accrue and become prohibitively expensive for the extended storage of multiple cell types or large volumes of a specific cell population. While there are active measures to reduce the incidence of contamination and to improve the long-term genetic and phenotypic stability of cultured cells [77, 79], this will not significantly improve the economics of in vitro culture relative to the other preservation strategies.

In addition to the economic constraints of in vitro culture, this preservation technology places a number of limitations on product manufacturing and end use [4, 79]. Maintaining an adequate inventory of cells or engineered tissues to meet end-user demand can result in significant manufacturing costs. As ex vivo expansion of a cell population or the engineering of a tissue construct can require several weeks, the overproduction and subsequent loss of a significant amount of product may be required to ensure sufficient inventory to meet clinical demand. Just-in-time delivery can further complicate manufacturing and end use as sufficient time to rigorously assess the safety and quality of individual products may not be available [4]. For these reasons, alternative methods for the preservation of cells and engineered tissues are necessary.

3
Hypothermic Storage

Hypothermic preservation of cells, tissues and organs is based on the principle that biochemical events and molecular reactions can be suppressed by a reduction in temperature. In the context of biopreservation, hypothermic conditions are those in which the temperature is lower than normal physiologic temperature but higher than the freezing point of the storage solution. As chemical reaction rates are temperature-dependent, cooling below normal physiological temperatures inhibits metabolic processes that deplete critical cellular metabolites and accumulate injury. Through the exploitation of this beneficial effect of temperature, hypothermic preservation has been critical in the advancement of transfusion and transplant medicine by facilitating the extended storage of red blood cells [80], platelets [81], hepatocytes [82, 83], pancreatic islets [84], corneas [85, 86], native and engineered skin [87, 88] and solid organs [89, 90]. As changes in temperatures have significant effects on the physicochemical properties of aqueous systems, biochemical reaction rates and transport phenomena that will disrupt cell homeostasis [91, 92], understanding the biochemical and physiological implications of hypothermic exposure has led to the development of strategies to minimize hypothermia-related injury.

3.1
Hypothermia-Induced Injury

Hypothermia-induced cell injury can be attributed to a number of events, including membrane pump inactivation, disruption of calcium homeostasis, cell swelling and free-radical-induced apoptosis [92–94]. The hypothermia-induced inhibition of transmembrane pumps, such as the Na^+/K^+ ATPase and the mitochondrial electron transport system disrupts the ability of the cell to maintain the necessary ionic gradients and high-energy phosphates (i.e. ATP) required for normal metabolism. Accumulation of intracellular calcium owing to the effect of ATP depletion on Ca^{2+} transport and the release of sequestered Ca^{2+} can have detrimental effects on cell signalling pathways and cytoskeletal organization. The net diffusion of sodium chloride into the cell transiently increases the intracellular osmolality, resulting in cell swelling owing to the osmotic influx of water. Disruption of the electron transport system, the hydrolysis of ATP and the glycolytic production of lactate results in a marked decrease in intracellular pH. Iron released from intracellular protein stores and carriers as a result of a decreasing pH can catalyse the production of reactive oxygen species that can lead to the induction of apoptosis [95]. In addition to the disruptions in cellular metabolism, thermotropic membrane phase transitions [96] and temperature-induced denaturation of cytoskeletal elements [82] result in physical destabilization of cell membranes. While hypothermic storage can delay degradative cellular processes, without adequate steps to protect against the molecular and physicochemical effects of hypothermia, cell damage will occur.

3.2
Strategies for Hypothermic Storage of Cells, Tissues and Organs

The successful use of hypothermic temperatures for the preservation and storage of cells, tissues and organs has resulted from extensive efforts to minimize hypothermia-induced injury. Two different strategies have been developed for hypothermic preservation [89, 90, 94]. The first approach involves storage in specially formulated preservation solutions that modulate the physiological response to low temperatures. These solutions may contain elements that maintain ionic gradients, calcium homeostasis, buffer pH and/or scavenge free radicals. The second approach to hypothermic storage involves the continuous circulation of an oxygenated preservation solution through the organ or around the cells and tissues. Continuous hypothermic perfusion prevents ATP depletion and the accumulation of harmful metabolites.

Hypothermic storage allows red blood cells that are preserved and stored for up to 42 days at 4 °C to be used in the treatment of anemic patients. The ability to preserve the viability of red blood cells for extended periods has not only made it possible to bank and distribute blood, but more importantly, to meet the growing requirements for blood fractionation, cross-matching and transmissible disease testing. Red blood cell preservation is an excellent example of how an understanding of cell metabolism and hypothermia-related injury can lead to the development of improved preservation solutions [80, 97]. Current red blood cell preservation solutions contain sodium citrate to prevent coagulation, dextrose as a source of metabolic energy, sodium phosphate to maintain pH and adenine to sustain ATP levels. The collection, processing, storage and transfusion of more than 16 million units of red blood cells in the USA and Canada each year [98, 99] is a testament of the successful application of hypothermic storage in cell banking.

The hypothermic preservation and storage of human kidneys has been achieved by simple storage in specially formulated solutions and using continuous hypothermic perfusion. Studies of the metabolic and physicochemical response of organs to hypothermia [93] led to the development of a number of preservation solutions [100]. For example, the University of Wisconsin (UW) solution used in the preservation of livers, kidneys and pancreases uses the cell impermeant molecules potassium lactobionate, raffinose and hydroxyethyl starch to minimize cell swelling, adenosine to stimulate ATP production, glutathione to scavenge free radicals and potassium phosphate to maintain pH [89]. Flushing the kidney with UW solution allows for the hypothermic storage of canine kidneys for up to 72 h [101, 102] and human kidneys for approximately 24 h [103]. Machine perfusion of tissues and organs has been used to extend the hypothermic storage time by supplying the necessary oxygenated preservation solutions that allow the organ to continue to function aerobically. Using a modified UW solution, researchers have demonstrated successful hypothermic perfusion of canine kidneys for up to 7 days [104] and human kidneys for at least 32 h [103].

Hypothermic storage of engineered and native tissues has typically been achieved using static storage in culture media or commercial preservation solutions. However, the growing trend to use perfusion bioreactors in the manufacturing of engineered tissues [105–107] may allow for the future development of techniques for the long-term storage of these tissues using continuous hypothermic perfusion.

3.3
Limitations of Hypothermic Storage

Hypothermic storage in suitably designed preservation solutions is a relatively inexpensive method for the storage and transportation of cells and tissues. The commercial availability of quality-controlled hypothermic preservation solutions and universal access to refrigeration equipment does not place restrictive operating constraints on this technology. In contrast, hypothermic organ perfusion is a technically demanding procedure that requires specialized equipment and experienced personnel and as a result is relatively expensive. Efforts to development and market portable perfusion devices (i.e. LifePort; Organ Recovery Systems) will expand the number of medical centres capable of performing hypothermic organ perfusion. While cellular metabolism is slowed during hypothermic storage, it is not completely suppressed, and accumulating cell damage and cell death eventually result in a decrease in the biological activity of the system. Because of the limited shelf life of biological products that are stored at hypothermic temperatures, this technique is not currently a viable solution for long-term storage of engineered cells and tissues.

4
Cryopreservation

Cryopreservation is the process of preserving the biological structure and/or function of living systems by freezing to and storage at ultralow temperatures. As with hypothermic storage, cryopreservation utilizes the beneficial effect of decreased temperature to suppress molecular motion and arrest metabolic and biochemical reactions. Below $-150\,°C$ [108], a state of "suspended animation" can be achieved as there are very few reactions or changes to the physicochemical properties of the system that have any biological significance. To take advantage of the protective effects of temperature and to successfully store cells and engineered tissues for extended periods using cryopreservation techniques, damage during freezing and thawing must be minimized. Over the last century, enormous progress has been made in understanding the basic elements responsible for low-temperature injury in cellular systems and in the development of effective techniques to protect cells from this cryoinjury. As there are a number of excellent books [109–112] and recent review articles [2, 113, 114] that summarize the current understanding of the fundamental principles of cryoinjury and cryoprotection, only a brief synopsis will be presented here.

Cell injury is related to the nature and kinetics of the cellular response to the numerous physical and chemical changes that occur during freezing and thawing. Under normal physiological conditions, when a cell suspension is cooled below the freezing point of the suspending solution, ice will form first in the extracellular space. As the cell membrane serves as an effective barrier to ice growth [115], and the cytoplasm contains few effective nucleating agents [116, 117], intracytoplasmic ice formation does not immediately occur. Extracellular ice nucleation results in the concentration of solutes in the unfrozen fraction. The development of a chemical potential difference across the cell membrane provides the driving force for the efflux of water from the cell. With additional cooling, more ice will form extracellularly, and the cell will become increasingly dehydrated. If the cooling rate is sufficiently slow, the movement of water across the membrane will maintain the intracellular and extracellular composition close to chemical equilibrium. Injury during slow cooling has been correlated with excessive cell shrinkage [118–120] and toxicity owing to the increasing concentrations of solutes [121, 122].

As the permeability of the plasma membrane to water is temperature-dependent, when cells are cooled rapidly, the formation of ice in the external solution and the concentration of extracellular solutes occur much faster than the efflux of water from the cell. This results in the cytoplasm becoming increasingly supercooled with an associated increase in the probability of intracellular ice nucleation. While the mechanism by which intracellular ice formation occurs and the means by which it damages the cell have not yet been resolved [123], the current tenet is that intracellular ice formation in cells in suspension is an inherently lethal event that should be avoided [123–126].

As cryoinjury results when cells are cooled too slowly (owing to exposure to high concentrations of solutes) or too rapidly (owing to intracellular freezing), an optimal cooling rate can be determined for a specific cell type under known conditions [127]. Unfortunately, cryoinjury is associated with a wide variety of physical and chemical events that occur during freezing and thawing. Cytoplasmic supercooling [128], ice nucleation and ice crystal morphology and growth [125, 129–131], osmotic stress [119, 132, 133], solute-related stresses [121, 122, 134], thermal gradients [135, 136] and recrystallization [137] and/or devitrification [138] during rewarming all effect the post-thaw viability of cryopreserved cells. As these variables are interdependent, determining the optimal cooling and warming conditions is very difficult to resolve empirically. Identifying mechanistic and phenomenological models of cryoinjury has assisted in the development of mathematical models that predict the low-temperature response of cells [133, 138–142]. These models have hastened the optimization of the freezing and thawing process for cell suspensions.

While the principles governing cellular cryobiology have been extensively studied, a lack of understanding of the mechanisms responsible for tissue damage during freezing and thawing has limited the successful cryopreservation of tissues used in clinical or industrial applications [143, 144]. The quality of cryopreserved tissues is a function of the viability of the constituent cells and the continuance of an intact tissue structure. Loss of cell viability or damage to the extracellular matrix during freezing and thawing will result in a severe reduction in overall tissue function. Cryopreservation of tissues therefore requires knowledge of the individual and combined contributions of the cell and matrix components to the overall response of the tissue to freezing and thawing.

There are a number of unique elements that complicate the cryobiology of tissue systems [2, 143, 144]. The macroscopic size and defined geometry of tissues results in heat and mass transfer constraints that can lead to spatial variations in cryoprotectant concentration and in achieved cooling and warming rates. This effect is further compounded by the heterogeneity of cell types within tissues and the fact that each specific cell type has well-defined optimal cooling and warming rates and cryoprotectant requirements [124]. In addition, the characteristic cell–cell and cell–matrix interactions of tissue systems may act as critical targets for [145, 146] or mediators of cryoinjury [147, 148]. Finally, the formation of extracellular ice within a tissue (i.e. within the matrix and/or intravascular space) can result in significant injury. Excessive dehydration of the surrounding cellular components [149] and/or damage to the vascular network [150, 151] will compromise the integrity of the tissue. The combined effect is that there is an irregular distribution of damage sites in intact cryopreserved tissue that will be localized to different layers of the tissue or to specific cell types within the tissue [149, 152, 153]. Development of protocols for the cryopreservation of tissues must consider not only the in situ cellular function, but also the effects that tissue structure and composition have on the low-temperature response of the cells.

4.1
Cryopreservation: Freeze–Thaw and Vitrification

Damage to cells can be caused by both intracellular ice formation and exposure to high concentrations of solutes [127]. The successful cryopreservation
of a wide variety of cell types has been a result of the development of effective techniques to minimize these two factors affecting cell survival. During
slow cooling, a reduction in the extracellular ice formed can limit the concentration of extracellular solutes and hence the degree of damage. Similarly,
during rapid cooling, if intracellular ice formation can be inhibited or limited,
then the scale of damage done to the cells can be significantly reduced. Minimizing the detrimental effects of ice formation during freezing and thawing
has therefore been the focal point for the cryopreservation of cell and tissue systems. To eliminate the damaging effects of ice formation, two different
cryopreservation approaches have emerged: freeze—thaw preservation and
vitrification.

In 1949, Polge et al. [154] introduced the idea of using chemical compounds to enhance the survival of frozen biological material. With the addition of glycerol to their samples, they were able to demonstrate a significantly
greater proportion of viable avian spermatozoa after thawing from $-70\,°C$.
Over the next 50 years, a number of chemical compounds were shown to
protect against the damaging effects of freezing and thawing [155, 156]. Effective cryoprotectants are relatively nontoxic at high concentrations and can
be broadly classified into two groups on the basis of the permeability of the
cell membrane to these agents [157]. Permeable cryoprotectants are small,
nonionic molecules that function to decrease the amount of ice present at
a given temperature. By lowering the temperature at which a cell is exposed
to the increasing intracellular and extracellular solute concentrations, pen-
etrating cryoprotectants mitigate damage owing to excessive cell shrinkage
and/or solute toxicity (slow cooling injury). Nonpermeable cryoprotectants
are generally long-chain polymers that act by dehydrating the cell prior to
freezing, thereby reducing the amount of intracellular water and hence the
probability of intracellular ice formation (rapid cooling injury). While relatively nontoxic at physiological temperatures, the rapid addition or removal
of high concentrations of cryoprotectants can generate damaging cell volume
fluctuations [132] that can be exacerbated by the rapid fluctuations of the intracellular and/or extracellular concentration of cryoprotectants in unfrozen
compartments during freezing and thawing [118–122]. Innovative protocols
have therefore been developed to minimize cryoprotectant toxicity by delay-
ing cell exposure to high cryoprotectant concentrations to lower temperatures
and to avoid damaging osmotic volume excursions through gradient or step-
wise addition and removal processes.

The detrimental effects of ice formation can be eliminated if the formation
of ice is completely avoided. In the context of cryopreservation, vitrifica-

tion is the process by which an aqueous solution bypasses ice formation and becomes an amorphous, glassy solid. By preventing the formation of a crystalline solid (ice), and the corresponding intracellular and extracellular solute accumulation, this method provides a means to significantly reduce the damage done to cells and tissues during freezing [131, 158]. However, in order to vitrify a sample, high concentrations of cryoprotectants and/or ultrarapid cooling rates must be employed. Devitrification, or the formation of ice crystals in an amorphous sample [159], can occur during suboptimal storage or slow warming and can result in significant damage to vitrified biological systems [150, 160]. Investigation of methods to add and remove high concentrations of cryoprotectants [161, 162] and the identification of glass-forming agents with reduced toxicity [160, 163, 164] have reduced the requirement for ultrarapid cooling rates. Similarly, high hydrostatic pressure [158, 165], synthetic ice blocking agents [166, 167] and the use of natural antifreeze proteins [168, 169] have been used to promote vitrification at lower cryoprotectant concentrations or at lower cooling rates and to minimize devitrification.

While there are inherent differences in the two approaches used to cryopreserve cells and tissues, both freeze–thaw and vitrification methods have resulted in the successful preservation and long-term storage of a variety of cell and tissue types from numerous species. Critical to the application of these cryopreservation strategies has been the multidisciplinary effort to elucidate the mechanisms and conditions responsible for cell injury. The following examples demonstrate the challenges that continue to face the cryopreservation of cells and engineered tissues using the freeze–thaw and vitrification approaches.

4.2
Freeze–Thaw Cryopreservation

The freeze–thaw method has been the traditional approach used to cryopreserve cells. A generic freeze–thaw protocol would involve slow freezing in the presence of a moderate concentration (1 M) of a chemical cryoprotectant (typically glycerol or dimethyl sulfoxide), storage at or below – 80 °C and then rapid thawing. While seemingly straightforward, determining what constitutes slow freezing or rapid thawing or what concentration of cryoprotectant will result in optimal post-thaw survival for the cell of interest is not trivial. As cell biology affects the interdependence of the cooling rate, the warming rate, the cryoprotectant concentration and the rate of cryoprotectant addition and removal, a cryopreservation protocol that works for one cell type may not work for another. Measurement of the permeability of the cell membrane to water and cryoprotectants, the incidence of intracellular ice formation as a function of the cooling rate and the toxicity limits of cryoprotectants has allowed researchers to use mathematical models [133, 138–142] to derive the freeze–thaw parameters for a specific cell line. These parameters would then be experimentally validated. Optimized freeze–thaw techniques have dramatically improved the quality and utility of cryopreserved cells as exemplified by the more than 75% post-thaw survival of red blood cells stored for more than 37 years used in transfusion medicine [170], the extended storage of bovine spermatozoa used in cattle breeding [171] and the use of cryopreserved plant germplasm in agriculture [172].

While freeze–thaw cryopreservation of single cells is a routine technique used in a wide variety of industries there is a significant interest in advancing current methods so as to improve the post-thaw recovery of viable cells. Existing freeze–thaw techniques result in between 30 and 90% of the cells being viable following preservation and storage [173]. In most applications, damage resulting from cryopreservation can be compensated for by ensuring that there is a significant margin between the minimum cell concentration required to achieve a result and the number of viable cells recovered following freezing and thawing. However, in many instances, it has become more and more critical that the number of viable cells following cryopreservation equals the number of cells cryopreserved (i.e. 100% recovery). This is clearly the case in umbilical cord blood banking [174] where there is no opportunity to increase the volume or number of $CD34^+$ cells collected from a donor, so minimizing damage to cord blood progenitor cells during cryopreservation is much more important. In addition, rigorous assessments are now revealing that the recovery of viability may be too low to guard against cell selection based on genetic or phenotypic attributes [78]. This is exemplified in the renewed interest to improve the recovery of human and animal spermatozoa after cryopreservation [175–177].

To improve the post-thaw recovery of functionally viable cells, cryobiologists are actively exploring a number of potential strategies. As existing freeze–thaw techniques are based on minimizing the biophysical response to ice formation and cryoprotectant addition and removal, improved mathematical models to predict ice growth [138, 178] and the cell osmotic response [179–182] will allow development and more accurate simulation of novel crypopreservation protocols. New approaches to cryoprotection such as the use of intracellular sugars [183–185] or intracellular ice [186, 187] will lead to better methods to preserve cell viability. Experimental and theoretical research is uncovering new models of cell cryoinjury that will lead to new methods to protect cellular systems [188, 189]. Finally, advances in molecular biology and biochemistry have created an intense interest in the molecular response of cells to freezing and thawing. As damage to critical subcellular structures during freezing and thawing may not immediately result in physical cell injury, the activation of molecular-based events may lead to cell necrosis or the induction of apoptosis [190, 191]. The identification of these molecular triggers and the development of effective inhibitors will improve the post-thaw recovery of cryopreserved cells [192–194]. By integrating a molecular-based understanding of the cellular response to freezing and thawing with existing physico-chemical-based models, a more comprehensive foundation for the development and improvement of freeze–thaw cryopreservation protocols will result.

As the physical structure of tissues creates unique challenges for freeze–thaw cryopreservation [143], only marginal success has been made in translating cell-based preservation methods to tissue systems. Even among those tissues being routinely cryopreserved, it is becoming increasingly apparent that the quality of cryopreserved tissues needs to be improved. The freeze–thaw cryopreservation of pancreatic islets used to treat diabetic patients has been routinely used since the late 1970s [195]. However, significant post-thaw cell loss, abnormal insulin secretion and delayed loss of function result in a decline in the long-term function of transplanted human islets [196]. Similarly, freeze–thaw cryopreserved skin [88, 197], heart valves [198, 199] and vascular tissue [200, 201] are routinely stored in tissue banks for use in transplant medicine even though the cellular constituents of these tissues may not be adequately preserved. This lack of successful application of cell-based freeze–thaw cryopreservation methods to tissue systems has prompted many researchers to question the underlying assumptions used to develop tissue-based cryopreservation protocols [2, 144, 189].

Recent efforts to develop protocols for the cryopreservation of articular cartilage serve as an excellent example of the challenges facing the preservation of tissue systems. Articular cartilage consists of an extracellular matrix composed of collagen fibres and other large molecules within which chondrocytes are embedded in well-defined regions. While the freeze–thaw preservation of isolated chondrocytes can result in more than 80% of the cells remaining viable, application of the same technique to intact tissues gives very poor in vitro and in vivo results [202, 203]. As viable chondrocytes in cryopreserved articular cartilage are localized to the periphery of the tissue [204, 205], efforts have been taken to better understand the permeability of isolated chondrocytes [206] and intact cartilage to cryoprotectants [207] and to characterize the formation of ice within intact cartilage [149, 204]. Damaged chondrocytes in the intermediate layer of cryopreserved cartilage are detached from the extracellular matrix and appear significantly shrunken [208]. This research has led to the hypothesis that the structure of articular cartilage affects ice growth and solute transport, resulting in significant mechanical and osmotic stresses that lead to chondrocyte injury [149, 189]. These biophysical events are further complicated by recent work examining the molecular response of chondrocytes to cryopreservation and the role that cell–cell matrix interactions have on cell proliferation [209]. The relevant questions in the cryobiology of articular cartilage are therefore similar to those for other tissues—heat and mass transfer in three-dimensional porous structures, ice nucleation and growth in complex systems and the role of cell–cell and cell–matrix interactions. Resolving these issues will be essential if current efforts to design freeze–thaw cryop-reservation protocols for the preservation of native and engineered tissues are to be successful.

4.3
Vitrification of Cells and Tissues

While the idea of using vitrification as a means to preserve cells, tissues and organs has been around since the 1930s [210], practical ice-free cry-opreservation was not achieved until 1985 when the vitrification of mouse embryos was successfully demonstrated [211]. To realize the high con-centrations of cryoprotectant needed to vitrify at readily attainable cool-ing rates, tolerable concentrations of multiple cryoprotectants with good glass-forming characteristics were combined. This first-generation vitrifica-tion solution, termed VS1, contained 20.5% dimethyl sulfoxide, 15.5% ac-etamide, 10% propylene glycol and 6% poly(ethylene glycol) and could be vitrified when cooled at $20\,°C/min$ and rapidly warmed to avoid devitri-fication [158, 211]. Vitrification was rapidly adopted as a practical replace-ment for the freeze–thaw cryopreservation of spermatozoa, oocytes and em-bryos used in reproductive medicine and animal husbandry owing to the increased simplicity, cost-effectiveness and speed of the preservation proced-ure [212–214]. In addition to the preservation of mammalian reproductive cells, vitrification solutions and cooling and warming protocols have been de-veloped for the ice-free cryopreservation of a variety of cell types, including monocytes [215], Drosophila embryos [216, 217], cell lines [218] and plant germplasm [219].

Preventing the physical damage resulting from the formation of ice in frozen tissues and organs has been the motivation for the development of ice-free cryopreservation methods for these complex systems [220]. The initial challenge facing tissue and organ cryopreservation was the development of carrier media and delivery methods for the equilibration of tissues and or-gans with the high concentrations of cryoprotectant required for vitrification at readily achievable cooling rates [221]. Significant advances in understand-ing cryoprotectant toxicity and in developing improved vitrification solu-tions [160, 222–224] have resulted in a number of tissues being successfully vitrified, including cornea [225], skin [226], islets [227], liver slices [228, 229], ovaries [230] and veins [231]. The perfusion of kidneys with a vitrifiable con-centration of cryoprotectants has been successfully demonstrated as a first step towards the vitrification of a whole organ [232].

As efforts to identify new cryoprotectants and to reduce the concentration of existing vitrification solutions using molecular ice blockers and antifreeze proteins continue [166, 169], research has begun to focus on the issue of de-vitrification. If vitrified tissues and organs are not warmed fast enough, ice crystals will form when the solution is brought above the glass-transition temperature. Unlike cell suspensions, the size and structure of tissues and organs reduces heat transfer and results in significant nonuniform heating. Efforts to develop technologies that use electromagnetic heating to minimize devitrification during the warming of vitrified tissues and organs are under way [233–235]. The future utility of vitrification in tissue and organ preservation will benefit from the development of methods to reduce cryoprotectant toxicity by improving the glass-forming properties of the vitrification solutions and to uniformly cool and warm large samples.

4.4
Limitations of Cryopreservation

Although cryopreservation has been used extensively for the long-term storage of various clinically important cells and native and engineered tissues, significant areas for improvement exist. First, the high concentrations of the chemical cryoprotectants used in freeze–thaw cryopreservation and vitrification can adversely affect transplant patients therefore necessitating costly post-thaw removal [236]. Dimethyl sulfoxide used in the cryopreservation of hematopoietic progenitor cells, for example, has been linked to gastrointestinal [237] and cardiovascular [237, 238] side effects and must therefore be removed through a complex series of washes that have been shown to reduce the absolute number and viability of the recovered cells [239]. Second, there are a number of commercial and clinically important cell types from different species that have limited function after cryopreservation, including platelets [240, 241], hepatocytes [242], granulocytes [243], spermatozoa [175, 244] and oocytes [245]. Moreover, the increased complexity of tissues compared with isolated cells limits the ability to use freeze–thaw techniques to preserve cell function at low temperatures [2, 143]. Finally, the cryopreservation process itself is costly and requires highly trained technicians and specialized equipment for processing, storage and distribution to end users, making it logistically prohibitive for routine use in large-scale or remote operations. With continued research, freeze–thaw and ice-free cryopreservation may overcome some of these limitations.

5
Desiccation and Dry Storage

In natural systems, desiccation is used as a strategy to preserve biological activity through times of extreme environmental stress. Termed anhydrobiosis, the ability to survive in a dry state for extended periods has been identified in a variety of diverse organisms, including plants, bacteria, yeasts, nematodes, fungi and crustaceans [246, 247]. Studies of these organisms have revealed a series of complex molecular and physiological adaptations that permit survival despite water loss exceeding 99%. Natural mechanisms of protection during desiccation include scavenging of reactive oxygen species [248, 249], downregulation of metabolism [250, 251] and the accumulation of amphiphilic solutes [249, 252], proteins [249, 253] and disaccharides [254]. These naturally occurring protective processes are being used as the foundation for the development of methods for the preservation of desiccation-sensitive biological material.

5.1
Adaptive Protection from Reactive Oxygen Species

The transient formation of partially reduced and activated forms of reactive oxygen contributes to the oxidative destruction of phospholipids, DNA and proteins in desiccation-sensitive cells. Exposure to toxic intermediates and products of oxygen metabolism is a result of an impairment of the electron transport chain in desiccated cells. Desiccation-tolerant organisms have developed a number of strategies to protect against oxidative damage, including the synthesis of antioxidants and the elimination of oxygen from the cells [249, 255].

The synthesis of specific antioxidant molecules has been shown to protect anhydrobiotic plants from the damaging effects of reactive oxygen [248, 255, 256]. The overproduction of enzymatic antioxidants such as lipoxygenase inhibitors [257] and ascorbate peroxidase [258] and the alteration of molecular antioxidants (ascorbate, tocopherol [256, 259]) occur during dehydration. As the regulation and role of specific antioxidants in desiccation tolerance is not yet resolved [249, 256], continued research to identify and characterize new antioxidants and reactive oxygen scavenging systems in anhydrobiotic organisms is needed.

An alternative strategy used by anhydrobiotes to protect against the damaging effects of oxygen is to reduce the intracellular concentration of oxygen during drying. As the formation of reactive oxygen species is dependent on the oxygen concentration, the synthesis of oxygen-binding proteins and the production of an oxygen-impermeable extracellular envelope protects cyanobacteria from reactive oxygen species [260]. In plants, a controlled increase in cytoplasmic viscosity has been suggested as another means by which the intracellular oxygen concentration is reduced during drying [261].

To minimize the damaging effects of oxygen a number of the strategies used by anhydrobiotes have been adapted for use with desiccation-sensitive cells. The protective effects of storage and rehydration under a vacuum [262–264], in nitrogen environments [265, 266] and in the absence of light [263, 267] have been investigated. The recovery of desiccated human foreskin fibroblasts [263] and mesenchymal stem cells [264] has been shown to significantly improve when they are stored under a vacuum and in the dark. This work suggests that protecting desiccation-sensitive cells from reactive oxygen species may have an important role in improving the desiccation tolerance of these cells.

5.2
Intracellular Sugars and Desiccation Tolerance

The best-characterized adaptation used by anhydrobiotes to protect biological structures during dehydration and dry storage has been the synthesis of intracellular and extracellular disaccharides. Sugars are believed to play a major role in the stabilization of membranes, proteins and other key cellular structures. The mechanism of sugar protection is an active area of research that includes the role of the glassy state in long-term stabilization [268, 269], and the interaction of sugars with biological molecules and supramolecular structures to afford stabilization [270–272]. By incorporating sugars into preservation media, freeze-drying (or lyophilization) has been used successfully for the dehydration and storage of pharmaceutical agents [273, 274], bacteria [260, 275–277], yeasts [278–280] viruses [281] and liposomes [282]. Current efforts are focusing on the use of sugars in the desiccation of mammalian cells.

It has been found that for a sugar to be maximally effective at protecting against the damaging effects of dehydration it must be present on both sides of the plasma membrane. The successful freeze-drying of liposomes [282], bacteria [277] and yeasts [279] has been shown to be dependent on the presence of intracellular and extracellular sugars. As mammalian cells do not naturally synthesize desiccation-important disaccharides, nor is the plasma membrane permeable to them, getting the sugars into the cells has been a major research effort. Trehalose has been shown to be one of the commonest disaccharides found in anhydrobiotes [246, 247] and has been the most popular disaccharide used in studies to induce desiccation tolerance in mammalian cells [283, 284]. Techniques used for the intracellular accumulation of trehalose have included viral transfection of the trehalose synthase genes [263, 285–287], thermal [185, 263, 288] and osmotic shock [263], microinjection [183], electroporation [289], and the use of a metal-actuated switchable membrane pore [290, 291]. While each technique suffers from one or more practical limitations [292], these methods have been successfully shown to enhance the intracellular concentration of sugars.

The presence of intracellular trehalose alone is insufficient to preserve mammalian cells during drying. Other factors that have been shown to modulate the effect of intracellular sugars and improve the survival of desiccated mammalian cells have included the concentration of intracellular sugar [290, 293], the rate of drying [263], the final moisture content [290, 291], the presence of oxygen and/or light [263], the storage temperature [290] and the rate of rehydration [294]. For example, intact murine fibroblasts reversibly permeabilized using a switchable membrane pore and dried in 0.4 M trehalose solutions were recovered after 90-day storage at − 20 °C [290]. More than 70% of fibroblasts dried to 10% residual moisture levels in the presence of isotonic trehalose solutions (0.2 M) have been shown to grow and divide following rehydration [291]. Similarly, the presence of intracellular trehalose in human fibroblasts [263, 285], embryonic kidney cells [285] and mesenchymal stem cells [264] has been shown to enhance cell recovery following air-drying and storage when held under a vacuum. These initial successes would strongly suggest that continued efforts to characterize and understand the critical parameters affecting the use of intracellular sugars for the stable storage of desiccated mammalian cells are warranted.

The freeze-drying of human platelets has been one of the most recent successful demonstrations of the protective effect of intracellular sugars. Platelet concentrates currently stored at 22 °C for up to 5 days are used for the clinical treatment of thrombocytopenia. Clinical preservation of platelets has been particularly challenging owing to the activation of platelets when they are exposed to temperatures below 20 °C [240, 295]. Using a temperature-sensitive endocytotic pathway to accumulate intracellular trehalose, Crowe and co-workers have shown that platelets freeze-dried in 150 mM trehalose and 5% albumin can be successfully rehydrated following 22-month storage at room temperature [288, 294]. Rehydrated platelets exhibited normal morphology and responded normally to clot-inducing agonists. While a demonstration of the clinical efficacy of freeze-dried platelets has yet to be performed, this work clearly indicates the future possibilities for using intracellular trehalose to increase the desiccation tolerance of mammalian cells and engineered tissues.

5.3
Quiescence and Diapause

The controlled entry into a hypometabolic or developmentally arrested state is another protective mechanism used by natural systems that undergo seasonal exposure to environmental stress. Quiescence and diapause are two naturally occurring methods by which organisms can enter a dormant state and have been shown to increase the tolerance of organisms to desiccation, freezing and anoxia [250, 251, 296]. Unlike quiescence, where entry into a hypometabolic state is induced by the environmental conditions, diapausing organisms enter this state prior to the environmental stress when growth conditions are optimal [250]. The systematic downregulation of energy-producing and energy-consuming processes [251], the reduction in macromolecular degradation [297] and the production of protective solutes [246, 249, 298] maintain the viability of the cell during the environmental insult. While the dormant state can last for months or years, the length of survival has been shown to be directly proportional to the degree of metabolic depression [299].

The coordinated downregulation of metabolism and entry into a dormant state in response to (quiescence) or in anticipation of (diapause) an environmental stress has been demonstrated in a majority of the animal phyla [247, 250, 251]. Resurrection plants [258] and seeds from a variety of plant species [300–302] display a significant reduction in respiration and photosynthetic activity prior to or during desiccation. Annual killifish survive seasonal droughts owing to the desiccation tolerance of their diapausing embryos. By depressing their oxidative metabolism by up to 90% and their protein synthesis by more than 93%, killifish embryos are remarkably resistant to anoxia and dehydration [250]. The best studied and most dramatic demonstration of the protective effects of metabolic depression is seen in the embryos of brine shrimp (*Artemia franciscana*) that survive severe anoxia in hypersaline lakes [297, 303, 304]. Entry into anaerobic quiescence occurs under fully hydrated conditions and is accompanied by a major depression in the respiration rate and protein synthesis that allows *Artemia* cysts to lie dormant until environmental conditions are conducive for normal growth and development [304]. As the signals responsible for the induction of hypometabolism are still relatively unknown, there has been an intense interest in better understanding the biochemistry and molecular pathways involved in quiescence and diapause. Changes in cytosolic adenylate sources [305], intracellular pH [299, 306], protein synthesis and phosphorylation [297, 299], and mitochondrial involvement [299, 307] remain active areas of investigation.

As researchers identify the molecular mechanisms responsible for entry into and exit from hypometabolic and developmentally arrested dormant states, the possibility of metabolically and/or genetically engineering mammalian cells to mimic this behaviour becomes possible. Efforts to modulate mammalian cell proliferation and growth by regulating cell size, shape, interaction with other cells and the extracellular matrix, and the addition or removal of soluble factors represent a first attempt to use native pathways [13, 14]. The potential benefit that induced quiescence and/or diapause would have in protecting mammalian cells from desiccation injury warrants expanding current efforts to exploit data emerging from studies of natural systems.

5.4
Future of Desiccation and Dry Storage

Although a relatively new technology, dry storage provides a long-term preservation strategy that alleviates many of the problems associated with other preservation technologies. First, dried storage is the only preservation method that permits the ambient temperature, long-term storage of biological molecules. This simplifies the distribution of the therapeutic product and reduces the need for stringent storage requirements at the manufacturing site and the end user's facility. Second, as the process of drying removes a substantial proportion of the sample water, the resultant product is much smaller and lighter than conventionally preserved products. This can significantly increase the storage capacity of a facility with a resulting increase in inventory levels. Third, the general acceptance of drying as a suitable method for the manufacturing of therapeutic products by regulatory authorities [308] makes this technology an appealing option for biopreservation. Finally, the significantly lower concentration of stabilizers used in dry storage compared with that used in standard freeze–thaw or vitrification protocols may eliminate the need for the costly and time-consuming removal of these agents before injection, transfusion or transplantation.

While the potential benefits are significant, the technical and scientific challenges facing the development of clinical and commercial methods for the desiccation and dry storage of mammalian cells are formidable. As natural anhydrobiosis is likely a result of multiple adaptive mechanisms, several different approaches may be needed to enhance the desiccation tolerance of mammalian cells. While the introduction of intracellular sugars has been shown to improve the survival of cells following desiccation, little is known about the mechanisms of damage, nor is there a consensus on the protective effects of sugars. Careful examinations of the molecular and biophysical effects of drying and the stabilizing effects of sugars in the cellular microenvironment are needed. The basic science of desiccation and dry storage of mammalian cells is only now emerging and many of the issues involved in the translation of this technology to the clinical and industrial preservation of cells and engineered tissues have not yet been addressed. Fortunately, many of the scale-up and processing issues involved in the desiccation and dry storage of biological material have been developed for the pharmaceutical and food science industries [309, 310]. Continued interdisciplinary research efforts are required to further develop this rapidly emerging area of biopreservation.

6
Conclusion

Preserving cell viability and function is an essential component in the translation of engineered cells and tissues from the bench to the bedside. As a number of different strategies are being used for the preservation and storage of native cells, tissues and organs, efforts are being made to apply this technology to engineered cells and tissues. With the number of different preservation choices available, selecting a technology that is appropriate for a specific engineered cell or tissue will depend on the intended clinical application, the logistics surrounding its manufacturing and distribution and, ultimately, the length of preservation and storage that can be successfully achieved in the system of interest. While there are a number of challenges facing the application of each technology, through interdisciplinary research efforts significant progress has been made to overcome these limitations. As the need for effective preservation technologies in cell and tissue engineering will be the motivation for more concerted efforts in the biopreservation sciences, there are encouraging prospects for the future of this science.

References

1. Langer R, Vacanti JP (1993) Science 260:920
2. Karlsson JOM, Toner M (2000) In: Lanza RP, Langer R, Vacanti JP (eds) Principles of tissue engineering. Academic, New York, p 293
3. Griffith LG, Naughton G (2002) Science 295:1009
4. Naughton GK (2002) Ann NY Acad Sci 961:372
5. Toner M, Kocsis J (2002) Ann NY Acad Sci 961:258
6. Freshney RI (2000) Culture of animal cells: a manual of basic technique. Wiley-Liss, Toronto
7. Iyer VR, Eisen MB, Ross DT, Schuler G, Moore T, Lee JCF, Trent JM, Staudt LM, Hudson J, Boguski MS, Lashkari D, Shalon D, Botstein D, Brown PO (1999) Science 283:83
8. Andrews RG, Briddell RA, Appelbaum FR, McNiece IK (1994) Curr Opin Hematol 1:187
9. Gumbiner BM (1996) Cell 84:345
10. Zieske JD (2001) Curr Opin Ophthamol 12:237
11. Mooney DJ (1992) J Cell Physiol 151:497
12. Koller MR, Papoutsakis ET (1995) In: Hjortso MA, Roos JW (eds) Cell adhesion fundamentals and biotechnological applications. Dekler, New York, p 61
13. Nelson CM, Chen CS (2002) FEBS Lett 514:238
14. Chen CS, Mrksich M, Huang S, Whitesides GM, Ingber DE (1997) Science 276:1425
15. Bhatia SN, Balis UJ, Yarmush ML, Toner M (1999) FASEB 13:1883
16. Hsiung GD (1989) Yale J Biol Med 62:79
17. Falkenberg FW (1998) Res Immunol 149:542
18. Aglietta M, Bertolini F, Carlo-Stella C, De Vincentiis A, Lanata L, Lemoli RM, Olivieri A, Siena S, Zanon P, Tura S (1998) Haematologica 83:824
19. Moyer MP (1989) Med Prog Technol 15:83
20. Lipman J, Flint O, Bradlaw J, Frazier J, McQueen C, Green C, Acosta D, Harbell J, Klaunig J, Resau J, Borenfreund E, Mehta R, Van Buskirk RG, Ekwall B, Ham R, Barnes D, Hay R, Schaeffer W (1992) Cytotechnology 8:129
21. McKeehan WL, Barnes D, Reid L, Stanbridge EJ, Murakami H, Sato GH (1990) In Vitro Cell Dev Biol 26:9
22. Lancini G, Lorenzetti R (1993) Biotechnology of antibiotics and other bioactive microbial metabolites. Plenum, New York
23. Finkelstein DB, Ball C (1992) Biotechnology of filamentous fungi: technology and products. Butterworth-Heinemann, Toronto
24. Johri BN, Satyanarayana T, Olsen J (2001) Thermophilic moulds in biotechnology. Kluwer, Dordrecht
25. Muhlbach HP (1998) Biotechnol Annu Rev 4:113
26. Phillipson JD (1990) In: Charlwood BV, Rhodes MJC (eds) Secondary products from plant tissue culture. Clarendon, Oxford, p 2
27. Pörtner R, Schilling A, Lüdemann I, Märkl H (1996) Bioprocess Eng 15:117

28. Zhou WC, Rehm J, Europa A, Hu WS (1997) Cytotechnology 24:99
29. Xie L, Wang DIC (1997) Trends Biotechnol 15:109
30. Dowd JE, Weber I, Rodriguez B, Piret JM, Kwok KE (2000) Biotechnol Bioeng 63:484
31. Kurkela R, Fraune E, Vihko P (1993) Biotechniques 15:674
32. Kratje RB, Reimann A, Hammer J, Wagner R (1994) Biotechnol Prog 10:410
33. Vunjak-Novakovic G, Searby N, De Luis J, Freed LE (2002) Ann NY Acad Sci 974:504
34. Schwarz RP, Goodwin TJ, Wolf DA (1992) J Tissue Cult Methods 14:51
35. Miller AO, Menozzi FD, Dubois D (1989) Adv Biochem Eng Biotechnol 39:73
36. Merten OW (1999) Dev Biol Stand 99:167
37. Hesse F, Wagner R (2000) Trends Biotechnol 18:173
38. Sinacore MS, Drapeau D, Adamson SR (2000) Mol Biotechnol 15:249
39. Schwartz MA, Assoian RK (2001) J Cell Sci 114:2553
40. Eliceiri BP, Cheresh DA (2001) Curr Opin Cell Biol 13:563
41. Bissell MJ, Barcellos-Hoff MH (1987) J Cell Sci 8:327
42. Folch A, Toner M (2000) Annu Rev Biomed Eng 2:227
43. Bhatia SN, Chen CS (1999) Biomed Microdevices 2:131
44. Ochoa ER, Vacanti JP (2002) Ann NY Acad Sci 979:10
45. Sorribas H, Padesta C, Tiefenauer L (2002) Biomaterials 23:893
46. Karlsson JOM (2002) Cryobiol 45:252
47. Lanza RP, Langer R, Vacanti JP (2000) Principles of tissue engineering. Academic, New York
48. Atala A, Lanza RP (2001) Methods of tissue engineering. Academic, New York
49. Prunieras M, Delescluse C, Regnier M (1976) J Invest Dermatol 67:58
50. Bell E, Ehrlich P, Buttle D, Nakatsuji T (1981) Science 211:1052
51. Coulomb B, Dubertret L (2002) Wound Repair Reg 10:109
52. Hamoen KE, Erdag G, Cusick JL, Rakhorst HA, Morgan JR (2002) Methods Mol Med 69:203
53. Tenchini ML, Ranzati C, Malcovati M (1992) Burns 18:S11
54. Breidahl AF, Judson RT, Clunie GJ (1989) Aust N Z J Surg 59:485
55. Bell E, B I, Merrill C (1979) Proc Natl Acad Sci USA 76:1274
56. Valyi-Nagy IT, Herlyn M (1991) Cancer Treat Res 54:85
57. Bessou-Touya S, Picardo M, Maresca V, Surlève-Bazeille JE, Pain C, Taieb A (1998) J Invest Dermatol 111:1103
58. Régnier M, Staquet MJ, Schmitt D, Schmidt R (1997) J Invest Dermatol 109:510
59. Rheinwald JG, Green H (1975) Cell 6:331
60. Michaeli D, McPherson M (1990) J Burn Care Rehabil 11:21
61. Eaglstein WH, Falanga V (1998) Adv Wound Care 11:1
62. Noordenbos J, Doré C, Hansbrough JF (1999) J Burn Care Rehabil 20:275
63. Le Panse R, Bouchard B, Lebreton C, Coulomb B (1996) Exp Dermatol 5:108
64. Grinnell F (2003) Trends Cell Biol 13:264
65. Werner S, Smola H (2001) Trends Cell Biol 11:143
66. Coulomb B, Lebreton C, Dubertret L (1989) J Invest Dermatol 92:122
67. Nielsen LK (1999) Annu Rev Biomed Eng 1:129
68. Douay L (2001) J Hematother Stem Cell Res 10:341
69. McNiece IK, Briddell RA (2001) Exp Hematol 29:3
70. Nakahata T (2001) Int J Hematol 73:6
71. Keil F, Elahi F, Greinix HT, Fritsch G, Louda N, Petzer AL, Prinz E, Wagner T, Kalhs P, Lechner K, Geissler K (2002) Transfusion 42:581
72. Gammaitoni L, Bruno S, Sanavio F, Gunetti M, Kollet O, Cavalloni G, Falda M, Fagioli F, Lapidot T, Aglietta M, Piacibello W (2003) Exp Hematol 31:261

73. Maurer AM, Liu Y, Caen JP, Han ZC (2000) Int J Hematol 71:203
74. Sasayama N, Kashiwakura I, Tokushima Y, Wada S, Murakami M, Hayase Y, Takagi Y, Takahashi TA (2001) Cytotherapy 3:117
75. Foley R, Tozer R, Wan Y (2001) Transfus Med Rev 15:292
76. Bordignon C, Carlo-Stella C, Colombo MP, De Vincentiis A, Lanata L, Lemoli RM, Locatelli F, Olivieri A, Rondelli D, Zanon P, Tura S (1999) Haematologica 84:1110
77. Lincoln CK, Gabridge MG (1998) Methods Cell Biol 57:49
78. Simione FP (1992) J Parenter Sci Technol 46:226
79. Spier RE (1997) Cytotechnology 23:113
80. Högman CF (1999) Vox Sang 76:67
81. Gulliksson H (2000) Transfus Med 10:257
82. Stefanovich P, Ezzell RM, Sheehan SJ, Tompkins RG, Yarmush ML, Toner M (1995) Cryobiol 32:389
83. Wigg AJ, Phillips JW, Berry MN (2003) Liver 23:201
84. Lakey JRT, Tsujimura T, Shapiro AM, Kuroda Y (2002) Transplantation 74:1809
85. Bourne WM (1991) Refract Corneal Surg 7:60
86. Basu PK (1995) Indian J Ophthalmol 43:55
87. Trent JF, Kirsner RS (1998) Int J Clin Pract 52:408
88. Bravo D, Rigley TH, Gibran N, Strong DM, Newman-Gage H (2000) Burns 26:367
89. Southard JH, Belzer FO (1995) Annu Rev Med 46:235
90. St Peter SD, Imber CJ, Friend PJ (2002) Lancet 359:604
91. Douzou P (1977) Cryobiochemistry: an tntroduction. Academic, New York
92. Taylor MJ (1987) In: Grout BWW, Morris GJ (eds) The effects of low temperatures on biological systems. Arnold, London, p 3
93. Belzer FO, Southard JH (1988) Transplantation 45:673
94. Fuller BJ (1987) In: Bowler K, Fuller BJ (eds) Temperature and animal cells. The Company of Biologists, Cambridge, p 460
95. Rauen U, de Groot H (2002) Biol Chem 383:477
96. Drobnis EZ, Crowe LM, Berger T, Anchordoguy TJ, Overstreet JW, Crowe JH (1993) J Exp Zool 265:432
97. Beutler E (1991) In: Rossi EC, Simon TL, Moss GS (eds) Principles of transfusion medicine. Williams & Wilkins, Baltimore, p 47
98. National Blood Data Resource Center (2003). Comprehensive report on blood collection and transfusion in the US in 2001. National Blood Data Resource Center, Bethesda, MD
99. Canadian Blood Services (2003) http://www.bloodservices.ca
100. Mühlbacher F, Langer F, Mittermayer C (1999) Transplant Proc 31:2069
101. Belzer FO, Ashby BS, Dunphy JE (1967) Lancet 2:536
102. Ploeg RJ, Goossens D, Vreugdenhil PK, McAnulty JF, Southard JH, Belzer FO (1988) Transplant Proc 20:935
103. D'Alessandro AM, Southard JH, Love RB, Belzer FO (1994) Surg Clin North Am 74:1083
104. McAnulty JF, Vreugdenhil PK, Lindell S, Southard JH, Belzer FO (1993) Transplant Proc 25:1642
105. Mironov V, Kasyanov V, McAllister K, Oliver S, Sistino J, Markwald R (2003) J Craniofac Surg 14:340
106. Carrier RL, Rupnick M, Langer R, Schoen FJ, Freed LE, Vunjak-Novakovic G (2002) Biotechnol Bioeng 78:617
107. Davisson T, Sah RL, Ratcliffe A (2002) Tissue Eng 8:807

108. Mazur P (1964) In: Timmerhaus KD (ed) Advances in cryogenic engineering, vol 9. Plenum, New York, p 28
109. Grout BWW, Morris GJ (1987) The effects of low temperatures on biological systems. Arnold, London
110. McGrath JJ, Diller KR (1988) Low temperature biotechnology: emerging applications and engineering contributions. American Society of Mechanical Engineers, New York
111. Benson E, Fuller BJ, Lane N (2003) Life in the frozen state. Taylor and Francis, London
112. Bowler K, Fuller BJ (1987) Temperature and animal cells. Society for Experimental Biology, Cambridge
113. Gao DY, Critser JK (2000) ILAR J 41:187
114. Pegg DE (2002) Semin Reprod Med 20:5
115. Mazur P (1965) Ann NY Acad Sci 125:658
116. Franks F, Mathias SF, Galfre P, Webster SD, Brown D (1983) Cryobiol 20:298
117. Rasmussen DH, Macaulay MN, MacKenzie AP (1975) Cryobiol 12:328
118. Meryman HT (1970) In: Wolstenholme GE, O'Connor M (eds) The frozen cell. Churchill, London, p 51
119. Steponkus PL, Wolfe J, Dowgert MF (1981) In: Morris GJ, Clarke A (eds) Effects of low temperatures on biological membranes. Academic, Toronto, p 307
120. Zade-Oppen AMM (1968) Acta Physiol Scand 73:341
121. Levitt J (1962) J Theor Biol 3:355
122. Lovelock JE (1957) Proc R Soc Lond Ser B 147:427
123. Karlsson JOM, Cravalho EG, Toner M (1993) CryoLett 14:323
124. Mazur P (1984) Am J Physiol 247:C125
125. Toner M (1993) In: Steponkus PL (ed) Advances in low temperature biology, vol 2. JAI, London, p 1
126. Muldrew K, McGann LE (1990) Biophys J 57:525
127. Mazur P, Leibo SP, Chu EHY (1972) Exp Cell Res 71:345
128. Diller KR (1975) Cryobiol 12:480
129. Karlsson JOM, Cravalho EG, Borel Rinkes IHM, Tompkins RG, Yarmush ML, Toner M (1993) Biophys J 65:2524
130. Ishiguro H, Rubinsky B (1994) Cryobiol 31:483
131. Luyet BJ, Gehenio PM (1940) Biodynamica 3:33
132. Meryman HT (1971) Cryobiol 8:489
133. Muldrew K, McGann LE (1994) Biophys J 66:532
134. Karow AM, Webb WR (1965) Cryobiol 2:99
135. Farrant J, Morris GJ (1973) Cryobiol 10:134
136. Lovelock JE (1955) Br J Haematol 1:117
137. Forsyth M, MacFarlane DR (1986) Cryo Lett 7:367
138. Karlsson JOM (2001) Cryobiol 42:154
139. Mazur P (1963) J Gen Physiol 47:347
140. Toner M, Cravalho EG, Karel M (1990) J Appl Phys 67:1582
141. Pitt RE (1992) In: Steponkus PL (ed) Advances in low-temperature biology, vol 1. JAI, London, p 63
142. Diller KR (1992) Adv Heat Transfer 22:157
143. Karlsson JOM, Toner M (1996) Biomaterials 17:243
144. Pegg DE (2001) CryoLett 22:105
145. Armitage WJ, Juss BK (1996) CryoLett 17:213
146. Hornung J, Muller T, Fuhr G (1996) Cryobiol 33:260

147. Acker JP, Elliott JAW, McGann LE (2001) Biophys J 81:1389
148. Berger WK, Uhrik B (1996) Experientia 52:843
149. Muldrew K, Novak K, Yang H, Zernicke R, Schachar NS, McGann LE (2000) Cryobiol 40:102
150. Pegg DE (1998) In: Pegg DE, Karow AM (eds) The biophysics of organ cryopreservation. NATO ASI series, vol 147. Plenum, New York, p 117
151. Rubinsky B, Lee CY, Bastacky J, Onik G (1990) Cryobiol 27:85
152. Zieger MAJ, Tredget EE, McGann LE (1997) CryoLett 18:126
153. Rajotte RV, Warnock GL, Bruch LC, Procyshyn AW (1983) Cryobiol 20:169
154. Polge C, Smith AU, Parkes AS (1949) Nature 164:666
155. Lovelock JE, Bishop MWH (1959) Nature 183:1394
156. Vos O, Kaalen CA (1965) Cryobiol 1:249
157. McGann LE (1978) Cryobiol 15:382
158. Fahy GM, MacFarlane DR, Angell CA, Meryman HT (1984) Cryobiol 21:407
159. MacFarlane DR (1986) Cryobiol 23:230
160. Fahy GM, Levy DI, Ali SE (1987) Cryobiol 24:196
161. Farrant J (1965) Nature 205:1284
162. Khirabadi BS, Fahy GM (1994) Cryobiol 31:10
163. Boutron P, Peyridieu J (1994) Cryobiol 31:367
164. Kuleshova LL, MacFarlane DR, Trounson AO, Shaw JM (1999) Cryobiol 38:119
165. MacFarlane DR, Angell CA, Fahy GM (1981) CryoLett 2:353
166. Wowk B, Leitl E, Rasch CM, Mesbah-Karimi N, Harris SB, Fahy GM (2000) Cryobiol 40:228
167. Wowk B, Darwin M, Harris SB, Russell SR, Rasch CM (1999) Cryobiol 39:215
168. O'Neil L, Paynter SJ, Fuller BJ, Shaw RW, DeVries AL (1998) Cryobiol 37:59
169. Sutton RL, Pegg DE (1993) CryoLett 14:13
170. Valeri CR, Ragno G, Pivacek LE, Cassidy GP, Srey R, Hansson-Wicher M, Leavy ME (2000) Vox Sang 79:168
171. Vishwanath R, Shannon P (2000) Anim Reprod Sci 62:23
172. Reed BM (2001) CryoLett 22:97
173. Morris CB (1995) Methods Mol Biol 38:179
174. Yang H, Acker JP, Hannon J, Miszta-Lane H, Akabutu JJ, McGann LE (2001) Cytotherapy 3:377
175. Critser JK, Mobraaten LE (2000) ILAR J 41:197
176. McLaughlin EA (2002) Hum Fertil (Camb) 5:S61
177. Nijs M, Ombelet W (2001) Hum Fertil (Camb) 4:158
178. Udaykumar HS, Mao L (2002) Int J Heat Mass Transfer 45:4793
179. Kleinhans FW (1998) Cryobiol
180. Jaeger M, Carin M, Medale M, Tryggvason G (1999) Biophys J 77:1257
181. Elliott JAW, McGann LE, Hakda S, Bannerman R (2002) Cryobiol 45:252
182. Elmoazzen HY, Elliott JAW, McGann LE (2002) Cryobiol 45:68
183. Eroglu A, Toner M, Toth TL (2002) Fertil Steril 77:152
184. Eroglu A, Russo MJ, Bieganski R, Fowler A, Cheley S, Bayley H, Toner M (2000) Nature Biotechnol 18:163
185. Beattie GM, Crowe JH, Lopez AD, Cirulli V, Ricordi C, Hayek A (1997) Diabetes 46:519
186. Acker JP, McGann LE (2002) Cell Transplant 11:563
187. Acker JP, McGann LE (2003) Cryobiol 46:197
188. Muldrew K, Acker JP, Wan R (2000) Cryobiol 41:337

189. Muldrew K, Acker JP, Elliott JAW, McGann LE (2003) In: Benson E, Fuller BJ, Lane N (eds) Life in the frozen state. Taylor & Francis, London
190. Baust JM, Vogel MJ, Van Buskirk RG, Baust JG (2001) Cell Transplant 10:561
191. Baust JM (2002) Cell Preserv Technol 1:17
192. Stroh C, Cassens U, Samraj AK, Sibrowski W, Schulze-Osthoff K, Los M (2002) FASEB 16:1651
193. Baust JM, Van Buskirk RG, Baust JG (2000) In Vitro Cell Dev Biol 36:262
194. Yagi T, Hardin JA, Valenzuela YM, Miyoshi H, Gores GJ, Nyberg SL (2001) Hepatology 33:1432
195. Rajotte RV (1999) Ann NY Acad Sci 875:200
196. Piemonti L, Bertuzzi F, Nano R, Leone BE, Socci C, Pozza G, Di Carlo V (1999) Transplantation 68:655
197. Zieger MAJ, Tredget EE, McGann LE (1996) Cryobiol 33:376
198. Kitagawa T, Masuda Y, Tominaga T, Kano M (2001) J Med Invest 48:123
199. Brockbank KGM, Bank HL (1987) J Card Surg 1:145
200. Song YC, Hunt CJ, Pegg DE (1994) Cryobiol 31:317
201. Zhang F, Attkiss KJ, Walker M, NBuncke HJ (1998) J Reconstr Microsurg 14:559
202. Schachar NS, McGann LE, Shrive N (1993) Curr Opin Orthop 4:90
203. Schachar NS, Novak K, Hurtig M, Muldrew K, McPherson R, Wohl G, Zernicke R, McGann LE (1999) J Orthop Res 17:909
204. Muldrew K, Hurtig M, Novak K, Schachar NS, McGann LE (1994) Cryobiol 31:31
205. Ohlendorf C, Tomford WW, Mankin HJ (1996) J Orthop Res 14:413
206. McGann LE, Stevenson M, Muldrew K, Schachar NS (1988) J Orthop Res 6:109
207. Muldrew K, Schachar NS, McGann LE (1996) CryoLett 17:331
208. Acker JP, Yang H, Studholme C, Muldrew K, Novak K, Zernicke R, Schachar NS, McGann LE (1995) Cryobiol 32:583
209. Muldrew K, Chung M, Novak K, Schachar NS, Zernicke R, McGann LE, Rattner JB, Matyas JR (2001) Osteoarthr Cartil 9:432
210. Luyet BJ (1937) Biodynamica 1:1
211. Rall WF, Fahy GM (1985) Nature 313:573
212. Liebermann J, Nawroth F, Isachenko V, Isachenko E, Rahimi G, Tucker MJ (2002) Biol Reprod 67:1671
213. Holt WV (1997) Reprod Fertil Dev 9:309
214. Vajta G (2000) Anim Reprod Sci 60-61:357
215. Takahashi T, Hirsh A, Erbe EF, Bross JB, Steere RL, Williams RJ (1986) Cryobiol 23:103
216. Steponkus PL, Myers SP, Lynch DV, Gardner L, Bronshteyn VL, Leibo SP, Rall WF, Pitt RE, Lin TT, MacIntyre RJ (1990) Nature 345:170
217. Mazur P, Cole KW, Hall JW, Schreuders PD, Mahowald AP (1992) Science 258:1932
218. Wusteman M, Pegg DE, Wang L, Robinson MP (2003) Cryobiol 46:135
219. Sakai A (2000) In: Engelmann F, Takagi H (eds) Cryopreservation of tropical plant germplasm. JIRCAS, Tsukuba, Japan, p 1
220. Armitage WJ, Rich SJ (1990) Cryobiol 27:483
221. Fahy GM, Saur J, Williams RJ (1990) Cryobiol 27:492
222. Fahy GM, Lilley TH, Linsdell H, Douglas MS, Meryman HT (1990) Cryobiol 27:247
223. Fahy GM (2002) Cryobiol 45:233
224. Wusteman MC, Pegg DE, Robinson MP, Wang LH, Fitch P (2002) Cryobiol 44:24
225. Armitage WJ, Hall SC, Routledge C (2002) Invest Ophthalmol Vis Sci 43:2160
226. Fujita T, Takami Y, Ezoe K, Saito T, Sato K, Takeda N, Yamamoto Y, Homma K, Jimbow K, Sato N (2000) J Burn Care Rehabil 21:304

227. Jutte NH, Heyse P, Jansen HG, Bruining GJ, Zeilmaker GH (1987) Cryobiol 24:403
228. Wishnies SM, Parrish AR, Sipes IG, Gandolfi AJ, Putnam CW, Krumdieck CL, Brendel K (1991) Cryobiol 28:216
229. de Graaf IAM, Koster HJ (2003) Toxicol In Vitro 17:1
230. Migishima F, Suzuki-Migishima R, Song SY, Kuramochi T, Azuma S, Nishijima M, Yokoyama M (2003) Biol Reprod 68:881
231. Song YC, Khirabadi BS, Lightfoot F, Brockbank KGM, Taylor MJ (2000) Nature Biotechnol 18:296
232. Khirabadi BS, Fahy GM (2000) Transplantation 70:51
233. Ruggera PS, Fahy GM (1990) Cryobiol 27:465
234. Marsland TP, Evans S, Pegg DE (1987) Cryobiol 24:311
235. Robinson MP, Wusteman M, Wang L, Pegg DE (2002) Phys Med Biol 47:2311
236. de la Torre JC (1983) Biological actions and medical applications of dimethyl sulfoxide. New York Academy of Sciences, New York
237. Zambelli A, Poggi G, Da Prada G, Pedrazzoli P, Cuomo A, Miotti D, Perotti C, Preti P, Robustelli della Cuna G (1998) Anticancer Res 18:4705
238. Martino M, Morabito F, Messina G, Irrera G, Pucci G, Iacopino P (1996) Haematologica 81:59
239. Yang H, Acker JP, Cabuhat M, McGann LE (2003) Bone Marrow Transplant
240. Blajchman MA (2001) Transfus Clin Biol 8:267
241. Arnaud F (1999) Cryobiol 38:192
242. Koebe HG, Muhling B, Deglmann CJ, Schildberg FW (1999) Chem Biol Interact 121:99
243. Yang H, Arnaud F, McGann LE (1992) Cryobiol 29:500
244. Johnson LA, Weitze KF, Fiser P, Maxwell WM (2000) Anim Reprod Sci 62:143
245. Wininger JD, Kort HI (2002) Semin Reprod Med 20:45
246. Crowe JH, Hoekstra FA, Crowe LM (1992) Annu Rev Physiol 54:579
247. Leopold AC (1986) Membranes, metabolism and dry organisms. Cornell University Press, Ithaca, NY
248. Ingram I, Bartels D (1996) Annu Rev Plant Physiol Plant Mol Biol 47:377
249. Oliver AE, Leprince O, Wolkers WF, Hincha DK, Heyer AG, Crowe JH (2001) Cryobiol 43:151
250. Hand SC, Podrabsky JE (2000) Thermochim Acta 349:31
251. Hand SC, Hardewig I (1996) Annu Rev Physiol 58:539
252. Rice-Evans CA, Miller NJ, Paganga G (1997) Trends Plant Sci 2:152
253. Blackman SA, Obendorf RL, Leopold AC (1995) Physiol Plant 93:630
254. Crowe LM (2001) Comp Biochem Physiol A
255. Hendry GA, Khan MM, Greggains V, Leprince O (1996) Biochem Soc Trans 24:484
256. Kranner I, Beckett RP, Wornik S, Zorn M, Pfeifhofer HW (2002) Plant J 31:13
257. Bianchi G, Gamba A, Murelli C, Salamini F, Bartels D (1992) Phytochemistry 31:1917
258. Farrant JMA (2000) Plant Ecol 151:29
259. Merritt DJ, Senaratna T, Touchell DH, Dixon KW, Sivasithamparam K (2003) Seed Sci Res 13:155
260. Potts M (1994) Microbiol Rev 58:755
261. Leprince O, Hoekstra FA (1998) Plant Physiol 118:1253
262. Becker MJ, Rapoport AI (1987) Adv Biochem Eng Biotechnol 35:127
263. Puhlev I, Guo N, Brown DR, Levine F (2001) Cryobiol 42:207
264. Gordon SL, Oppenheimer SR, Mackay AM, Brunnabend J, Puhlev I, Levine F (2001) Cryobiol 43:182

265. Matsuo S, Toyokuni S, Osaka M, Hamazaki S, Sugiyama T (1995) Biochem Biophys Res Commun 208:1021
266. Poirier I, Marechal P-A, Richard S, Gervais P (1999) J Appl Microbiol 86:87
267. Seel W, Hendry G, Atherton N, Lee J (1991) Free Radical Res Commun 15:133
268. Crowe JH, Carpenter JF, Crowe LM (1998) Annu Rev Physiol 60:73
269. Wolfe J, Bryant G (1999) Cryobiol 39:103
270. Gaber BP, Chandrasekhar I, Pattabiraman N (1986) In: Leopold AC (ed) Membranes, metabolism and dry organisms. Cornell University Press, Ithaca, NY, p 231
271. Crowe JH, Crowe LM, Carpenter JF (1993) Biopharm 4:28
272. Crowe JH, Crowe LM, Carpenter JF (1993) Biopharm 5:40
273. Franks F (1999) Biotechnol Genet Eng Rev 16:281
274. Kreilgaard L, Frokjaer S, Flink JM, Randolph TW, Carpenter JF (1998) Arch Biochem Biophys 360:121
275. Billi D, Wright DH, Helm RF, Prickett T, Potts M, Crowe JH (2000) Appl Environ Microbiol 66:1680
276. Israeli E, Shaffer BT, Lighthart B (1993) Cryobiol 30:519
277. Leslie SB, Israeli E, Lighthart B, Crowe JH, Crowe LM (1995) Appl Environ Microbiol 61:3592
278. Leslie SB, Teter SA, Crowe LM, Crowe JH (1994) Biochim Biophys Acta 1192:7
279. Cerrutti P, Segovia de Huergo M, Galvagno M, Schebor C, del Pilar Buera M (2000) Appl Microbiol Biotechnol 54:575
280. Slaughter JC, Nomura T (1992) Enzyme Microb Technol 14:64
281. Bieganski RM, Fowler A, Morgan JR, Toner M (1998) Biotechnol Prog 14:615
282. Crowe LM, Crowe JH, Rudolph AS, Womersley C, Appel L (1985) Arch Biochem Biophys 242:240
283. Crowe LM, Reid DS, Crowe JH (1996) Biophys J 71:2087
284. Crowe JH, Crowe LM, Oliver AE, Tsvetkova N, Wolkers WF, Tablin F (2001) Cryobiol 43:89
285. Guo N, Puhlev I, Brown DR, Mansbridge J, Levine F (2000) Nature Biotechnol 18:168
286. Garcia de Castro A, Tunnacliffe A (2000) FEBS Lett 487:199
287. Lao G, Polayes D, Xia JL, Bloom FR, Levine F, Mansbridge J (2001) Cryobiol 43:106
288. Wolkers WF, Walker NJ, Tablin F, Crowe JH (2001) Cryobiol 42:79
289. Mussauer H, Sukhorukov VL, Zimmermann U (2001) Cytometry 45:161
290. Chen T, Acker JP, Eroglu A, Cheley S, Bayley H, Fowler A, Toner M (2001) Cryobiol 43:168
291. Acker JP, Fowler A, Lauman B, Cheley S, Toner M (2002) Cell Preserv Technol
292. Acker JP, Chen T, Fowler A, Toner M (2003) In: Benson E, Fuller BJ, Lane N (eds) Life in the frozen state. Taylor & Francis, London
293. Tunnacliffe A, Garcia de Castro A, Manzanera M (2001) Cryobiol 43:124
294. Wolkers WF, Walker NJ, Tamari Y, Tablin F, Crowe JH (2003) Cell Preserv Technol 1:175
295. Winokur R, Hartwig JH (1995) Blood 85:1796
296. Crowe JH, Clegg JS (1973) Anhydrobiosis. Dowden, Hutchinson and Ross, Stroudsburg
297. van Breukelen F, Hand SC (2000) J Comp Physiol B 170:125
298. Clegg JS, Jackson SA, Liang P, MacRae TH (1995) Exp Cell Res 219:1
299. Hand SC (1998) J Exp Biol 201:1233
300. Leprince O, Buitink J, Hoekstra FA (1999) J Exp Bot 50:1515
301. Rogerson NE, Matthews S (1977) J Exp Bot 28:304
302. Leprince O, Hendry GAF, McKersie BD (1993) Seed Sci Res 3:231

303. Clegg JS, Jackson SA, Warner AH (1994) Exp Cell Res 212:77
304. Clegg JS (1997) J Exp Biol 200:467
305. Podrabsky JE, Hand SC (1999) J Exp Biol 202:2567
306. Busa WB, Crowe JH (1983) Science 221:366
307. Eads BD, Hand SC (2003) J Exp Biol 206:577
308. Franks F (1998) Eur J Pharm Biopharm 45:221
309. Levine H (2002) Amorphous food and pharmaceutical systems. Royal Society of Chemistry, Cambridge
310. Rahman S, Rahman MS (1999) Handbook of food preservation. Dekker, New York

Adv Biochem Engin/Biotechnol (2006) 103: 189–205
DOI 10.1007/10_010
© Springer-Verlag Berlin Heidelberg 2005
Published online: 20 December 2005

Fabrication of Three-Dimensional Tissues

Valerie Liu Tsang[1] · Sangeeta N. Bhatia[2,3] (✉)

[1]Department of Bioengineering, University of California, San Diego, La Jolla, CA 92093, USA

[2]Harvard – MIT Division of Health Sciences & Technology / Department of Electrical Engineering and Computer Science, Massachusetts Institute of Technology, Cambridge, MA 02139, USA

[3]Brigham and Women's Hospital, Boston, MA 02115, USA
sbhatia@mit.edu

Abstract The goal of tissue engineering is to restore or replace the lost functions of diseased or damaged organs. Ideally, engineered tissues should provide nutrient transport, mechanical stability, coordination of multicellular processes, and a cellular microenvironment that promotes phenotypic stability. To achieve this goal, many engineered tissues require both macro- ($\sim$ cm) and micro- ($\sim 100\,\mu$m) scale architectural features. In recent years, techniques from the manufacturing world have been adapted to create scaffolds for tissue engineering with defined three-dimensional architectures at physiologically relevant length scales. This chapter reviews three-dimensional fabrication techniques for tissue engineering, including: acellular scaffolds, cellular assembly, and hybrid scaffold/cell constructs.

Keywords Micropatterning · Hydrogels · Poly(ethylene glycol) · Scaffolds · Tissue engineering

Abbreviations
PLGA   Poly(DL-lactic-*co*-glycolic) acid
PDMS   Polymethylsiloxane
FDM   Fused deposition molding
SLS   Selective laser scintering
3-DP   Three-dimensional printing
PAM   Pressure-assisted microsyringe
PEG   Poly(ethylene glycol)
CAD   Computer-aided design

1
Introduction

The goal of tissue engineering is to restore or replace the lost functions of diseased or damaged organs. Tissue constructs are typically assembled in vitro by combining cells and biomaterial scaffolds. The expectation is that the cells will provide key tissue functions and the scaffold will provide mechanical support; however, in many cases, the mere addition of cells to a porous scaffold is insufficient for the reproduction of normal tissue function. In vivo, tissues are composed of repeating three-dimensional units on the scale of $100-1000\,\mu m$ (e.g., nephron, islet) [1]. The three-dimensional architecture of these repeating tissue units underlies the coordination of multicellular processes, emergent mechanical properties, and integration with other organ systems via the microcirculation. Another key element of tissue structure in vivo is the local cellular environment. The "microenvironment" ($\sim 10\,\mu m$) presents biochemical, cellular, and physical stimuli that orchestrate cellular fate processes such as proliferation, differentiation, migration, and apoptosis. Thus, successful fabrication of fully functional tissues must be addressed at two levels: (1) At the microscale, cells must be presented with an appropriate environment for cell survival and function, and (2) at the macroscale, the tissue construct must facilitate coordination of multicellular processes, provide adequate transport of nutrients, and possess suitable mechanical properties.

Traditional scaffold fabrication techniques have involved the production of porous polymer constructs as substrates for cell attachment [2]; however, complex architectures with tunable micro- and macro-scale features are difficult to achieve. In recent years, CAD-based manufacturing technologies have therefore been applied toward the fabrication of tissues. These three-dimensional fabrication approaches offer several potential opportunities in tissue engineering. First, the independent control of micro- and macro-scale features may enable the fabrication of multicellular structures that are required for complex tissue function. Second, fabrication of three-dimensional vascular beds would allow support of larger tissue constructs than could be otherwise achieved. Third, the combination of clinical imaging data with three-dimensional fabrication techniques may offer the capability to create tissue engineered constructs that are customized to the shape of the defect or injury. Fourth, such fabrication technology may allow "mass production" of many identical tissue constructs for use in drug discovery or fundamental studies. Indeed, in two-dimensional model systems, fabrication ("micropat-

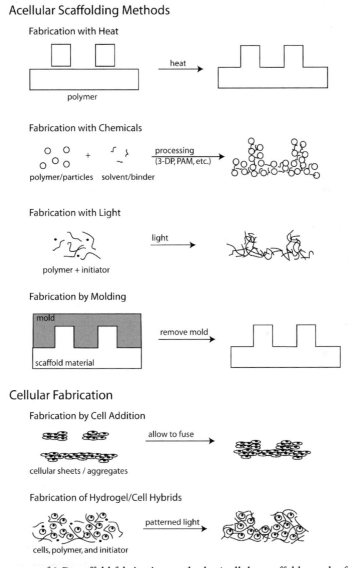

Fig. 1 Summary of 3-D scaffold fabrication methods. Acellular scaffolds can be fabricated using various techniques, such as heat (FDM), chemicals (3-DP), light (SLA), and molding. Cells themselves can be incorporated in the fabrication process by cellular addition or by photopatterning of hydrogels

terning") technologies have previously led to insights on the effect of cell–cell and cell–matrix interactions on hepatocyte and endothelial cell fate [1]. Thus, we anticipate that 3-D fabrication techniques may also prove useful for studying structure/function relationships in model tissues.

In this chapter, we review and compare various three-dimensional fabrication methods with regard to their resolution, stage of development, and relative utility for tissue engineering. Three general approaches are described schematically in Fig. 1 and include: (1) fabrication of acellular scaffolds, (2) techniques for cellular assembly, and (3) hybrid cell/scaffold systems.

2
Fabrication of Three-Dimensional Acellular Scaffolds

"Traditional" scaffold fabrication techniques such as solvent casting/particulate leaching allow definition of both scaffold shape and pore size; however, neither the internal scaffold architecture nor the connectivity of the void space can be designed a priori. In contrast, rapid prototyping technologies, developed in the realm of manufacturing engineering, provide exquisite spatial control over polymer architecture. Therefore, in recent years, various CAD-based techniques have been adapted to fabricate three-dimensional polymer scaffolds for tissue engineering applications. We have classified the various scaffold fabrication techniques by their modes of assembly as seen in Fig. 1: fabrication with heat, fabrication with binders, fabrication with light, and fabrication by molding. A summary of such techniques is presented below – detailed reviews on solid freeform fabrication are available elsewhere [3–5].

2.1
Fabrication with Heat

Heat-based fabrication generally relies on the application of heat energy to fuse layers of material to each other by bringing the biopolymer above its glass transition temperatures and applying pressure. One of the simplest examples of this technique is *sheet lamination*, in which shapes are cut out of polymer sheets using a laser, and are then sequentially fused together by applying heat and pressure [5]. In its current stage of development, the resulting prototype is extremely dense (i.e., low void volume) and may not be practical for construction of highly cellular tissues.

Membranes with both higher porosity and smaller features can also be laminated to create more-intricate scaffolds. Borenstein et al. created biodegradable membranes containing small channels (20 μm in diameter) for vascular tissue engineering by casting thin films of poly(DL-lactic-*co*-glycolic) acid (PLGA) onto microfabricated silicon wafers (Fig. 2c) [6]. By

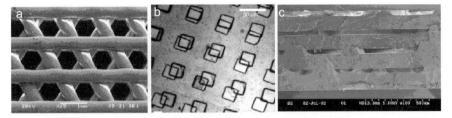

Fig. 2 Fabrication using heat. **a** Fused deposition molding. Molten biomaterials are extruded through a nozzle to build 3-D scaffolds layer by layer [4]. **b,c** Molded lamination. Membranes of the biodegradable polymer PLGA are cast from PDMS [7] (**b**) or silicon (**c**) molds and then laminated to create 3-D scaffolds. In **c**, layers of PLGA are fused together to form microfluidic channels for vascular tissue engineering (photo courtesy of Jeff Borenstein and Kevin King, Draper Laboratory) [7]

laminating patterned PLGA membranes to each other, channels were formed between the layers. Bhatia and coworkers used a similar method to create porous tissue engineering scaffolds using *soft lithography* techniques (Fig. 2b) [7]. A mold made of the elastomeric polydimethylsiloxane (PDMS) is cast from a microfabricated silicon master [8]. A solution of PLGA is cast onto the PDMS mold and then baked, forming a PLGA layer containing microstructures equivalent to those on the silicon master (20–30 µm resolution). Micropores are incorporated into the PLGA by solvent casting and particulate leaching. A 3-D structure is then constructed by lamination of the patterned, porous PLGA membranes.

Fused deposition molding (FDM) is another heat-based manufacturing technology that has been applied toward the building of 3-D scaffolds. Molten plastics or ceramics are extruded through a nozzle and deposited layer by layer to form the 3-D scaffold. Hutmacher and coworkers have used FDM to fabricate bioresorbable scaffolds of poly(ε-caprolactone) (PCL) with feature sizes of approximately 250–700 µm [9]. Furthermore, their group has demonstrated primary human fibroblast proliferation and extracellular matrix production on the PCL scaffolds [3]. Other groups have also explored the use of FDM for scaffold production using bioceramic or polymer materials (Fig. 2a) [4]. One limitation of this method is that the height of the pores created is predetermined by the size of the polymer filament extruded through the nozzle. In addition, the materials that can be used for this method are limited by the melting points and processing conditions involved.

Selective laser scintering (SLS) also uses heat to fuse polymers into desired shapes and layers. This technique involves the use of a laser beam to raise the local surface temperature of the powder bed in order to fuse the particles, forming patterned structures within each layer. The resolution of SLS is limited by the laser beam diameter used in this system, which currently is approximately 400 µm [5]. Because of the nature of the process,

Fig. 3 3-D plotting. Heated liquid agar solidifies into a 3-D hydrogel scaffold when deposited into a cooled medium [11]

unfused powders within the structures also result in a higher porosity than with other methods such as FDM. Lee and Barlow have demonstrated the use of this method with polymer-coated calcium phosphate powders to fabricate scaffolds, and have reported bone tissue ingrowth over several weeks using dog models [10]. In addition to ceramic/polymer blends, others are also working on ways to improve the SLS process and to apply it to biopolymers [4].

As a rule, the materials used in heat-based fabrication methods must retain their desired properties (degradation, biocompatibility, etc.) after exposure to elevated temperatures, limiting the choice of materials primarily to synthetic polymers. A few natural biomaterials have also been shown to be compatible with heat-based free-form fabrication by exploiting phase transitions from liquid to hydrogel. Mulhaupt and coworkers used 3-D plotting technology to deposit heated agar and gelatin solutions (90 °C) into a cooled plotting medium (10–15 °C) to create 3-D hydrogel scaffolds (Fig. 3) [11]. The scaffolds were then sterilized in ethanol, coated with fibrin, and seeded with a human osteosarcoma cell line or with mouse fibroblasts, which subsequently adhered to the surfaces.

2.2
Fabrication with Binders

Another application for bioengineering is *three-dimensional printing* (3-DP). This method involves the use of an ink-jet printer to deposit a binder solution onto a biomaterial powder bed to fabricate structures (approximately 200–500 μm) one layer at a time (Fig. 4) [12]. Because the fusion of particles is performed with solvents and adhesives rather than heat and pressure, the range of biomaterials that can be used is greater than with the heat-based fabrication methods. Griffith and coworkers fabricated porous PLGA scaffolds using 3-DP and particulate leaching, and demonstrated the attachment

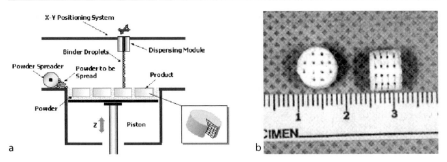

Fig. 4 3-D printing. Ink-jet technology is used to print a binder solution onto a bed of polymer powder. An additional layer of powder is then deposited, and the process is repeated to form 3-D scaffolds [12]

of cocultures of rat hepatocytes and nonparenchymal cells [13]. Zeltinger et al. expanded upon this work and explored its limitations by examining cell attachment, growth, and matrix deposition on 3-D printed scaffolds with various pore sizes and cell types [14].

Like 3-DP, fabrication of scaffolds using the *pressure- assisted microsyringe* (PAM) method involves layer-by-layer deposition, with the solvent acting as the binding agent. However, rather than printing the binder onto a bed of powder, the microsyringe method involves the deposition of a solution of a polymer in solvent through a syringe fitted with a $10-20\,\mu m$ glass capillary needle [7]. The size of the polymer stream deposited can be varied by controlling the syringe pressure, solution viscosity, syringe tip diameter, and motor speed. This method is similar to FDM, but requires no heat, and is able to produce structures with greater resolution. While the resolution of PAM is greater than most of the other fabrication methods, micropores cannot be incorporated using particulate leaching due to the syringe dimensions.

2.3
Fabrication with Light

Photopolymerization involves the use of light energy to initiate a chain reaction, resulting in the formation of a solid from the original liquid polymer solution. *Stereolithography* is a manufacturing photopolymerization technique that can be used for fabrication of tissue engineering scaffolds. Spatial patterning of the material is achieved by directing the position of the light using a laser beam or by exposing certain areas of an entire layer through a photomask. The stage is then lowered and the process repeated to form additional layers. Cooke et al. reported the use of stereolithography technology for generating biodegradable 3-D polymer scaffolds of diethyl fumarate, poly(propylene fumarate), and the photoinitiator bisacylphosphine oxide (Fig. 5) [15]. Structures fabricated using stereolithography typically

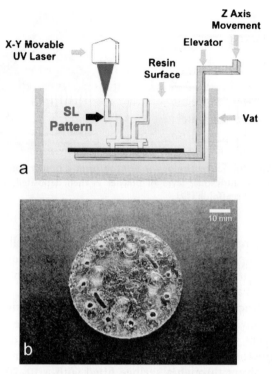

Fig. 5 Stereolithography. **a** UV light is used to crosslink the material in specific regions of a layer. The elevator is then lowered to reveal a new layer of polymer, and the process is repeated to create the desired shape. **b** A prototype scaffold designed using SLA [15]

contain features as small as 250 μm, but certain systems using small-spot lasers have been shown to produce 70 μm features [5].

The use of light to pattern polymers can be applied not only to solidify rigid polymers such as in stereolithography but also to fabricate *hydrogel polymer scaffolds* using photolithographic techniques. Hydrogels are crosslinked networks of hydrophilic polymers that swell with water. They are becoming an increasingly popular material for tissue engineering due to their high water content and mechanical properties that resemble those of living tissues. Yu et al. reported a photolithographic method of patterning 2-hydroxyethyl methacrylate which is then dried, and can later be rehydrated for cell adhesion [16]. This method has been used to create single-layer structures, though theoretically it could be adapted for multilayer fabrication. However, some patterning resolution may be lost during the rehydration process. Photopatterning of hydrogels will be discussed in greater detail in a later section.

2.4
Fabrication by Molding

The techniques described above were utilized to fabricate scaffolds for cell adhesion and tissue ingrowth; however, the same methods can also be used to indirectly fabricate scaffolds by using the prototypes as molds. Molding allows for the creation of patterned scaffolds from materials that are not compatible with the fabrication processes. For example, stereolithography was used by Orton and coworkers to create a negative epoxy mold of the desired scaffold design (Fig. 6a) [17]. A hydroxyapatite/acrylate suspension was then cast onto the mold and cured with heat, and then placed in a furnace to burn out the mold and acrylate binder, resulting in a 3-D hypoxyapatite scaffold. They also reported bone ingrowth 9 weeks after implantation into minipig mandibles [18]. Like the 3-D printing method described above, ink-jet printing technology can also be used to fabricate molds by depositing wax or other low melting point compounds (Solidscape). Hollister and coworkers have used this technique to create molds that have then been used to cast

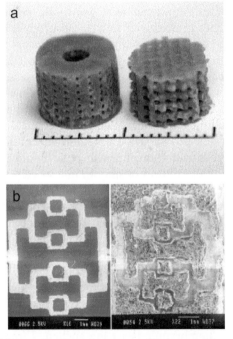

Fig. 6 Molded scaffolds. **a** Hydroxyapatite was cast into a negative epoxy mold (manufactured using stereolithography) and then cured by heat. The scaffold was then placed in a furnace to burn out the mold. [17] **b** The extracellular matrix compound collagen was cast onto a negative mold that was printed using ink-jet technology. The mold was then dissolved away with ethanol, leaving a patterned collagen scaffold [21]

hypoxyapatite, poly(L)lactide, and polyglycolide scaffolds [19, 20], and have also created micropores within the polymer scaffolds by combining the technique with particulate leaching. The wax molds are then removed by melting and washing with solvents. Sachlos et al. also used the same ink-jet printing system to create molds for casting of the extracellular matrix component collagen (Fig. 6b) [21]. In this case, the mold is removed by dissolving in ethanol, resulting in patterned collagen scaffolds with features as small as 200 μm.

3
Fabrication of Cellular Structures

In contrast to the scaffold fabrication techniques described in the previous section, several groups are taking an orthogonal approach to tissue engineering by directly constructing layers of live cells. By building with cells directly, researchers hope to alleviate the problem of insufficient and heterogeneous cellular engraftment that can occur upon seeding cells on acellular scaffolds. Yamoto and coworkers have proposed the construction of 3-D tissues by layering of cell sheets that have been grown in culture and then released as a contiguous layer [22]. Cardiomyocytes were grown on temperature-responsive culture surfaces (dishes grafted with poly(N-iso-propylacrylamide)) and allowed to proliferate. Lowering the temperature by 5 degrees results in hydration of the grafted polymer, causing the cells to lift off as a sheet. Multiple sheets of cardiomyocytes can then be layered to create an in vitro myocardial tissue construct. The cell layering method may allow for a small degree of patterning by varying the composition of each layer, but it is not a means for creating more complex three-dimensional patterned structures.

Auger and coworkers also designed a method of creating blood vessels by "rolling" sheets of cells and culturing with pulsatile flow to form an engineered blood vessel [23]. Cultures of smooth muscle cells formed sheets that were then wrapped around a tubular support, allowing endothelial cells to be seeded within the lumen. After culture, the tissue-engineered blood vessels demonstrated excellent mechanical properties and the cells exhibited several markers of native vessels. While this technique can be used for simple tubular structures such as a blood vessel, it cannot be adapted for fabricating arbitrary 3-D tissue architectures.

Others are attempting to develop methods to directly "plot" living cells into 3-D structures by depositing cells layer by layer and allowing them to fuse spontaneously [24–26]. Mironov et al. demonstrated that the printing of cell aggregates and embryonic heart mesenchymal fragments resulted in fusion into a tube-like structure when placed in a three-dimensional collagen or thermosensitive gel [25]. Such technology, if successful, could allow cells to be placed into precise locations within a three-dimensional construct. Odde et al. have also developed methods to directly plot cells using laser guidance. A stream of cells is "written" onto a surface in a specified pattern using optical trapping forces to guide cells [27]. Because these emerging technologies rely to some extent on cellular self-assembly (itself an evolving field), future progress will be required to establish which tissues will be amenable to assembly by this approach.

4
Fabrication of Hybrid (Cell/Scaffold) Constructs

In general, we note that acellular scaffolds provide excellent mechanical integrity but may be difficult to populate with cells. Similarly, cellular constructs provide excellent tissue density but may be mechanically unstable. Thus, to address the challenges of providing structural support and high tissue density while maintaining an in vivo-like environment for cells, hydrogel polymers have been explored for tissue engineering [28].

4.1
Cell-Laden Hydrogels

Biological hydrogels such as fibrin and collagen have been explored to entrap cells. Hubbell and coworkers have expanded the functionality of fibrin gels by incorporating genetically engineered bioactive sites for cell adhesion and proteolytic remodelling [29–31]. Desai and coworkers have used collagen gels to deposit patterned structures containing cells by using microfluidic molding methods [32]. This method would be useful to fabricate some model tissues; however, it may be difficult to generalize due to the constraints of the microfluidic network on a flat surface.

Synthetic polymer hydrogels have also been utilized. Many of the chemistries rely on photocrosslinking of polymer solutions around entrapped cells. This allows a homogenous distribution of cells at relatively high density throughout a hydrogel network with tunable transport properties. Poly(ethylene glycol) (PEG)-based hydrogels are of particular interest because of their biocompatibility, hydrophilicity, and ability to be customized by changing the chain length or chemically adding biological molecules [33]. PEG-based hydrogels have been used to homogeneously immobilize various cell types including chondrocytes [34, 35], vascular smooth muscle cells [36], osteoblasts [37], and fibroblasts [38, 39] that can attach, grow, and produce matrix. PEG-based hydrogels have been customized by incorporation of adhesion domains of extracellular matrix proteins to promote cell adhesion, growth factors to modulate cell function, and degradable linkages [36, 39–45]. Clearly, photopolymerization of hydrogels for tissue engineering is a field that is rapidly growing because of its chemical flexibility to be customized and tissue-like physical properties.

4.2
Three-Dimensional Photopatterning of Cell-Laden Hydrogels

Typically, the shape of a cell-laden hydrogel is defined by the container used for photocrosslinking. For example, Fig. 7 depicts disc-like structures formed by casting in a cylindrical vial. One property of photosensitive hydrogel systems that had not been exploited was the potential to localize photocrosslinking with light, thereby forming defined hydrogel features containing living cells. Elsewhere, in non-biological systems, photolithographic patterning has indeed been applied to pattern hydrogel microstructures [46] and valves within flow systems [47]. In experiments designed to create cell-based sensors, Pishko and coworkers used photolithography to create a thin layer of hydrogel microstructures containing cells by spinning polymer and cells onto a silicon surface and exposing to UV light through a mask [48].

Hence, the application of photolithographic methods toward hydrogel tissue engineering may enable the construction of three-dimensional tissues. We have recently combined photolithographic techniques with existing PEG-based cell encapsulation methods to build structural features within a 3-D cell/hydrogel network (Fig. 8) [49]. The method allows localized photoimmobilization of live mammalian cells in a controlled hydrogel architecture. Cells are suspended in a polymer solution and photoimmobilized in multiple cellular domains. Uncrosslinked polymer (and cells) are rinsed away and the process can be repeated. Notably, different cell types can be introduced in each step—represented in Fig. 8 by cells that have been labeled fluorescently (red, green, blue). By increasing the height of the photocrosslinking chamber in between steps, additional layers can be added to create a 3-D cellular hydrogel tissue construct. Thus far, hydrogel features as small as 50 μm containing cells have been achieved. Furthermore, using this method, structures up to three layers have been fabricated (Fig. 8b). In complementary experiments, we have

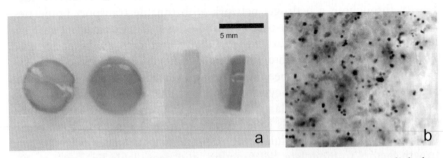

Fig. 7 PEG-based hydrogels containing cells. **a** PEG-based hydrogels are crosslinked to form the shape of the container (dye added for clarity). **b** Living cells are suspended within the crosslinked hydrogel (MTT stain for viability) (Photos by courtesy of Jennifer Elisseeff, Johns Hopkins University)

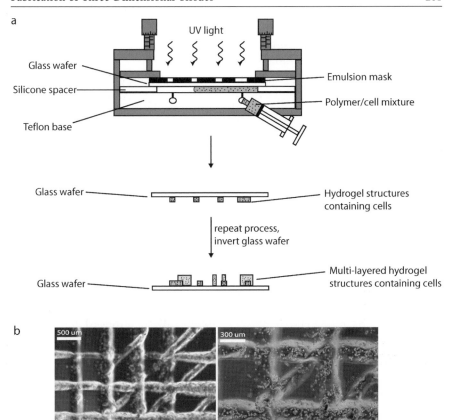

Fig. 8 3-D photopatterning of hydrogels. **a** Photopatterning method. Polymer solution and cells are introduced into a chamber. The unit is exposed to 365 nm light through an emulsion mask, causing crosslinking of the polymer in the exposed areas and trapping the cells within these regions. The uncrosslinked polymer solution and cells are then washed away, and the process is repeated with thicker spacers and a new mask to create 3-D cellular hydrogel structures. Each layer may contain the same type of polymer/cell mixture, or can be composed of different polymer properties or different cell types. **b** Three-layered hydrogel structure containing cells [49]

also developed a tool to specify cellular location within the prepolymer solution (as opposed to random dispersal) using electromagnetic fields (data not shown). In conjunction with hydrogel technologies being explored by other groups (bioactive materials, incorporation of adhesion peptides and growth factors), photopatterning of hydrogels for tissue engineering may lead to the development of tissue constructs that can be customized spatially, physically, and chemically to improve upon existing polymer scaffold systems.

Table 1 Comparison of 3-D scaffolding methods

	Resolution (μm)	Advantages	Disadvantages
ACELLULAR 3-D SCAFFOLDS			
		use of well-established fabrication methods, usually automated	must seed cells post-processing, less control in cell placement and
Fabrication with Heat			
Micro Molding [6, 7]	20–30	simple; reusable molds	limited to thin membranes, each layer must be contiguous structure, manual alignment required
Fused Deposition Modeling [3, 4, 9]	250–700	no trapped particles or solvents, automated	high temperatures during processing
Selective Laser Sintering [4, 5, 10]	400	high porosity, automated	high temperatures during process, powder may be trapped
3-D Plotting [11]	1000	use of hydrogel materials (agar, gelatin), automated	limited resolution
Fabrication with Chemicals			
3-D Printing [12–14]	200–500	versatile; high porosity, automated	limited choice of materials (e.g. organic solvents as binders); difficult to reduce resolution below polymer particle size
Pressure Assisted Micro-syringe [7]	10	high resolution, not subject to heat, automated	viscosity dependent, no inclusion of particles
Fabrication with Light			
Stereolithography [5, 15]	70–250	ease of use, easy to achieve small features, automated	limited choice of materials – must be photosensitive and biocompatible; exposure of material to laser
Fabrication by Molding			
Matrix Molding [21]	200	use of biological matrix materials (collagen), mold fabrication can use automated methods (above)	features must be inter-connected, weaker mechanical properties

Table 1 (continued)

	Resolution (µm)	Advantages	Disadvantages
CELL-LADEN 3-D SCAFFOLDS			
		Precise placement of cells throughout construct, ability to place multiple cell types arbitrarily	limited fabrication conditions (sterility, temperature, pH), still in earlier phases of development
Cellular Addition			
Organ Printing [24–26]	100	incorporation of cell aggregates or tissue explants, precise cell placement, automated	lack of structural support, dependence on self assembly
Laser-Guided Deposition [27]	< 1	precise single cell placement, automated	has yet to be extended to 3-D structures, lack of structural support
Cell / Biopolymer Hybrids			
Hydrogel Photo-patterning [49]	100	incorporation of living cells within scaffold, leverages existing hydrogel chemistry (incorporation of peptides, degradation domains), versatile	not yet automated, exposure of cells to ultraviolet light, diffusion of large molecules limited by hydrogel pore size

5
Summary

Technological advances in scaffold fabrication have led to the development of architecturally complex tissue engineering constructs. We have reviewed several methods that have been developed or adapted from manufacturing to create scaffolds with arbitrary 3-D architectures. In general, the approaches that researchers have taken in creating implantable cellular therapies are to create acellular, cellular, or hybrid constructs. Table 1 summarizes various techniques that have been developed for 3-D tissue fabrication. Technologies are compared with regard to their spatial resolution and relative merits and limitations. Ultimately, the utility of each technique for a given tissue will depend on many specific design criteria—mechanical stability, chemical composition, degradation, cellular organization, nutrient requirements, and so on. In the future, fundamental studies of structure/function relationships may also help to specify the best approach for fabricating a particular tissue.

6
Future Directions in Three-Dimensional Tissue Fabrication

What started as a simple idea of populating synthetic polymer scaffolds with living cells to "grow" tissues has become a much more complex endeavor. As new developments are made, the breadth and depth of the tissue engineering field continues to grow. While it was known from the start that cell signaling, extracellular matrix components, and mechanical properties would play a role in the development of engineered tissues, it has become clear that a deeper understanding of the biology underlying fundamental structure-function relationships is necessary. Towards that end, emerging technologies are enabling the design of experiments in three dimensions that better mimic the in vivo environment. These same technologies are being applied to the fabrication of increasingly complex tissue engineered constructs.

The first generation of synthetic tissue-engineered scaffolds were limited in size, structure, and chemistry; however, great progress has been made in the past few years. Free-form fabrication technologies now allow the customization of scaffold shapes and the construction of larger scaffolds by incorporating channels for transport. The ability to pattern hydrogel polymers into arbitrary structures provides a way to create 3-D scaffolds embedded uniformly with cells. New insights in polymer chemistry and extracellular matrix chemistry are contributing to the customization of tissue engineered constructs based on the cell types and functions. There is still much progress to be made before tissue-engineered constructs can be used effectively for whole or partial replacement of human organs. Further studies in structure-function relationships, biomaterials, and developmental biology will ensure a bright future for positively impacting human health.

Acknowledgements We would like to thank the Whitaker Foundation (V.L.), American Association of University Women (V.L.), NIH NIDDK, NSF CAREER, David and Lucile Packard Foundation, and NASA for their generous support.

References

1. Bhatia SN, Chen CS (1999) Biomed Microdevices 2:131
2. Langer R, Vacanti JP (1993) Science 260:920
3. Hutmacher DW (2001) J Biomater Sci Polym Ed 12:107
4. Leong KF, Cheah CM, Chua CK (2003) Biomaterials 24:2363
5. Yang S, Leong KF, Du Z, Chua CK (2002) Tissue Eng 8:1
6. Borenstein JT, Terai H, King KR, Weinberg EJ, Kaazempur-Mofrad MR, Vacanti JP (2002) Biomed Microdevices 4:167
7. Vozzi G, Flaim C, Ahluwalia A, Bhatia S (2003) Biomaterials 24:2533
8. Chen CS, Mrksich M, Huang S, Whitesides GM, Ingber DE (1997) Science 276:1425
9. Zein I, Hutmacher DW, Tan KC, Teoh SH (2002) Biomaterials 23:1169

10. Lee G, Barlow J, Fox W, Aufdermorte T (1996) Proc Solid Freeform Fabrication Symp, Austin, TX, p 15
11. Landers R, Hubner U, Schmelzeisen R, Mulhaupt R (2002) Biomaterials 23:4437
12. Park A, Wu B, Griffith LG (1998) J Biomater Sci Polym Ed 9:89
13. Kim SS, Utsunomiya H, Koski JA, Wu BM, Cima MJ, Sohn J, Mukai K, Griffith LG, Vacanti JP (1998) Ann Surg 228:8
14. Zeltinger J, Sherwood JK, Graham DA, Mueller R, Griffith LG (2001) Tissue Eng 7:557
15. Cooke MN, Fisher JP, Dean D, Rimnac C, Mikos AG (2003) J Biomed Mater Res 64B:65
16. Yu T, Chiellini F, Schmaljohann D, Solaro R, Ober CK (2000) Polymer Preprints 41:1699
17. Chu TM, Hollister SJ, Halloran JW, Feinberg SE, Orton DG (2002) Ann NY Acad Sci 961:114
18. Chu TM, Orton DG, Hollister SJ, Feinberg SE, Halloran JW (2002) Biomaterials 23:1283
19. Taboas JM, Maddox RD, Krebsbach PH, Hollister SJ (2003) Biomaterials 24:181
20. Hollister SJ, Maddox RD, Taboas JM (2002) Biomaterials 23:4095
21. Sachlos E, Reis N, Ainsley C, Derby B, Czernuszka JT (2003) Biomaterials 24:1487
22. Shimizu T, Yamato M, Kikuchi A, Okano T (2003) Biomaterials 24:2309
23. L'Heureux N, Paquet S, Labbe R, Germain L, Auger FA (1998) Faseb J 12:47
24. EnvisionTec (2003) www.envisiontec.de
25. Mironov V, Boland T, Trusk T, Forgacs G, Markwald RR (2003) Trends Biotechnol 21:157
26. Sciperio I (2003) www.sciperio.com
27. Odde DJ, Renn MJ (2000) Biotechnol Bioeng 67:312
28. Nguyen KT, West JL (2002) Biomaterials 23:4307
29. Halstenberg S, Panitch A, Rizzi S, Hall H, Hubbell JA (2002) Biomacromolecules 3:710
30. Lutolf MP, Lauer-Fields JL, Schmoekel HG, Metters AT, Weber FE, Fields GB, Hubbell JA (2003) Proc Natl Acad Sci USA 100:5413
31. Sakiyama SE, Schense JC, Hubbell JA (1999) Faseb J 13:2214
32. Tan W, Desai TA (2003) Tissue Eng 9:255
33. Peppas NA, Bures P, Leobandung W, Ichikawa H (2000) Eur J Pharm Biopharm 50:27
34. Elisseeff J, McIntosh W, Anseth K, Riley S, Ragan P, Langer R (2000) J Biomed Mater Res 51:164
35. Bryant SJ, Anseth KS (2002) J Biomed Mater Res 59:63
36. Mann BK, Gobin AS, Tsai AT, Schmedlen RH, West JL (2001) Biomaterials 22:3045
37. Behravesh E, Zygourakis K, Mikos AG (2003) J Biomed Mater Res 65A:260
38. Gobin AS, West JL (2002) Faseb J 16:751
39. Hern DL, Hubbell JA (1998) J Biomed Mater Res 39:266
40. Kao WJ, Hubbell JA (1998) Biotechnol Bioeng 59:2
41. Koo LY, Irvine DJ, Mayes AM, Lauffenburger DA, Griffith LG (2002) J Cell Sci 115:1423
42. Alsberg E, Anderson KW, Albeiruti A, Rowley JA, Mooney DJ (2002) Proc Natl Acad Sci USA 99:12025
43. Schmedlen RH, Masters KS, West JL (2002) Biomaterials 23:4325
44. Sawhney AS, Pathak CP, van Rensburg JJ, Dunn RC, Hubbell JA (1994) J Biomed Mater Res 28:831
45. Nuttelman CR, Henry SM, Anseth KS (2002) Biomaterials 23:3617
46. Revzin A, Russell RJ, Yadavalli VK, Koh WG, Deister C, Hile DD, Mellott MB, Pishko MV (2001) Langmuir 17:5440
47. Beebe DJ, Moore JS, Bauer JM, Qing Y, Liu RH, Devadoss C, Byung-Ho J (2000) Nature 404:588
48. Koh WG, Revzin A, Pishko MV (2002) Langmuir 18:2459
49. Liu VA, Bhatia SN (2002) Biomed Microdevices 4:257

Adv Biochem Engin/Biotechnol (2006) 103: 207–239
DOI 10.1007/b137206
© Springer-Verlag Berlin Heidelberg 2006
Published online: 5 January 2006

Engineering Skin to Study Human Disease –
Tissue Models for Cancer Biology and Wound Repair

Jonathan A. Garlick

Division of Cancer Biology and Tissue Engineering Department of Oral and Maxillo-
facial Pathology, Tufts University, 55 Kneeland Street, Room 116,
Boston, Massachusetts 02111, USA
Jonathan.Garlick@Tufts.edu

Abstract Recent advances in the engineering of three-dimensional tissues known as skin equivalents, that have morphologic and phenotypic properties of human skin, have provided new ways to study human disease processes. This chapter will supply an overview of two such applications – investigations of the incipient development of squamous cell cancer, and studies that have characterized the response of human epithelium during wound repair. Using these novel tools to study cancer biology, it has been shown that cell-cell interactions inherent in three-dimensional tissue architecture can suppress early cancer progression by inducing a state of intraepithelial dormancy. This dormant state can be overcome and cancer progression enabled by altering tissue organization in response to tumor promoters or UV irradiation or by modifying the interaction of tumor cells with extracellular matrix proteins or their adjacent epithelia. By adapting skin equivalent models of human skin to study wound reepithelialization, it has been shown that several key responses, including cell proliferation, migration, differentiation, growth-factor responsiveness and protease expression, will mimic the response seen in human skin. In this light, these engineered models of human skin provide powerful new tools for studying disease processes in these tissues as they occur in humans.

Keywords Tissue engineering · Human skin equivalents · Intraepithelial neoplasia · Wound repair · Squamous cell carcinoma

Abbreviations

IE   Intraepithelial
ECM   Extracellular matrix
β-gal β-Galactosidase
BM   Basement membrane

1
Introduction

Engineered human tissue models designed to advance our understanding of human disease processes require the ability to fabricate tissues that faithfully mimic their in vivo counterparts. In vitro studies using human cells have often been limited by their inability to simulate the characteristic morphology, phenotype and behavior found in vivo. For example, monolayer cultures of human skin keratinocytes demonstrate limited stratification, partial differentiation and hyperproliferative growth [1], and have been of limited use when studying the complex cellular responses seen in intact tissues due to their lack of three-dimensional (3-D) tissue architecture. Biologically-meaningful signaling pathways, mediated by the linkage of adhesion and growth, function optimally when cells are spatially organized in 3-D tissues, but are uncoupled and lost in two-dimensional culture systems [2–4]. It is therefore essential to generate 3-D cultures that display the architectural features seen in vivo in order to engineer tissue models that will allow the study of human diseases in their appropriate tissue context.

During the last decade, the development of tissue-engineered models that mimic human skin, known as skin equivalents (SE), have provided novel experimental systems to study the behavior of normal and altered human stratified squamous epithelium. The SE is cultured at an air-liquid interface on a collagen matrix populated with dermal fibroblasts to generate 3-D, organotypic tissues that demonstrate in vivo-like epithelial differentiation and morphology, as well as rates of cell division similar to those found in human skin [5, 6]. Organotypic, SE tissue models have previously been adapted to study epithelial and skin biology on a variety of connective tissue substrates that have served as dermal equivalents [7, 8]. A well-stratified epithelium was seen when cultures were grown on dermal equivalents fabricated as Type I collagen gels which were populated with fibroblasts [5, 9, 10]. Porous membranes seeded with fibroblasts or coated with extracellular matrix proteins have been used to generate skin-like organotypic cultures [11]. Alternatively, fibroblasts have been incorporated into a three-dimensional scaffold, where these cells could secrete and organize an extracellular matrix [12]. While organotypic cultures of stratified epithelium have been shown to express basement membrane components in organotypic culture [13–17], limited success has been achieved in attaining structured basement membrane [8, 18, 19]. Since it is known that basement membrane components play a functional role in the regulation of epidermal growth and differentiation [20], it is important to generate cultures that have a well-structured basement membrane. Furthermore, it has previously been shown that the correct spatial organization and polarity of basal cells was associated with functional hemidesmosomes and basement membrane integrity [21].

Since the goal in the fabrication of SEs of human skin has been to fabricate and maintain a stratified epithelium that demonstrates in vivo-like features of epidermal morphology, growth and differentiation, it is critical to optimize these features in 3-D cultures. This has been accomplished by combining the three components thought to be critical in epidermal normalization – keratinocyte stem cells with high proliferative potential, viable dermal fibroblasts and structured basement membrane. Epidermal stem cells have been shown to be present in SE cultures as transplants of these cultures have shown the persistence of genetically-marked progenitor cells in these tissues up to one year after grafting. Dermal fibroblasts are required to stimulate epithelial growth and to promote its stratification while the presence of pre-existing basement membrane components were required to initiate and promote the rapid assembly of structured assembly BM in SE cultures [6], which is needed to sustain keratinocyte growth and to optimize epithelial architecture. Our laboratory has optimized the growth and differentiation of skin-like, organotypic cultures by growing keratinocytes on an acellular, human dermal substrate (AlloDerm) that was repopulated with human fibroblasts to generate SEs with a high degree of tissue normalization forming a structured, mature basement membrane (Fig. 1). This human tissue model recapitulates the morphology of skin to a large degree and has facilitated further clarification of the contributions made by basement membrane components and dermal fibroblasts to normal epidermal morphogenesis.

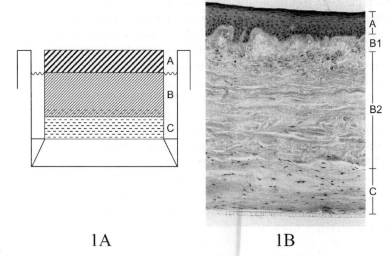

1A   1B

Fig. 1 Appearance of skin equivalents grown in organotypic culture at an air–liquid interface. Skin equivalent cultures of normal human keratinocytes were grown at the air–liquid interface for ten days on a deepidermalized human dermis (AlloDerm) containing basement membrane components (layer B). The epithelium generated (layer A) demonstrated in vivo-like tissue architecture, characterized by the presence of all morphologic strata and epithelial rete pegs at the interface with the connective tissue. The contracted collagen gel (layer C) containing dermal fibroblasts on which the AlloDerm was layered is seen at the bottom of the tissue, where fibroblasts have migrated into the lower part of the AlloDerm

Such optimally-engineered human tissues can be adapted to study a variety of human disease processes that simulate events that occur in human skin and other stratified squamous epithelia. As examples of how SEs can be adapted to study human disease, this chapter will describe how SE cultures have been used to characterize the response of human skin-like tissues following wounding and during early cancer development in a premalignant tissue. First, studies that have utilized these 3-D tissues to demonstrate the critical role of tissue architecture and cell–cell interactions during the earliest stages in the development of cancer in stratified squamous epithelium will be reviewed. Secondly, the response of human tissue models designed to mimic the in vivo reepithelialization of wounded human skin, from the initiation of keratinoctye activation until restoration of epithelial integrity. This will be accomplished by reviewing previous studies that have defined key response parameters, such as growth, migration, differentiation, growth-factor responsiveness and protease expression after wounding SEs. These applications demonstrate the utility of these engineered, human epithelial tissues in the study of responses that characterize the switch from a normal to a regenerative epithelium during wound reepithelialization and during the earliest, intraepithelial stage of carcinogenesis. It is hoped that by describing human tissue models that recapitulate tissue regeneration and carcinogenesis in vivo, further study of the nature of these processes will be facilitated.

2
Engineered human tissue models used to study early cancer progression in stratified squamous epithelium

Squamous cell cancer is initiated as a small nest of aberrant, dysplastic cells that expand to dominate a tissue and form a macroscopic tumor. During the earliest, intraepithelial (IE), premalignant stages of cancer progression, before the onset of cancer cell invasion, premalignant lesions demonstrate dysplastic cell foci, with abnormal nuclear and cytoplasmic morphologic features, that are initially surrounded by normal, undisturbed tissue [22, 23]. However, the role of 3-D tissue architecture, as characterized by interactions between potentially neoplastic cells and their normal neighbors, in the progression of human cancer during these early IE stages is not known. While it has been shown that normal cells can alter the phenotype of transformed cells in vitro [24–28], these studies were performed in conventional 2-D cultures that do not account for the role that tissue architecture plays in cancer development [29, 30]. Furthermore, the role of normal cell context in controlling the growth of cells with malignant potential has been difficult to study in vivo. Most studies of in vivo carcinogenesis, including transgenic models, follow the progression of cells with malignant potential that are surrounded by cells manifesting similar properties of transformation [31]. This does not accurately reflect the early progression of spontaneous tumors in stratified epithelial tissues, since cells with neoplastic potential are usually in contact with normal cellular neighbors during the incipient stages of tumor development.

In recent years, evidence has been accumulating that cancer is a disease of altered tissue organization. The view that cancer development and progression is a consequence of altered interactions between tumor cells and their immediate tissue microenvironment has recently been named the "tissue organization field theory of carcinogenesis" [32]. This theory proposes that cancer development disrupts normal interactions between adjacent cells or stroma, which dramatically modifies the ability of cells to sense normal regulatory signals that are inherent in tissue architecture. A corollary of this theory is that cells with neoplastic potential can be reprogrammed to behave like normal cells if found in normal tissue context. In this light, investigation of the role of cell–cell interactions in early neoplastic progression requires the capacity to detect and characterize small numbers of cells with malignant potential in the context of a 3-D network of more normal cells. Engineered human tissues that display 3-D human tissue architecture are therefore essential tools for accomplishing this goal.

Over the last decade, novel tissue models have been engineered to study early neoplastic progression in stratified squamous epithelium in which normal cell context is respected and cells with malignant potential are marked to study their fate and phenotype. To adapt these 3-D cultures to simulate premalignant disease as it occurs in human tissues, SEs with varying degrees of dysplasia were fabricated by mixing normal keratinocytes with tumor cells. The tumor cells used were a cell line (HaCaT-II-4) that was de-

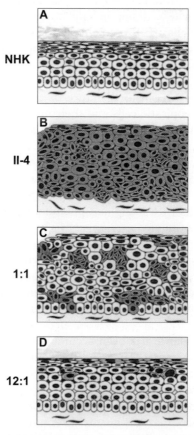

Fig. 2 A human tissue model for premalignant disease of stratified squamous epithelium. Normal keratinocytes (NHK) form a well-stratified epithelium with normalized tissue architecture when grown in skin-equivalent culture (**A**) , while II-4 cells that were labeled with the gene for β-galactosidase generate a disorganized and dysplastic tissue (**B**). When these two cell types are mixed in a 1 : 1 ratio, expanded clusters of β-gal-positive cells are randomly distributed among normal cells (**C**). The presence of such large numbers of II-4-gal cells has disrupted normal tissue architecture. When II-4 cells are mixed in a 12 : 1 ratio (NHK : II-4 cells), β-gal-positive intraepithelial tumor cells do not expand and remain as individual cells in the context of the well-preserved tissue architecture of normal cells (**D**)

rived by transfection of the spontaneously-immortalized human keratinocyte line (HaCaT) [33] with an activated c-Harvey-ras oncogene [34]. These cells have previously been shown to display severe dysplasia in organotypic culture and low-grade malignant behavior after in vivo transplantation [35]. By genetically-marking these potentially-malignant keratinocytes with a retroviral vector encoding β-glactosidase (β-gal) and mixing them at varying ratios with normal keratinocytes, epithelial tissues with varying degrees of dysplasia were generated to directly study the intraepithelial dynamics between NHK and adjacent, intraepithelial tumor cells (Fig. 2). These SEs were then transplanted into the dorsa of nude mice as surface transplants to allow the earliest events in neoplastic progression to be studied in vivo, by generating a stratified epithelium which evolved from a preinvasive, focally dysplastic tissue to one demonstrating tumor cell invasion into the connective tissue.

2.1
Cell–cell interactions inherent in 3-D tissue architecture suppress early cancer progression by inducing a state of intraepithelial dormancy

Using the approach described above to engineer precancerous lesions in human tissues, it was found that normal tissue architecture acted as a dominant suppressor of early cancer progression in stratified epithelium [36]. This occurred as interactions with adjacent normal keratinocytes induced intraepithelial tumor cells to withdraw from cell cycle and undergo terminal differentiation. These findings showed that the signaling network inherent in cell interactions in stratified epithelia was effective for tumor control and that a higher level of tissue organization, such as that seen in intact 3-D tissues, could predominate over cellular genotype in early cancer progression.

This was first shown in premalignant tissues that were generated as mixtures of NHK and II-4 cells, in which normal cells were the predominant cell type (12 : 1, NHK : II-4), that were clinically and morphologically normal four weeks after grafting to nude mice and showed no β-gal positive. This suggested that cells with malignant potential were eliminated from the tissue at this 12 : 1 mixing ratio. II-4 cells were only detected in 1 : 1 mixtures after grafting and they demonstrated larger foci of dysplastic cells that invaded into the underlying connective tissue. These findings suggested that a relatively high, critical number of cells with malignant potential needed to be present for these cells to persist, undergo clonal expansion and form a focally dysplastic and early invasive tumor in vivo. To explain the observation that II-4 cells failed to persist in the tissue when a greater number of normal cells were present in the mixture, the fate and distribution of cells with malignant potential was studied after in vitro growth in organotypic cultures that were generated before transplantation. Tissues constructed by mixing cells at 1 : 1 and 12 : 1 ratios were analyzed by immunohistochemical staining to determine if cells with malignant potential were undergoing changes in their biologic behavior due to interactions with adjacent, normal cells that could explain their subsequent loss from the tissue after grafting. Tissue dynamics were studied after growth for one week in organotypic culture by assessing the proliferation and differentiation of II-4 cells in these mixtures. Double immunofluorescent staining of 12 : 1 mixed cultures for β-gal and BrdU demonstrated that no proliferation was seen in the individual β-gal positive cells, which were only present above the basal cell layer, suggesting that II-4 cells were growth-suppressed when surrounded by normal cells. In contrast, 1 : 1 mixtures demonstrated numerous, suprabasal β-gal positive clusters which were BrdU-positive, showing that these cells were able to continue to proliferate and expand. This demonstrated that clusters containing larger numbers of II-4 cells continued to proliferate and that cell growth was not affected when the number of contiguous II-4 cells was sufficiently high. In addition, when mixed with normal keratinocytes in a ratio of 12 : 1, II-4-gal cells underwent terminal differentiation as evidenced by the colocalization of β-gal and filaggrin, a marker of keratinocyte terminal differentiation. This showed that II-4 cells were being normalized by adjacent normal keratinocytes which were undergoing terminal differentiation. In contrast, II-4 cells grown in cultures at a ratio of 1 : 1 were not induced to express filaggrin by neighboring cells. Furthermore, since most II-4 cells mixed with NHK at growth-supressive ratios were detected as individual cells in the suprabasal layers, it was important to determine how this sorting occurred. When the distribution of II-4 cells was examined in 12 : 1 mixtures (NHK : II-4) shortly (16 hours) after seeding, a monolayer of keratinocytes was seen on collagen gels that contained a small number of basal β-gal-positive cells amidst a large number of NHK. Within two days, all β-gal-positive, II-4 cells had been displaced to a position above the basal cell layer. It appeared that NHK could actively compete with II-4 cells for basal position and displace them as they preferentially attached to this Type I collagen matrix. This supports the view that the suprabasal distribution of II-4 cells was due to an active sorting process through which these cells were displaced from their initial basal position, leading to their ultimate loss from the tissue.

These findings are shown schematically in Fig. 3. When tumor cells were mixed in a 1 : 1 ratio with normal cells, IE clusters were able to form in vitro (Fig. 3A) and tumor cells were able to invade into the connective tissue after in vivo transplantation (Fig. 3C). In contrast, tumor cells grown in the context of a majority of normal cells demonstrated individual, differentiated β-gal-positive cells that were growth-suppressed in vitro (Fig. 3B) and desquamated from the tissue after grafting in vivo (Fig. 3D). By using 3-D tissues that mimic the early stages of epithelial cancer in humans, these

In Vitro   In Vivo

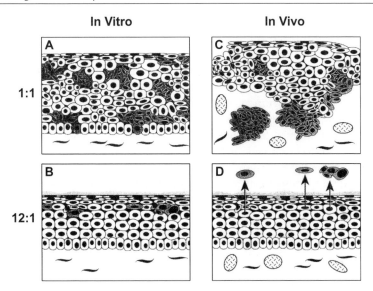

Fig. 3 Cancer progression can only occur when intraepithelial dormancy is overcome through the presence of elevated numbers of tumor cells in the tissue. Skin equivalent cultures were grown as mixtures of II-4 and normal keratinocytes at ratios of either 1 : 1 (**A**) or 12 : 1 (**B**) and grafted to nude mice. 1 : 1 mixtures demonstrated clusters of II-4 cells that that underwent intraepithelial expansion in vitro (**A**) and invaded into the un-derlying connective tissue after transplantation to nude mice (**C**). However, when 12 : 1 mixtures were grown (**B**), tissues demonstrated individual β-gal-positive cells that did not expand while grown in vitro and underwent desquamation of II-4 cells from the tissue after grafting (**D**). This demonstrated that a state of intraepithelial dormancy was induced by surrounding normal cells that suppressed the growth of II-4 cells and prevented their persistence within the tissue at this suppressive ratio. Only when tumor cell clusters of sufficiently large size were present, as in the 1 : 1 mixtures, were II-4 cells able to evade this local growth-suppression and invade into the connective tissue

findings demonstrated a novel mechanism for elimination of cells with ma-lignant potential that could suppress early neoplastic progression, namely a tissue-based growth control induced by interactions between adjacent cells that leads to normalization and elimination of sufficiently small numbers of potentially malignant cells. The size of an initiated clone is therefore crucial in determining its survival potential during development of early neoplasia, and progression to malignancy from a premalignant state requires the pres-ence of a number of potentially malignant cells above a critical threshold. In larger tumor cell clusters, greater numbers of cells can interact and can escape this environmental growth suppression. When individual tumor cells were present, they entered a quiescent state known as "intraepithelial dormancy", in which the full, neoplastic potential of the tissue was not realized. These cells are therefore held in a conditionally-suppressed state and can undergo one of two ultimate fates – either being eliminated from the tissue together with adjacent normal cells, or overcoming this dormant state when tumor cells interact with each other in sufficiently large clusters. Further proof of this "conditional" state of growth suppression was shown by determining whether II-4 cells could resume growth when this tissue was disaggregated and single cells regrown in submerged monolayer culture. To isolate only suprabasal II-4 cells from mixtures, SEs were grown in low-calcium media in order to strip all suprabasal cell layers while leaving basal cells attached to the collagen matrix. Suprabasal cells detached as a sheet, which was then trypsinized and single cells grown at clonal density in submerged culture on a 3T3 feeder layer. Expanded colonies of II-4 cells were seen throughout these cultures, proving that growth-inhibited II-4 cells had only transiently withdrawn from the cell cycle when contacting NHK in 12 : 1 mixtures in organotypic culture.

These findings supported earlier observations made in the "classical", two-stage theory of skin carcinogenesis in experimental animals, which had shown that application of a tumor promoter to previously initiated skin produces tumors, while "subcarcinogenic" initiation alone results in no tumors. Initiated skin must therefore contain altered cells that cannot be identified microscopically since they do not form discrete foci and are "operationally normal" in the absence of promotion and in the appropriate cell microenvironment [36]. It has been theorized that these "repressed single mutant cells" are held in a nonproliferating state by feedback from normal, differentiated cells [37]. An epithelium exposed to subcarcinogenic levels of initiating effects may therefore contain large numbers of repressed, initiated cells that may never progress to neoplasia. This may help explain why premalignant lesions such as actinic keratosis of skin, cervical dysplasia, oral leukoplakia and lobular carcinoma of the breast can contain initiated or dysplastic cells that do not always advance to invasive cancer.

These studies on the role of 3-D tissue architecture in cancer development directly implicate tissue architecture and the cellular milieu as dominant regulators of the neoplastic phenotype, and support the "tissue organization field theory of carcinogenesis" [32]. In this light, maintenance of normal tissue architecture can constrain potentially malignant tumor cells in a conditionally-suppressed state and this is sufficient to abrogate cancer progression through an intrinsic, tissue-based elimination of tumor cells by normal cell neighbors. Using a 3-D culture system, it has been shown that this phenomenon also occurs in human breast tumorigenesis, where interactions between extracellular matrix (ECM) proteins and their receptors could normalize tissue architecture and revert malignant cells to a normal phenotype [38]. Neoplastic progression could occur only if the microenvironment was changed to allow growth of initiated cells so that this suppressive tissue-barrier could be overcome. Clearly, studies on the role of 3-D tissue architecture in tumor biology would not have been feasible without the capacity to adapt engineered human SEs to mimic early cancer progression.

2.2
Factors altering cell–cell
and cell–matrix interactions abrogate the microenvironmental control
on intraepithelial tumor cells and promote cancer progression

Since the signaling network inherent in cell–cell interactions plays an important role in tumor control, one may ask how potentially malignant cells can overcome this restrictive microenvironment of the normal cell context that was found to limit intraepithelial tumor cell expansion in vitro and prevent clonal persistence in vivo. Several studies were designed in which 3-D tissue architecture was perturbed to determine whether the state of intraepithelial dormancy could be abrogated and tumor progression promoted.

2.2.1
The tumor promoter TPA enables expansion
of intraepithelial tumor cells

12-O-Tetradecanoylphorbol-13-acetate (TPA) is a phorbol ester tumor promoter that is thought to selectively stimulate proliferation of distinct subpopulations of skin keratinocytes, leading to the clonal expansion of initiated cells [39, 40]. Direct studies of whether exposure of TPA to premalignant tissues generated using the SE model described above would permit foci of intraepithelial tumor to undergo expansion and abrogate the growth suppression induced by adjacent normal cells were performed [41]. TPA (0.001 ug/ml) was added to organotypic cultures containing mixtures of NHK : II-4 cells at varying ratios to determine whether this agent could selectively stimulate clonal expansion of II-4 cells in mixtures previously found to be growth-suppressive. When 12 : 1 and 4 : 1 cell mixtures (NHK : II-4) were exposed to 0.001 ug/mL TPA, expanded β-gal-positive clusters were visualized in these mixtures when compared to control cultures not treated with TPA (Fig. 4). To study the association of such expansion with proliferation, SEs were double-stained by immunofluorescence for BrdU and β-gal. Colocalization of β-gal and BrdU in II-4 cells in TPA-treated 12 : 1 mixtures (Fig. 4H) showed that the TPA-associated increase in clonal expansion of II-4 cells was accompanied by their proliferation, while no such BrdU-positive nuclei were seen in β-gal-positive cells in non-TPA-treated 12 : 1 or 4 : 1 cultures (Figs. 3C and D). Staining for the proliferation revealed a dramatic decrease in BrdU-positive nuclei in pure NHK cultures treated with TPA (Figs 4F, H, I and J) while untreated NHK cultures continued to show large numbers of proliferative basal cells (Figs. 4 A, C, D and E). In contrast, II-4 cell proliferation was similar in the presence (Fig. 4G) and absence (Fig. 4B) of TPA. Since it was found that proliferation of II-4 cells was not significantly altered by TPA, it appeared that intraepithelial expansion of II-4 cells occurred as a result of the growth advantage of these cells in relation to the suppressed normal cells rather than being caused by the direct stimulation of II-4 growth.

Fig. 4   TPA induces clonal expansion and proliferation of II-4 cells in 12 : 1 and 4 : 1 ▶
mixtures, which is associated with the decreased proliferation of NHK. Double-immuno-
fluorescence stain demonstrating superimposed fluorescent signals for β-galactosidase
(FITC channel, *green*) and bromodeoxyuridine (Brdu) (Texas red channel, *red*).
(**A**) Normal human keratinocyte (NHK) cultures grown without TPA demonstrating
Brdu-positive nuclei limited to the basal layer and no β-gal expression. All other nuclei counterstained with DAPI, *blue channel*. (**B**) Pure II-4 cell cultures grown without
TPA demonstrating that all II-4 cells express β-gal and the presence of Brdu-positive nu-
clei in basal and suprabasal layers. (**C**) Mixture of NHK : II-4 cells (12 : 1) demonstrating
individual β-gal cells in a suprabasal position and Brdu-positive nuclei limited to the
normal basal keratinocytes. Individual β-gal cells lack colocalization of β-gal and Brdu
demonstrate withdrawal of II-4 cells from cell cycle in the absence of TPA. (**D**) Mixture of
NHK : II-4 cells (4 : 1) demonstrating individual β-gal cells in a suprabasal position and
Brdu-positive nuclei limited to the normal basal keratinocytes. (**E**) Mixture of NHK : II-4
(1 : 1) showing large β-gal positive II-4 cell clusters and Brdu-positive nuclei limited to
the normal basal keratinocytes. (**F**) NHK cultures grown with 0.001 ug/ml TPA demonstrating decreased numbers of Brdu-positive nuclei limited to the basal layer and no
β-gal expression. (**G**) Pure II-4 cell cultures grown with 0.001 ug/ml TPA demonstrating continued proliferation as evidenced by presence of Brdu-positive nuclei in all strata.
(**H**) Mixture of NHK : II-4 cells grown with 0.001 ug/ml TPA demonstrating increased size
of β-gal positive II-4 cells. Note positive cell cluster which is also Brdu-positive (*arrow*) as
seen by the *yellow* nucleus and *green* cytoplasm (*arrow*) in the center of expanding cluster. (**I**) Mixture of NHK : II-4 (4 : 1) grown with 0.001 ug/ml TPA demonstrating increased
numbers of β-gal-positive II-4 cells (*arrow*). (**J**) Mixture of normal keratinocytes : II-4
(1 : 1) showing large clusters of β-gal cells, an occasional BrdU-positive NHK nucleus
(*dotted arrow*) and a BrdU positive II-4 cell nucleus (*solid arrow*). The dermal–epithelial
interface is marked with the *white dotted line*

These results demonstrated that the induction of II-4 cell expansion by
TPA in mixed cultures could be explained by the differential regulation of
proliferation of normal and II-4 cells in these cultures, that allowed tumor cells to circumvent the suppressive effect of neighboring, normal keratinocytes. TPA enabled clonal expansion of IE tumor cells by altering the rate
of growth and differentiation potential of normal cells, and not through direct
alteration of potentially malignant cells. TPA thus stimulates the early stages
of neoplastic progression in human stratified epithelium by creating a microenvironment conducive for clonal expansion of previously suppressed,
potentially malignant cells, by permitting them to overcome the growth suppressive effects of normal cell context.

2.2.2
Immortalization of adjacent epithelial cells
cannot induce intraepithelial dormancy of tumor cells

Since IE neoplasia arises in the context of cells with variably transformed
phenotypes, it was important to determine the extent to which the state of
transformation and stage of neoplastic progression of neighboring cells could

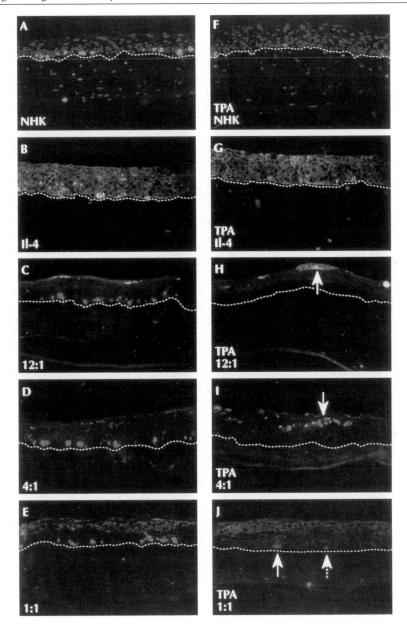

control the malignant phenotype in stratified epithelium [42]. To accomplish this, the distribution of genetically marked II-4 cells was determined after mixing with either NHK or with their immortalized, but nontumorigenic, parental HaCaT line. At all mixing ratios, II-4 cells demonstrated larger β-gal-positive clusters when surrounded by HaCaT cells than when grown in the

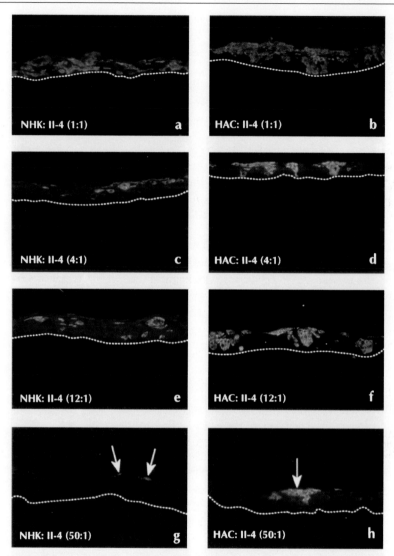

Fig. 5 Potentially malignant II-4 keratinocytes undergo clonal expansion in the context of HaCaT cells but not when surrounded by normal keratinocytes. II-4 keratinocytes were mixed with either normal human keratinocytes (NHK) at ratios of 1 : 1 (**a**), 4 : 1 (**c**), 12 : 1 (**e**) and 50 : 1 (**g**) or with HaCaT (HAC) keratinocytes at ratios of 1 : 1 (**b**), 4 : 1 (**d**), 12 : 1 (**f**) and 50 : 1 (**h**). Mixed cultures were grown for seven days in skin equivalent culture and stained for β-galactosidase expression. Individual β-gal-positive cells are indicative of induction of intraepithelial dormancy and failure of II-4 cells to proliferate while expansion of β-gal-positive clusters shows that growth suppression was lost. In 1 : 1 mixtures, II-4 cells grown with HAC cells demonstrate larger β-gal-positive clusters (**b**) than those grown with NHK (**a**). This was most dramatically seen for 4 : 1, 12 : 1 and 50 : 1 mixtures of II-4 and HAC cells, demonstrating that the nature of the cells adjacent to II-4 cells could determine whether the microenvironment was permissive for intraepithelial expansion. *Scale bar* = 100 μm

context of NHK (Fig. 5). This was most striking for the 4 : 1, 12 : 1 and 50 : 1 ratios, which demonstrated single II-4 cells in NHK context (Figs. 5C,E and G) and expanded β-gal-positive II-4 clusters when mixed with HaCaT cells (Figs. 5D,F and H).

Double immunofluorescence staining for BrdU incorporation and β-gal-positive cells in these mixtures showed the continued proliferation of II-4 cells when grown with HaCaT cells in a 12 : 1 ratio, while II-4 cells in 12 : 1 (NHK : II-4) mixtures were growth suppressed. Similarly, 12 : 1 mixtures showed that II-4 cells surrounded by HaCaT cells did not express filaggrin while those contacting NHK expressed this protein, showing that contact with HaCaT cells could not induce differentiation in II-4 cells as normal cell neighbors could. These findings showed that immortalized HaCaT cells could not limit the growth of these II-4 cells as the normal cell environment was able to, and that the capacity to induce IE dormancy was considerably reduced in the context of an immortalized cell line. The distribution and behavior of low-grade malignant cells was therefore dependent on the state of transformation of adjacent keratinocytes, indicating that alterations in the cellular microenvironment are central to the induction of clonal expansion and early neoplastic progression in 3-D, stratified epithelium. This shows that it is not sufficient for a cell to have malignant potential, but that it must be in a permissive environment in order to abrogate normal cell control of early cancer progression.

2.2.3
UV-B Irradiation is permissive for tumor cell expansion by inducing a differential apoptotic and proliferative response between tumor cells and adjacent normal cells

Solar irradiation in the ultraviolet (UV) range is known to be associated with the development of non-melanoma skin cancers in humans [43, 44]. UV-B irradiation has been linked to mutations in the p53 gene which are found in a large majority of cutaneous squamous cell carcinomas [44], premalignant lesions such as actinic keratosis, and in normal, sun-damaged skin [45]. In recent years, it has been shown that p53-mutated keratinocytes are arranged in clonal patches in normal human skin and may involve as much as 4% of the epidermis [45]. It was theorized that further UV-B exposure is more likely to induce cell death in normal cells that do not harbor p53 mutations, thereby allowing the expansion of individual p53 mutant cells into a niche left by the death of neighboring, normal cells [46]. This suggested that in addition to its role as a mutagen, UV-B irradiation can act as a tumor promoter by enabling the clonal expansion of p53-mutated cells. Since ethical reasons have limited the ability to directly study the effects of UVB irradiation in the skin of human volunteers, the use of skin-like, 3-D SE tissue models containing human keratinocytes offers an attractive alternative for studying the effects of UV-B irradiation on human skin. Using the 3-D tissue models for premalignant disease described above, it has recently been determined that biologically meaningful UVB exposure enables the intraepithelial expansion of II-4 tumor cells by inducing a differential apoptotic and proliferative response between these cells and adjacent normal keratinocytes [47]. To mimic the effects of sunlight on engineered human, premalignant tissues, UV-B was administered to mixed cultures of normal keratinocytes and II-4 cells using an FS20 sunlamp at doses between 0–50 mJ/cm^2. It was found that when SE mixtures were exposed to a UVB dose of 50 mJ/cm^2, intraepithelial tumor cells underwent a significant degree of proliferative expansion when compared to nonirradiated cultures. When mixed organotypic cultures (12 : 1, NHK : II-4) were not irradiated, the cultures demonstrated small numbers of individual β-gal-positive cells in the middle and upper spinous layers of the epithelium that occupied 2% of the tissue. In contrast, mixed cultures irradiated with UVB at 50 mJ/cm^2 demonstrated significant expansion of II-4 cells, as seen by the presence of large clusters of green β-gal-positive II-4 cells in the upper spinous layer of the epithelium. These clusters had expanded to occupy roughly 28% of the tissue, demonstrating that the number of II-4 cells in IE tumor cell clusters had increased. This showed that UV-B irradiation could induce the IE expansion of II-4 tumor cells and it allowed these cells to escape the growth control of adjacent NHK. To understand how UVB irradiation enabled the expansion of II-4 cells, irradiated and nonirradiated, mixed organotypic cultures were double-stained by immunofluorescence to measure cell proliferation. Only irradiated mixed cultures demonstrated colocalization of BrdU-positive nuclei and were β-gal-positive, proving that UVB-induced expansion of IE tumor cells was associated with the active proliferation of II-4 cells.

In addition to the effects of UV irradiation on cell growth, UV-B-associated apoptosis in normal keratinocytes and II-4 cells was measured by TUNEL assay. Numerous TUNEL-positive cells, making up roughly 14% of the cells in this tissue, were seen in normal keratinocytes, but II-4 cells did not demonstrate any TUNEL-positive cells following irradiation. This suggested that p53-mutant, II-4 cells were resistant to apoptosis and that the observed expansion of II-4 cells when the two cell populations were mixed and irradiated with UV-B was apparently due to the differential induction of apoptosis of NHK relative to II-4 cells. Thus, the differential sensitivity of normal keratinocytes and the early-stage II-4 tumor cells to induction of growth arrest and apoptosis was associated with the expansion of apoptosis-resistant IE tumor cells; their escape had been mediated from a growth-suppressed, dormant state previously induced by adjacent normal cells. The use of these 3-D tissue models to model the effects of UV-B-induced sun-damage in skin thus supports current theories for malignant progression in UV-damaged skin that propose that UV-B exposure can preferentially induce apoptotic cell death in cells that are not resistant to apoptosis, while adjacent tumor cells that harbor p53 mutations can undergo clonal expansion. Since it is known that sun-exposed skin contains thousands of clones of p53-mutant keratinocytes, histologically-normal sun-damaged cells are likely to be at risk for the early progression of skin carcinogenesis. UV-B can promote the clonal expansion of IE tumor cells by altering the behavior of normal keratinocytes in the epithelium rather than by directly altering the phenotype of the potentially malignant cells, thereby enabling UV-damaged clones to overcome the growth-suppressive effects exerted by normal cells. This suggests that clonal expansion and the earliest stages of UV-associated non-melanoma skin cancer are driven by the promoting effects of UV irradiation that are not necessarily due to the acquisition of additional mutations in previously sun-damaged skin, thus creating a microenvironment conducive to clonal expansion of cells with neoplastic potential.

2.2.4
Basement membrane proteins promote progression of early cancer by rescuing tumor cells from intraepithelial dormancy through their selective adhesion to laminin 1 and Type IV collagen and subsequent expansion

Fabrication of 3-D SEs have provided new opportunities to study the contribution of stromal components to the early stages of cancer progression in epithelial tissues. These tissue models have facilitated study of the dynamic and reciprocal effects of both diffusible factors [8] and structural matrix components [12, 18] that can be incorporated into SEs. For example, it has previously been shown that cell–ECM interactions can alter the malignant phenotype of potentially neoplastic breast cells [38] and restrict inappropriate cell growth by inducing apoptosis [48]. In stratified epithelia, only those keratinocytes interacting with basement membrane are division-competent, while keratinocytes distant from the ECM undergo apoptosis [49]. Proliferation is normally restricted to these matrix-attached cells [21, 50], and it is thought that this anchorage-dependent regulation of growth may influence the degree of epithelial dysplasia seen in the tissue [51]. It has previously been shown that when mixtures of transformed and normal tracheal cells were inoculated into the lumen of a denuded trachea, transformed cells survived only if they contacted the connective tissue substrate [52]. Contact with the connective tissue interface appears to be essential for allowing premalignant cells to persist, expand and eventually invade.

Since the progression from IEN to invasive cancer is associated with the migration of tumor cells through the basement membrane barrier (BM) after its proteolytic modification [53], it was important to understand the role of adhesive interactions between intraepithelial tumor cells and BM proteins, during the earliest, intraepithelial stage of cancer progression. To study this directly, 3-D tissue models were used to analyze the stromal contribution of

ECM and BM proteins to the progression of IE (Fig. 6). This was achieved by culturing mixtures of NHK and II-4 cells on a variety of connective tissue substrates including: 1) contracted Type I collagen gel containing human dermal fibroblasts; 2) a substrate containing pre-existing BM components present on a de-epidermalized dermis derived from human cadaver skin (Alloderm, LifeCell Corp., Branchburg, NJ, USA); this dermis was treated to remove the surface epithelium and stromal cells, while preserving the BM proteins Types IV and VII collagen and laminin 1 on its surface [54]; 3) polycarbonate membranes coated with purified BM proteins laminin 1 and Type IV collagen or ECM components not found in BM, such as fibronectin, Type I collagen or a mixture of fibronectin and Type I collagen (Becton Dickinson, Billerica, MA, USA). AlloDerm and polycarbonate membranes were layered on the contracted collagen gel to enable fibroblasts to repopulate the Alloderm from below and to support the growth of cells on the polycarbonate membrane (seen schematically in Fig. 6). II-4 cells and NHK were seeded as mixtures at ratios previously shown to result in loss of attachment of II-4 cells to their ECM substrate [36] (12 : 1 and 4 : 1/NHK : II-4) in order to determine whether BM or specific ECM proteins could overcome IE dormancy by rescuing tumor

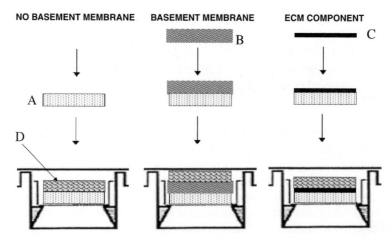

Fig. 6 Human three-dimensional tissue models developed to study the role of BM proteins in the progression of intraepithelial neoplasia. Keratinocytes were grown in the absence of pre-existing BM proteins by seeding cells directly on a contracted Type I collagen gel containing dermal fibroblasts (**A**). Cells were cultured in the presence of BM components (AlloDerm) by seeding keratinocytes on a deepidermalized dermal substrate (**B**) which was layered on the collagen gel (**A**) to allow it to be repopulated with fibroblasts. Cultures were grown on individual BM or extracellular matrix proteins by layering a polycarbonate membrane coated with these purified proteins (**C**) on the contracted Type I collagen gel. All cultures resulted in a well-stratified squamous epithelium (**D**) when grown at an air–liquid interface

cells from growth suppression by allowing their attachment and persistence in the tissue.

When mixed SE cultures were grafted to nude mice, tissues grown on a Type I collagen substrate that lacked BM components exhibited clinically (Fig. 7A) and histologically (Fig. 7C) normal epidermal structures. In contrast, mixtures grown in SE culture on AlloDerm in the presence of pre-existing BM components showed a thickened, white lesion (Fig. 7B) with microscopic evidence of persistence of IE tumor cells that were beginning to develop early invasive structures (Fig. 7D). These findings demonstrated that the presence of BM proteins was a permissive signal for the persistence of IE tumor cells, leading to the formation of dysplastic, premalignant lesions in vivo. Thus, in mixtures of tumor cells and normal keratinocytes (4 : 1/normal : II-4), tumor cells selectively attached, persisted and proliferated at the dermal–epidermal interface in vitro and generated dysplastic tissues only when grown in the presence of the AlloDerm substrate that contained BM proteins. It was found that this selective tumor cell attachment

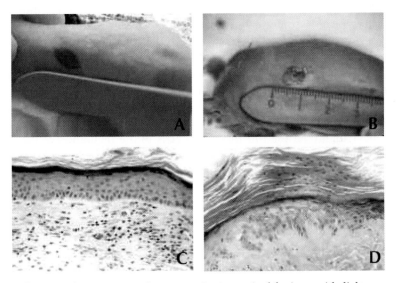

Fig. 7 Adhesion to basement membrane proteins is required for intraepithelial tumor cells to overcome intraepithelial dormancy and persist in the tissue to form a dysplastic, premalignant lesion in vivo. Skin equivalent cultures (4 : 1 mixtures, NHK : II-4) were grown in the absence or presence of BM components and grafted to nude mice for eight weeks. In the absence of pre-existing BM proteins, grafts appeared to be clinically normal skin (**A**) and showed a well-differentiated epithelium with normal architecture (**C**). In contrast, in the presence of BM proteins (AlloDerm), grafts fabricated from 4 : 1 mixtures of normal and II-4 cells showed a thickened, white lesion (**B**) that demonstrated moderate dysplasia, characterized by disorganization and altered polarity of basal and suprabasal layers, a hyperparakeratotic surface and an irregular epithelial–stroma interface (**D**)

was mediated by the rapid assembly of structured BM only in the presence of pre-existing BM components found in AlloDerm, which created a microenvironment in which tumor cells could compete with normal cells for attachment to BM. To determine which particular ECM or BM components enabled tumor cell attachment, mixtures were than grown on polycarbonate membranes coated with individual ECM or BM components, as described schematically in Fig. 6. Selective attachment and significant intraepithelial expansion occurred only on laminin 1 and Type IV collagen-coated membranes, and not on membranes coated with fibronectin or Type I collagen. Thus, intraepithelial progression towards premalignant disease is dependent on the selection of IE tumor cells with malignant potential to BM proteins that provide a permissive template for their persistence and expansion.

All of these studies show that the state of IE dormancy can be overcome through the attachment of tumor cells to BM proteins that serve as a selective template for early cancer progression by providing a permissive environment for IE tumor cells to adhere to the dermal–epidermal interface, persist in a basal position and proliferate to dominate the tissue. By allowing adherence, IE tumor cells gain a selective growth advantage and an increased risk of progression to invasive cancer. In this way, the BM microenvironment modulates the phenotype of IE tumor cells by enabling their selection for further cancer progression and directing their escape from IE growth control.

Until 3-D SE cultures were adapted to study early cancer development, most studies on neoplastic progression focused on the genotypic changes that induce the transformed state. Our findings suggest that it is not enough for a cell to have malignant potential, but it must be in a permissive environment for malignant progression to occur. This confirms previous studies which have identified the role of normal cells in the regulation of the transformed phenotype in 2-D monolayer cultures, and extends these observations to stratified epithelia with intact 3-D tissue architecture. Premalignant cells which cannot compete for attachment to BM can be displaced and become candidates for growth suppression dictated by normal cells in their immediate cellular environment. In this way, stratified epithelium may intrinsically eliminate small numbers of cells that have undergone the early stages of malignant change. In addition, factors such as UV-B irradiation, the stage of transformation of adjacent cells and the application of chemical promoters of tumorigenesis can abrogate NHK-induced growth control of clonal expansion by modifying the nature of the cells in direct contact with them. Thus, the positional and behavioral fates of potentially malignant cells is dependent on the nature of cell–cell and cell–matrix interactions, and this suggests that the release of a potentially neoplastic keratinocyte from a repressed state through modifications to its microenvironment is central to the induction of clonal expansion and early neoplastic change. By facilitating clonal expansion, cells are now more likely to acquire additional genetic changes, thereby leading to future malignant progression.

3
Three-dimensional skin-equivalent tissue models to study wound reepithelialization of human stratified epithelium

In vitro studies of wound reepithelialization have often been limited by their inability to simulate wound repair in the same way as it occurs in humans. For example, wound models using skin explants [55, 56] or monolayer submerged keratinocyte cultures [57] demonstrate limited stratification, partial differentiation and hyperproliferative growth. These culture systems have aided studies of keratinocyte migration in response to wounding, [58] but have been of limited use when studying the complex nature of keratinocyte response during wound repair, as they do not provide the proper tissue architecture to study wound response as it occurs in vivo.

SEs adapted to study wound repair in human epithelium have been found to simulate the chronology of events that occur during reepithelialization in human skin and have advanced our understand of the healing of wounds in human skin and other stratified epithelia [60]. This tissue model has allowed direct determination of the key response parameters of wounded epithelium including cell proliferation, migration, differentiation, growth-factor response and protease expression. Studies that have defined these responses of keratinocytes that are mobilized during reepithelialization will be reviewed in this chapter. These applications of tissue engineering technology demonstrate the utility of these human-like tissues in studying phenotypic responses that are characteristic of the switch from a normal to a regenerative tissue during wound repair.

3.1
Morphology of wounded skin equivalents

Immediately after keratinocyte injury, epithelium at the wound edge undergoes a sequence of coordinated temporal and spatial events that prepare these cells for new tissue formation. These initial phenotypic changes characterize a preparative phase that precedes migration into the wound site. This phase of wound response has been defined as the stage of "keratinocyte activation", during which the injured epithelium responds to wound injury by reprogramming patterns of gene expression to prepare for reepithelialization [65]. This "activated" keratinocyte undergoes a shift from a program of differentiation to one leading to directed and sustained migration and proliferation, which is then followed by stratification and differentiation.

The morphologic appearance of wounded SEs that occur during these events have been studied by several investigations [59–61]. Figure 8 demonstrates a schematic of the construction of SEs that describes how these cultures have been adapted to study wound healing, and Fig. 9 demonstrates the appearance of a wounded SE four days after wounding. The generation

Fig. 8 Construction of a skin-equivalent wounding model. **A** Schematic of stratified keratinocyte sheet growing on a contracted collagen matrix containing fibroblasts (layer II). This organotypic culture rests on a semipermeable membrane that is nourished with media from below at an air–liquid interface. An incisional wound is formed by cutting through the epithelium and collagen matrix (*dotted line*). **B** The wounded culture is then transferred onto a second contracted, collagen matrix (layer I). **C** The resulting cultures consisted of two layers of contracted matrix and one layer of epithelium. Wound margins are seen at the transition zone from layer I to layer II and are noted with arrows. **D** Reepithelialization occurs as wounded keratinocytes migrate onto the collagen in layer II, and is then followed by stratification of the tissue to reconstitute a fully stratified epithelium that covers the wound bed

Fig. 9 Appearance of skin equivalent culture four days after wounding. A wounded skin equivalent was transferred to a second collagen gel after an incisional, elliptical wound was generated. The incised skin equivalent demonstrates a degree of reepithelialization from the wound edges

of this model has demonstrated wound response that recapitulates many of the morphologic events known to occur during cutaneous reepithelialization in vivo, and thus provides an opportunity to study the appearance of wound

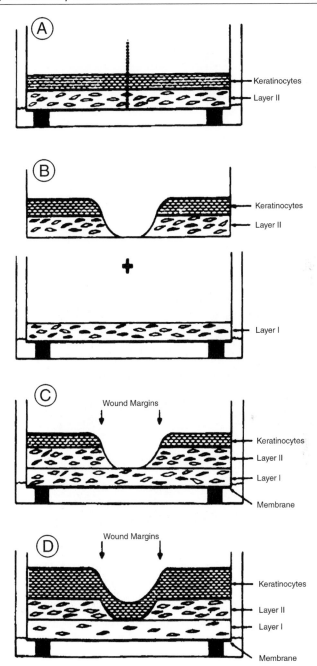

keratinocytes during the various stages of reepithelialization and epithelial reconstitution (Fig. 10). At 8 h, a wedge-shaped epithelial tongue 2–3 cells in thickness was seen at the edge of the wound margin (Fig. 10) in a tissue in

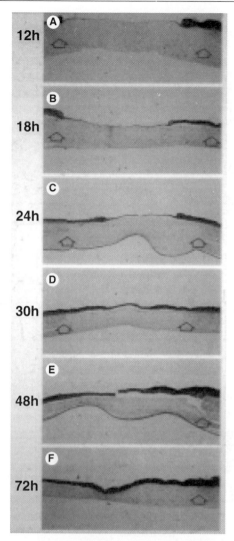

Fig. 10 Morphology of skin equivalents at various points after wounding. Wounded skin equivalents were stained with hematoxylin and eosin at **A** 12 hours, **B** 18 hours, **C** 24 hours, **D** 30 hours **E** 48 hours and **F** 72 hours after wounding. Two phases of epithelial response can be seen. The first phase extends from the earliest migration of epithelium **A, B** until the wound is completely covered by a thin epithelium (**C**). The second phase of stratification begins at 30 hours (**D**) and is complete when the tissue is of similar thickness to that of the unwounded epithelium at the wound margins (**E**). *Open arrows* demarcate the wound margins. (Original magnification ×10)

which wound margins could be seen at the transition from a double layer of collagen matrix to a single layer (Fig. 10). By 24 h, the wound floor was completely covered with a monolayer or bilayer of keratinocytes in the center and a more stratified epithelium towards the wound margins (Fig. 10). This showed that the stage of reepithelialization was complete and that stratification could now occur adjacent to the wound margins. Tissue stratification continued at 48 h, and a multilayer epithelium was generated by 4 d post-wounding. At 6 d after wounding, the reepithelialized surface demonstrated a fully stratified epithelium (Fig. 10) that was similar to nonwounded epithelium. This chronology of events during reepithelialization was therefore very similar to those reported in earlier in vivo studies. Geer and coauthors have shown that cells in wounded SEs initiated migration as an epithelial tongue by 48 hours after wounding and were completely reepithelialized within 72 hours [61]. Falanga et al. showed that wounded SEs initiated migration after 12 hours, reepithelialization was complete by 48 hours, and that a well-stratfied epithelium had reformed by 96 hours after wounding [59]. This consistency of epithelial response in these three studies demonstrated the ability to adapt a variety of SE-based models to study wound response in a manner that mimics that seen in vivo.

3.2
Proliferation in skin equivalents in response to wounding

Following the response of wounded SEs in organotypic culture provides an opportunity to directly measure the proliferation of keratinocytes during various stages of wound response. Previous studies have shown that the proliferative activity of SEs during reepithelialization had two distinct temporal and spatial phases. The first of these occurred during early reepithelialization, as the wound floor was being covered with the epithelial tongue. During this stage, proliferation was low in the migrating epithelial tongue, but it was elevated at the wound margins. The second phase of proliferative activity occurred after coverage of the wound was complete, as epithelium in the center of the wound underwent stratification. No BrdU labeling was seen in the tip of the wound edge at the earliest timepoint, 8 h after wounding, suggesting that these cells were nonproliferative and had assumed a migratory phenotype. This finding is in agreement with skin explant studies, which show that migration can be initiated as an active process independent of proliferation [57]. Proliferative activity in the epithelium of the wound margin peaked at 24 h after wounding as LIs were as high as 50% in epithelium of the wound margin and reached 80% in the center of wound epithelium. This burst of mitotic activity in the wound margin was transient, since the marginal epithelium returned to lower levels of growth activity at 48 h after wounding. These findings matched previous in vivo studies demonstrating a similarly delayed and transient increase in proliferative activity at the wound margin beginning after the initiation of migration. The finding that proliferation at the wound margin was higher than in the epithelium more distal to the margin suggested that cells were displaced from the wound margin onto the wound floor upon their proliferation. The movement of suprabasal cells at the edge of the wounded epithelium would leave that tissue somewhat denuded and would be analogous to the elimination of suprabasal cells by tape stripping. During the later phase of proliferation that occurred after wound coverage, mitotic activity continued to remain high in the wound epithelium at the 48 h and 4 d timepoints. This elevated mitotic activity led to stratification of the wound epithelium even after the proliferative rates in all other areas of the epithelium returned to baseline levels.

Proliferative response of SEs to wounding has also been evaluated using Ki67 as a marker of cell growth. Geer et al. found a similar pattern of growth to that described above, as was seen by a delayed and transient elevation of Ki67 expression at the wound edge [61]. In this study, Ki67 expression decreased significantly after wound closure as maturation and differentiation of the epithelium occurred. A similar pattern of Ki67 expression was seen in the study performed by Falanga et al. [59]. Using an innovative approach to characterize proliferative activity following SE wounding, these authors confirmed their findings on the spatial distribution of Ki67-positive cells in wounded SEs by assaying for the presence of S-phase nuclei using flow cytometry. To accomplish this, epithelial cells were separated from the underlying connective tissue in the SE by thermolysin treatment and were then disaggregated. It was found that SEs demonstrated a roughly twofold decrease in the fraction of S-phase cells after wounding. These findings are in agreement with other SE studies described above and with in vivo findings that have characterized the shift of wounded keratinocytes from a proliferative to a migratory phase shortly after wounding. The proliferative indices and rates of reepithelialization observed in these SE models are somewhat greater than those seen in wounds with scab formation in vivo, as the SE model presented here is analogous to a wet rather than dry healing environment. A wet environment is more conducive to an accelerated healing response and likely explains the more rapid healing response seen in wounded SEs compared to in vivo wound repair. Interestingly, the proliferation of wound epithelium is subject to environmental regulation as the growth response was shown to be sensitive to growth factor regulation, as described below.

3.3
Migration in skin equivalents in response to wounding

In vivo, wound response is known to alter the temporal and spatial patterns of integrin receptor expression and that of their ligands during reepithelialization. Several studies have characterized the distribution of these proteins during reepithelialization of wounded SEs. Geer et al. have developed an SE model for wound repair in which fibrin was incorporated as a substrate for reepithelialization by generating fibrin gels in the wound bed which contained physiologic concentrations of fibrinogen and thrombin [61]. The presence of fibrin was found to accelerate keratinocyte activation and it reduced the time of wound closure when compared to controls that did not contain fibrin. This promotion of reepithelialization was associated with the de novo synthesis of α5-integrin, which is not expressed in mature epithelium and is known to be upregulated during wound response in vivo. Integrin upregulation in response to wounding of SEs has also been determined in other SE wounding models [66]. It has also been shown that expression of proteins that serve as integrin ligands needed for cell migration as well as basement membrane components are also altered during reepithelialization of wounded SEs. Expression of laminin 1 has been found to be delayed somewhat during early reepithelialization, as migrating cells at the tip of the epithelial tongue did not express this protein and expression was seen closer to the wound margin. However, all basal cells expressed laminin 1 after wound closure, suggesting that keratinocytes can synthesize their own basement membrane proteins upon completion of the migratory phase of reepithelialization. These studies using wounded SEs have shown that following wounding, keratinocytes were activated to express an altered distribution of integrin receptors and their associated ligands to facilitate migration shortly after wounding and to stabilize the epithelium through the assembly of new basement membrane after wound coverage was complete.

3.4
Growth factor responsiveness and synthesis in wounded skin equivalents

SE models of wound repair facilitate the determination of the synthesis and response of surface keratinocytes to soluble growth factors, as it is possible to directly assay their effect on reepithelialization by adding these soluble factors to culture media. In vivo studies have previously shown that growth factors such as TGFβ-1 [63, 64] modulate reepithelialization and wound repair through autocrine or paracrine pathways [67]. Systemic administration of TGF-β1 is known to accelerate cutaneous wound healing [68] while topical administration inhibits epithelial regeneration at elevated doses [69] and can stimulate epithelial regeneration at low doses [70].

The temporal response of wounded SEs to TGFβ-1 was determined in the presence of levels of this growth factor that are known to be present in the in vivo environment shortly after wounding [67]. It was found that addition of 2.5 ng/ml TGF-β1 to cultures at the time of wounding could delay reepithelialization and reduce hyperproliferation. Twenty-four hours after wounding, untreated cultures had undergone reepithelialization while TGF-β1-treated cultures showed only a thin tongue of elongated cells moving onto the wound surface. Proliferation in this tongue, as detected by BrdU incorporation, was considerably lower than that seen in the wound epithelium not treated with TGF-β1. This delay in reepithelialization was shown to be transient as TGF-

β1-treated cultures had completely reepithelialized by 48 h after wounding. This TGF-β1-induced shift towards a migratory response and delayed reepithelialization was found to be dose-dependent, as a progressively greater delay in reepithelialization was observed 24 h after wounding at increased doses of TGF-β1. For example, complete reepithelialization and stratification had occurred within 48 h, even in the presence of 7 ng/ml TGF-β1, where epithelial proliferation was completely suppressed. This suggested that reepithelialization at this concentration was primarily due to the stimulation of keratinocyte migration. These studies confirmed in vivo studies on the effects of TGF-β1 on the wound environment, which have shown its ability to induce enhanced migration and a dose-dependent effect on epidermal regeneration.

These studies used either foreskin [64] or gingival [63] keratinocytes to demonstrate that the TGF-β1-induced delay in reepithelialization was due to a reduced hyperproliferative response at the wound margins. While proliferation was lowered two- to fourfold by the presence of 2.5 ng/ml TGF-β1, the level of proliferation was still considerably higher than that of unwounded control cultures treated with TGF-β1, demonstrating that wounded keratinocytes were refractory to the known TGF-β1-induced inhibition of proliferation [71]. This supports the view that keratinocytes activated after wounding are not subject to the same inhibitory effects that TGF-β1 has been shown to exert on unwounded cultures. Similarly, the addition of TGF-β1 enhanced the migratory phenotype of wounded SE cultures. This confirms previous findings from studies performed using simple monolayer culture systems, as it has been shown that migrating keratinocytes can upregulate their repertoire of migration-associated integrin receptors and their ECM ligands in response to TGF-β1 in order to facilitate their movement {292}. Thus, the use of SE wounding models has demonstrated that TGF-β1 alters reepithelialization by modifying both proliferation and migration. Keratinocyte activation following wounding is therefore thought to needed to facilitate the switch from a normal to an activated, regenerative epithelial cell phenotype [65]. TGF-β1 may then modulate this altered cell phenotype by enabling dose-dependent control of proliferation and migration in order to modulate the phenotype of the activated keratinocyte during different stages of reepithelialization.

Falanga et al. studied the production of growth factors and proinflammatory cytokines by RT-PCR after wounding of SEs, and found a sequential program of expression of these proteins [59]. Expression of the cytokines IL-1α, IL-1β, IL-6, IL-8, IL-11 and TNF-α was turned on shortly after wounding and peaked shortly thereafter. In contrast, levels of growth factors, such as insulin growth factor-2, TGF-β1 and PDGF-B increased following this point, at 48 to 72 hours after wounding. These levels of expression were closely correlated with protein levels of these soluble factors as determined by ELISA analysis of supernatants from wounded SEs. Taken together, these studies demonstrated the utility of SEs in determining the presence of and response to soluble factors in patterns that simulate wound repair events known to occur in vivo.

3.5
Matrix metalloproteinase activity in wounded skin equivalents

The matrix metalloproteinase (MMP) family of proteinases acts to degrade all components of the ECM [72]. During reepithelialization, such degradation directs tissue remodeling and facilitates removal of damaged tissue, thus paving the way for migration of keratinocytes over dermal connective tissue in the wound bed. In vivo, cutaneous wounds demonstrate the spatially and temporally coordinated expression of MMP-1 (Type I collagenase) [73, 74], while MMP-10 are expressed at the migrating edge of keratinocytes and MMP-3 (stromelysin 1) was found to be expressed just distal to these cells [75]. In migrating gingival keratinocytes, expression of MMP-9 (92 kd Type IV collagenase) was found to be elevated [76]. This distinct compartmentalization of these degradatory enzymes in response to wounding suggests they play specific functions during reepithelialization.

To determine if this in vivo pattern of MMP expression was also present upon wounding of SEs, MMP-1 RNA expression was assayed by in situ hybridization. Wounding of SEs showed that MMP-1 RNA was expressed only in keratinocytes that were actively repopulating the wound. At 8 h after wounding, only keratinocytes that had initiated reepithelialization and were in contact with the Type I collagen on the wound surface were positive for MMP-1. Similarly, at 24 h after wounding, expression of MMP-1 RNA was only detected in the center of the wound and not in unwounded keratinocytes at the wound margin. At later timepoints, MMP-1 expression was no longer seen in keratinocytes in the center of the wound or at wound margins.

These findings suggested that only keratinocytes that were in direct contact with the Type I collagen in the wound bed could activate MMP-1 expression, as this connective tissue interface did not contain the basement membrane components that were present under keratinocytes found at the wound margins. As these basement membrane components were not present on the substrate on which reepithelialization occurs, it appeared that activated keratinocytes turn on expression of MMP-1 to promote remodeling of extracellular matrix proteins as cells move over the wound bed. Expression of MMP-1 was terminated upon synthesis of new basement membrane components in the center of the wound, as initial assembly of basement membrane structure occurred at this site. In this light, the initiation of reepithelialization onto a Type I collagen substrate served as an activation signal to direct wound repair and activated protease expression and activity. Thus, the remodeling of the dermal–epidermal interface was an essential step in the restoration of epithelial attachment and renewed stabilization at the basement membrane zone.

3.6
Keratinocyte differentiation in wounded skin equivalents

The ability of wounded keratinocytes to alter their expression of markers of keratinocyte differentiation have been studied in wounded SEs [60]. Expression of keratins 1 and 10 (K1,K10) was seen at the edge of the epithelial tongue at 8 h after wounding, in both suprabasal and basal cells at the wound edge. In addition, cells in this position also expressed involucrin, supporting the view that cells initiating migration were already committed to terminal differentiation. This initial epithelial tongue was likely formed by suprabasal cells that had migrated over the basal cells beneath them to assume their position at the edge of the tongue. This suggested that cells initiating migration were suprabasal, differentiated cells at the wound edge that became displaced laterally in order to attach to the wound surface. This supports the "leap-frog hypothesis" of reepithelialization, wherein suprabasal cells roll over cells adjacent to them to reach the wound surface. It therefore appears that migration started as a multilayer cell sheet rather than the epidermal monolayer proposed by the "sliding model" of reepithelialization.

At 24 h post-wounding, the wound bed was covered by an epithelial sheet and cells expressing K1,K10 were no longer seen in a basal position. This suggested that migration was initiated by differentiated cells but was maintained by proliferating cells and their progeny as the epithelial tongue covered the wound floor. It is likely that these replicating cells originated from these proliferative cells at the wound margin described above and were displaced laterally onto the wound bed. No K1,K10 was seen in suprabasal cells in the epithelium at the center of the wound after 24 h, suggesting that its expression was delayed during stratification due to the hyperproliferative nature of the regenerating epithelium in this region. However, involucrin was correctly expressed immediately upon stratification in cells directly above the basal layer in the wound epithelium, even when only two cell layers were present. This demonstrates that cells became committed to terminal differentiation in the reepithelializing tissue shortly after covering the wound and as soon as stratification occurred.

In summary, the varied adaptations of SE technology described that serve as 3-D wounding models have shown similar tissue responses and demonstrate the broad adaptability of these tissue models to the study of human wound repair. As follows, these engineered tissue models using SEs recapitulate key events that occur in wound reepithelialization in vivo. The first cells to initiate migration were nonproliferating, terminally differentiated keratinocytes that form an epithelial tongue by moving laterally onto the wound floor. This loss of cells from a suprabasal position may be partially responsible for the proliferative response that occurs at the wound margins. At this stage, progeny cells from this proliferative edge displaced proliferative cells centrally into the elongating epithelial tongue that partially covers the wound floor. At the same time, nondividing cells continually migrated and together with the progeny of proliferating cells, advanced to completely cover the wound floor. These early changes mark a phenotypic switch from a proliferating, unwounded epithelium to one in which cell migration is predominant. Once reepithelialization is complete, the proliferative phenotype becomes dominant again as cell division induces stratification. As cells reform a multilayer tissue, keratinocyte differentiation of cells at the center of the wound floor lags behind that of the areas closer to the wound margins as cells first undergo terminal differentiation near the wound margins. Finally, proliferative activity continues to be high in the wound epithelium even after the wound margins return to baseline levels of mitotic activity. This allows the wound epithelium to stratify to a thickness similar to that of the unwounded epithelium. Keratinocyte activation following wounding is thought to be a prerequisite to sustaining cell proliferation and to enhancing cell migration in vivo [65]. These effects are coincident with the switch from a normal to an activated, regenerative epithelial cell phenotype that occurs following wounding, and they have been simulated and studied using SE technology.

This review has demonstrated the utility of SEs as biologically relevant models of epithelial wound response by demonstrating several important points. First, the proliferative, migratory and synthetic response of wounded, "activated" keratinocytes can be monitored in a "controlled" culture environment. Secondly, the phenotype of keratinocytes during the wound response in SEs closely mimics that seen upon cutaneous wounding. Third, the absence of in vivo factors such as the variety of mesenchymal cells in SEs allows the response of keratinocytes to wounding to be determined directly. Finally, the wound milieu can be easily modified in SEs in order to study the effects of agents or potential therapeutic agents that may alter the course of wound response.

References

1. Freeman AE, Eigel HJ, Herman BJ, Kleinfeld KL (1976) In Vitro 12:352
2. O'Brien LE, Zegers MM, Mostov KE (2002) Nat Rev Mol Cell Biol 2:748
3. Streuli CH, Bissell MJ (1990) J Cell Biol 110:1405
4. Bissell MJ, Radisky D (2001) Nat Rev Cancer 1:46
5. Bell E, Ehrich P, Butte DJ, Nakatsuji T (1981) Science 211:1052
6. Andriani F, Margulis A, Lin N, Griffey S, Garlick JA (2003) J Invest Dermatol 120:923
7. Boxman I, Lowik C, Aarden L, Ponec M (1993) J Invest Dermatol 101:316
8. Smola H, Thiekotter G, Baur M, Stark HJ, Breitkreutz D, Fusenig NE (1994) Toxicol In Vitro 8:641
9. Asselineau D, Bernard BA, Bailly C, Darmon M (1989) Dev Biol 133:322
10. Parenteau NL, Nolte CM, Bilbo P, Rosenberg M, Wilkins LM, Johnson EW, Watson S, Mason VS, Bell E (1991) J Cell Biochem 45:245

11. Rosdy M, Claus L-C (1990) J Invest Dermatol 95:409
12. Fleischmajer R, Utani A, MacDonald ED, Perlish JS, Pan TC, Chu ML, Nomizu M, Ninomiya Y, Yamada Y (1998) J Cell Sci 111(Pt 14):1929
13. Ohji M, SundarRaj N, Hassell JR, Thoft RA (1994) Invest Ophthalmol Vis Sci 35:479
14. Prunieras M, Regnier M, Fougere S, Woodley D (1983) J Invest Dermatol 81:74s
15. Bohnert A, Hornung J, Mackenzie IC, Fusenig NE (1986) Cell Tissue Res 244:413
16. Grinnell F, Takashima A, Lamke-Seymour C (1986) Cell Tissue Res 246:13
17. Contard P, Bartel RL, Jacobs L, Perlish JS, MacDonald ED, Handler L, Cone D, Fleischmajer R (1993) J Invest Dermatol 100:35
18. Marinkovich MP, Keene DR, Clytie SR, Burgeson RE (1993) Dev Dynamics 197:255
19. Zieske JD, Mason VS, Wasson ME, Meunier SF, Nolte CJ, Fukai N, Olsen BR, Parenteau NL (1994) Exp Cell Research 214:621
20. Stoker AW, Streuli CH, Martins-Green M, Bissell MJ (1990) Curr Opin Cell Biol 2:864
21. Dowling J, Yu QC, Fuchs E (1996) J Cell Biol 134:559
22. Farber E (1996) The step-by-step development of epithelial cancer: from phenotype to genotype. Advances in Cancer Research. Academic, London, p 21
23. Weinberg RA (1991) Oncogenes, tumor suppressor genes, cell transformation: trying to put it all together. In: Brugge J, Curran T, Harlow E, McCormick F (eds) Origins of human cancer: A comprehensive review. Cold Spring Harbor Laboratory Press, New York, p 1
24. Hennings H, Lowry DT, Robinson VA, Morgan DL, Fujiki H, Yuspa SH (1992) Mol Carcinog 13:2145
25. Hennings H, Lowry DT, Robinson VA (1991) Skin Pharmacol 4:79
26. Hennings H, Robinson VA, Michael DM, Pettit GR, Jung R, Yuspa SH (1990) Cancer Res 50:4794
27. Terzaghi-Howe M, McKeown C (1986) Cancer Res 46:917
28. Terzaghi-Howe M (1989) Carcinogenesis 10:967
29. Chow M, Rubin H (1999) Proc Natl Acad Sci USA 96:2093
30. Chow M, Rubin H (1999) Proc Natl Acad Sci USA 96:6976
31. Adams JM, Corey S (1991) Science 254:1161
32. Sonnenschein C, Soto AM (2000) Mol Carcinog 29:205
33. Boukamp P, Petrussevka RT, Breitkreutz D, Hornung J, Markham A, Fusenig NE (1988) J Cell Biol 106:761
34. Boukamp P, Stanbridge EJ, Yin-Foo D, Cerutti PA, Fusenig NE (1990) Cancer Res 50:2840
35. Fusenig NE, Boukamp P, Breitkreutz D, Hulsen A, Petrusevska S, Cerutti P, Stanbridge E (1990) Toxicol In Vitro 4:627
36. Javaherian A, Vaccariello M, Fusenig NE, Garlick JA (1998) Cancer Res 58:2200
37. Potter V (1980) Yale J Biol Med 53:367
38. Weaver VM, Petersen OW, Wang F, Larabell CA, Briand P, Damsky C, Bissell MJ (1997) J Cell Biol 137:231
39. Yuspa SH (1981) Prog Dermatol 15:1
40. Yuspa SH, Ben T, Hennings H, Lichti U (1982) Cancer Res 42:2344
41. Karen J, Wang Y, Javaherian A, Vaccariello M, Fusenig NE, Garlick JA (1999) Cancer Res 59:474
42. Vaccariello M, Javaherian A, Wang Y, Fusenig NE, Garlick JA (1999) J Invest Dermatol 113:384
43. Setlow RB (1974) Proc Natl Acad Sci USA 71:3363
44. Brash DE, Ziegler A, Jonason AS, Simon JA, Kunala S, Leffell DJ (1996) J Investig Dermatol Symp Proc 1:136

45. Jonason AS, Kunala S, Price GJ, Restifo RJ, Spinelli HM, Persing JA, Leffell DJ, Tarone RE, Brash DE (1996) Proc Natl Acad Sci USA 93:14025
46. Brash DE (1997) Trends Genet 13:410
47. Winter GD (1992) Epidermal regeneration studied in the domestic pig. In: Maibach HI, Rovee DT (eds) Epidermal wound healing. Year Book, Chicago, IL, p 71
48. Frisch SM, Francis H (1994) J Cell Biol 124:619
49. McCall C, Cohen J (1991) J Invest Dermatol 97:111
50. Pignatelli M, Stamp G (1995) Cancer Surv 24:113
51. Stoker M, O'Neill C, Berryman S, Waxman V (1968) Int J Cancer 3:683
52. Terzaghi M, Nettesheim P (1979) Cancer Res 39:4003
53. Liotta LA, Rao CN, Barsky SH (1983) Lab Invest 49:636
54. Wainwright DJ (1995) Burns 21:243
55. Hintner H, Fritsch PO, Foidart JM, Stingl G, Schuler G, Katz SI (1980) J Invest Dermatol 74:200
56. Stenn KS, Madri JA, Tinghitella T, Terranova VP (1983) J Cell Biol 96:63
57. Stenn KS, Milstone LM (1984) J Invest Dermatol 83:445
58. Woodley DT, O'Keefe EJ, Prunieras M (1985) J Am Acad Dermatol 12:420
59. Falanga V, Isaacs C, Paquette D, Downing G, Kouttab N, Butmarc J, Badiavas E, Hardin-Young J (2002) J Invest Dermatol 119
60. Garlick JA, Taichman LB (1994) Lab Invest 70:916
61. Geer DJ, Swartz DD, Andreadis ST (2002) Tissue Eng 8:787
62. O'Leary R, Arrowsmith M, Wood EJ (2002) Cell Biochem Funct 20:129
63. Garlick JA, Parks WC, Welgus HG, Taichman LB (1996) J Dent Res 75:912
64. Garlick JA, Taichman LB (1994) J Invest Dermatol 103:554
65. Coulombe PA (1997) Biochem Biophys Res Commun 236:231
66. O'Leary R, Arrowsmith M, Wood EJ (1997) Biochem Soc Trans 25:369S
67. Cromack DT, Sporn MB, Roberts AB, Merino MJ, Dart LL, Norton JA (1987) J Surg Res 42:622
68. Cromack DT, Pierce GF, Mustoe TA (1991) Prog Clin Biol Res 365:359
69. Mustoe TA, Pierce GF, Morishima C, Deuel TF (1991) J Clin Invest 87:694
70. Ksander GA, Ogawa Y, Chu GH, McMullin H, Rosenblatt JS, McPherson JM (1990) Ann Surg 211:288
71. Nathan C, Sporn M (1991) J Cell Biol 113:981
72. Brinckerhoff CE, Matrisian LM (2002) Nat Rev Mol Cell Biol 3:207
73. Saarialho-Kere UK, Chang ES, Welgus HG, Parks WC (1992) J Clin Invest 90:1952
74. Saarialho-Kere UK, Kovacs SO, Pentland AP, Olerud JE, Welgus HG, Parks WC (1993) J Clin Invest 92:2858
75. Saarialho-Kere UK, Pentland AP, Birkedal-Hansen H, Parks WC, Welgus HG (1994) J Clin Invest 94:79
76. Makela M, Salo T, Uitto VJ, Larjava H (1994) J Dent Res 73:1397

Adv Biochem Engin/Biotechnol (2006) 103: 241–274
DOI 10.1007/10_023
© Springer-Verlag Berlin Heidelberg 2006
Published online: 11 October 2006

Gene-Modified Tissue-Engineered Skin: The Next Generation of Skin Substitutes

Stelios T. Andreadis

Bioengineering Laboratory, Department of Chemical & Biological Engineering,
University at Buffalo, The State University of New York (SUNY), Amherst, NY 14260,
USA
sandread@eng.buffalo.edu

Abstract Tissue engineering combines the principles of cell biology, engineering and materials science to develop three-dimensional tissues to replace or restore tissue function. Tissue engineered skin is one of most advanced tissue constructs, yet it lacks several important functions including those provided by hair follicles, sebaceous glands, sweat glands and dendritic cells. Although the complexity of skin may be difficult to recapitulate entirely, new or improved functions can be provided by genetic modification of the cells that make up the tissues. Gene therapy can also be used in wound healing to promote tissue regeneration or prevent healing abnormalities such as formation of scars and keloids. Finally, gene-enhanced skin substitutes have great potential as cell-based devices to deliver therapeutics locally or systemically. Although significant progress has been made in the development of gene transfer technologies, several challenges have to be met before clinical application of genetically modified skin tissue. Engineering challenges include methods for improved efficiency and targeted gene delivery; efficient gene transfer to the stem cells that constantly regenerate the dynamic epidermal tissue; and development of novel biomaterials for controlled gene delivery. In addition, advances in regulatable vectors to achieve spatially and temporally controlled gene expression by physiological or exogenous signals may facilitate pharmacological administration of therapeutics through genetically engineered skin. Gene modified skin substitutes are also employed as biological models to understand tissue development or disease progression in a realistic three-dimensional context. In summary, gene therapy has the potential to generate the next generation of skin substitutes with enhanced capacity for treatment of burns, chronic wounds and even systemic diseases.

1
Introduction

Tissue Engineering applies the principles and methods of engineering and the life sciences toward the development of tissue substitutes to restore, maintain or improve tissue function [18, 71, 97]. The field of tissue engineering is motivated by the tremendous need for transplantation of human tissue. In particular, the large number of patients with severe burns (13 000 per year, with 1000 of these involving more than 60% of the body surface), diabetic ulcers (about 600 000 per year), venous ulcers ($\sim$ 1 million per year) and pressure sores (about 2 million per year), creates a pressing need for artificial skin substitutes [117]. In addition to providing an alternative to autologous transplantation, engineered tissues have great potential as realistic biological models to obtain fundamental understanding of the structure-function relationships under normal and disease conditions and as toxicological models to facilitate drug development and testing.

To engineer tissues in the laboratory, cells must grow on three-dimensional scaffolds that provide the right geometric configuration, mechanical support and bioactive signals that promote tissue growth and differentiation. The cells may come from the patient (autologous), another individual (allogeneic) or a different species (xenogeneic). Cell sourcing may be overcome by use of adult or embryonic stem cells that have the capacity for self-renewal and can differentiate into multiple cell types, thus providing an unlimited supply of cells for tissue and cellular therapies. Application of stem cells in tissue engineering requires control of their differentiation into specific cell types, which in turn depends on fundamental understanding of the factors that affect stem cell self-renewal and lineage commitment [162, 173].

2
Tissue Engineering of Skin

2.1
Skin Structure and Physiology

The skin has two distinct layers, the dermis (D) and the epidermis (E) (Fig. 1). The dermis is the inner thicker layer that provides mechanical strength and elasticity. The main cells of the dermis are fibroblasts that synthesize extracellular matrix, endothelial cells organized in small vessel capillaries (VC) and other cellular structures such as hair follicles, sebaceous glands and sweat glands. Individual cells and cellular structures are interspersed in a network of collagen and elastin fibers of varying diameter depending on the distance from the epidermis. The dermal zone right underneath the epidermis (papillary dermis) contains small-diameter fibers and the lower dermal compartment (reticular dermis) contains collagen and elastin fibers of thicker diameter. Interestingly, the diameter of vessel capillaries follows a similar distribution pattern.

Epidermis is the outer layer that provides a barrier to infection and water loss. It is separated by the dermis with a basement membrane and is comprised of multiple layers of keratinocytes that form a stratified squamous epithelium. From the innermost to the outermost, these layers are the basal layer (BL), spinous layer (SL), granular layer (GL) and stratum corneum (SC) (Fig. 1). The epidermis undergoes continuous self-renewal

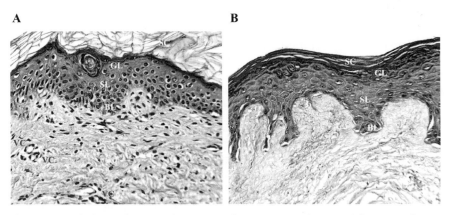

Fig. 1 A Morphology of mouse skin tissue. Skin tissue was harvested from an athymic mouse and processed for histology. **B** Bioengineered skin was prepared by culture of neonatal human keratinocytes on acellular dermis at the air-liquid interface for 7 days. Paraffin embedded tissue sections were stained with hematoxylin and eosin following standard protocols (magnification 40×). BL: basal layer; SL: suprabasal layer; GL: granular layer; SC: stratum corneum; VC: vessel capillary

through proliferation of the basal cells, the only cell compartment with the ability to proliferate.

The epidermis contains cells with different growth potential and at different stages of differentiation: slowly dividing stem cells that continue to proliferate for the lifetime of the tissue; transit amplifying cells that divide fast but are limited to a finite number of cell divisions before their progeny must commit to differentiate; and cells that are committed to differentiation along a certain lineage, which will eventually reach full maturity and die. Stem and transit amplifying cells are located in the basal layer of the epidermis. Periodically the transit amplifying cells leave the basement membrane and move upwards as they undergo a process of terminal differentiation that results in the anucleate cells of the stratum corneum. In the last stages of differentiation, cells extrude lipids into the intercellular space to form the permeability barrier and break down their nuclei and other organelles as they form the highly cross-linked protein envelope immediate beneath their cellular membranes. This envelope is connected to a network of keratin filaments that provide much of the physical strength of the epidermis. The cells of the stratum corneum are eventually sloughed off and replaced with new cells coming from the lower layers. The entire renewal process takes approximately 30 days.

2.2
Tissue-Engineered Skin

Research in tissue-engineered skin has produced two types of skin substitutes: biomaterials that act as synthetic dermal and epidermal analogs and can serve as temporary skin dressings and engineered tissues that contain skin cells, which provide the basic functions of the skin and may actively stimulate tissue regeneration and wound healing.

2.2.1
Biomaterial Dressings

Naturally derived and synthetic dressings, including Alloderm, Xenoderm and Integra [129], have been approved by the food and drug administration. Alloderm is acellular matrix from cadaver skin, which is processed to remove the cells of the epidermis and dermis. The processed acellular dermis is immunologically inert and retains an intact basement membrane [105]. When transplanted onto the wound bed, Alloderm is covered with a split thickness autograft to form a functional epidermis while the dermis is infiltrated by cells of the host and populated by new blood vessels [40, 41]. Xenoderm is very similar to Alloderm but is obtained from porcine skin. Finally, Integra contains an artificial dermis composed of bovine collagen and chondroitin-6-sulfate and an artificial epidermis composed of a disposable silicone sheet [24, 169–172]. A few weeks after transplantation the collagen-glycosaminoglycan matrix is vascularized by the host and the silicone layer is removed and replaced by a split thickness autograft to form the epidermis. Although both Alloderm and Integra are immediately available, immunologically inert and have been used successfully in the treatment of burns and wounds, the lack of a functional epidermis necessitates a second surgery to implant a split thickness autograft from a neighboring site.

2.2.2
Cell-based Skin Substitutes

Tissue-engineered skin substitutes with epidermal and dermal components have been designed to provide the lost cellular functions of the epidermis and dermis, respectively. The main types of skin substitutes employ three-dimensional biomaterials that provide the scaffolds for cell attachment, growth and differentiation to form functional tissues. To date three commercially available products have obtained FDA approval: Transcyte and Dermagraft, which are produced by Advanced Tissue Sciences, and Apligraf, which is produced by Organogenesis.

Transcyte is a dermal substitute that is composed of allogeneic human fibroblasts cultured in a nylon mesh for 4–6 weeks to form a dense cellular tissue and secrete a plethora of growth factors and extracellular matrix molecules. This cellular construct is used for treatment of burns after rendering the cells non-viable by a freeze-thaw process. Transcyte has been reported to provide considerable relief from pain, reduce scarring and prevent conversion of partial thickness burns into more serious full thickness injuries [107].

Dermagraft is also a dermal analog that is approved by the FDA for use in diabetic foot ulcers. It is composed of human fibroblasts cultured in a biodegradable polyglactin matrix, where they form a dense three-dimensional tissue containing extracellular matrix and growth factors. Dermagraft is used alone or as a base for the meshed autografts or possible epidermal cultures that provide barrier function. One of its main advantages is that it possesses considerable angiogenic activity, which is enhanced by the process of cryopreservation used to store the product [104, 127].

Apligraf is a living skin equivalent that contains both dermal and epidermal components and is approved by the FDA for treatment of venous ulcers. The dermal component consists of human fibroblasts embedded in type I collagen and cultured for a few weeks until the cells contract the matrix. At this point epidermal keratinocytes are added to overlay the matrix and cultured to the air-liquid interface to promote complete differentiation and stratification [15, 16, 124, 165]. Apligraf resembles human skin histologically and biochemically and possesses limited barrier function [124, 165]. Surprisingly, transplanted skin equivalents are not rejected by the host possibly due to the absence of endothelial cells, suggesting that engineered tissues from allogeneic cells are appropriate for transplantation in humans [23]. This is an important consideration as it suggests that tissue-engineered products can be produced from a limited number of donors and stored to provide immediately available, off-the-shelf tissues for transplantation.

Other tissue-engineered skin equivalents are also available or under development in industrial or academic laboratories for clinical applications or toxicological testing [9, 77, 132, 160]. Several studies have used human acellular dermis as a matrix for culture of epidermal keratinocytes [111–113]. Acellular dermis retains the biochemical components of the basement membrane (e.g., collagen IV, VII and laminin), the microtopology of human dermis (rete-ridge pattern) and dermal porosity that promotes ingrowth of fibroblasts and blood vessels along the pathway of preexisting vascular conduits. When keratinocytes are seeded on the basement membrane side of the dermis and raised to the air-liquid interface they differentiate to form a fully stratified epidermis with basal, suprabasal, granular and cornified cell layers exhibiting barrier function [6, 65]. In contrast to skin equivalents with collagen gels, the dermis retains the mechanical strength and elasticity of human skin, and therefore it is easy to handle during transplantation.

Others have used fibrin as a biomaterial for growth of tissue engineered skin. Fibrin is particularly attractive because it is a natural biomaterial that acts as a scaffold for tissue regeneration during wound healing and has been widely used as an adhesive in plastic and reconstructive surgery. Fibrin was found to maintain the stem cell phenotype and the proliferative potential of epidermal keratinocytes, while improving the "take-rate" of epidermal grafts onto massive full thickness burns [126]. Others used fibrin as a scaffold for fibroblast growth to re-create the dermal component of the skin before addition of epidermal keratinocytes [110, 131]. In combination with novel methods that have been developed to incorporate peptides and growth factors [65, 123, 135–137] fibrin formulations may be ideal for cell, growth factor and gene delivery to accelerate the healing response.

2.3
Limitations of Current Technologies

Although substantial progress has been made, several drawbacks must be overcome to increase the clinical success of tissue-engineered skin substitutes. Current skin substitutes lack several skin cells including mast cells, Langerhans cells and adnexal structures. Although some of these structures may not be necessary for patient survival, they are important for restoration of normal skin functions such as sensation and sweating. Despite successful attempts to add other cell types into tissue-engineered skin e.g., melanocytes [19, 103, 154], engineering the full complexity of the skin tissue may be much more challenging.

Part of the challenge may be addressed by appropriate design of bioactive scaffolds that provide the appropriate molecular signals and mechanical environment to guide cellular infiltration and function. Alternatively, cellular function may be directed by molecular engineering at the most fundamental level, the genome. Gene delivery can be applied in tissue engineering in order to impart new functions or enhance existing cellular activities in tissue substitutes. This is achieved by genetic modification of cells that will be part of the implant or gene transfer to the site of injury to facilitate in situ tissue regeneration. Cells can be genetically engineered to express a variety of molecules including growth factors that induce cell growth/differentiation or cytokines that prevent an immunologic reaction to the implant. Therefore, gene delivery has the potential to improve the quality of skin substitutes by altering the genetic basis of the cells that make up the tissues.

3
Gene Therapy in Tissue Engineering of Skin

3.1
Delivery Vehicles

The skin is an attractive target for gene therapy because it is easily accessible and shows great potential as an ectopic site for protein delivery in vivo. The cells that are primarily used to recreate skin substitutes are epidermal keratinocytes and/or dermal fibroblasts, which can be genetically modified with viral or non-viral vectors [42, 118]. Genetically modified cells are then used to engineer three-dimensional skin equivalents, which when transplanted in vivo can act as in vivo "bioreactors" to produce and deliver the desired therapeutic proteins either locally or systemically. Local delivery of proteins may be used for treatment of genetic diseases of the skin or wound healing of burns or injuries, while systemic delivery may be used for correction of systemic diseases like hemophilia or diabetes.

3.1.1
Gene Delivery Vehicles

Gene delivery vehicles can be broadly classified in two categories: viral and non-viral.

The genome of recombinant viruses has been modified by deletion of some or all viral genes and replacement with foreign therapeutic or marker genes. Recombinant viruses that are currently used in gene therapy include retrovirus, lentivirus, adenovirus and adeno-associated virus. Since viruses have evolved to infect cells, they display significantly higher gene transfer efficiency than non-viral systems. Recombinant retrovirus is the most commonly

Table 1 Physicochemical and biological properties of the most common gene transfer technologies *

Properties	RT	LT	AV	AAV	Plasmid DNA
Titer	10^5–10^7	10^5–10^7	10^7	10^{12}	10^7
Integrates into host genome?	Yes	Yes	No	Yes	No
Persistence of gene expression	Years	Years	Months	Years	Weeks
Stability	No	No	Yes	Yes	Yes
Maximum transgene size	7–8	7–8	36	4–5	Unlimited
Immunogenicity	No	No	Yes	Yes	No
Gene transfer to non-dividing cells	No	Yes	Yes	Yes	Yes
Potential for gene transfer to stem cells	Yes	Yes	No	Yes	No

* RT = Retrovirus; AV = Adenovirus; AAV = Adeno-associated virus; LT = Lentivirus

used vehicle for gene transfer to epidermal keratinocytes and skin substitutes. Lentivirus, adenovirus and non-viral gene transfer technologies have also been used but to a much lesser extent.

Non-viral methods include delivery of DNA using physical and chemical means. Physical methods such as particle acceleration (gene gun) facilitate entry into target cells and may be useful in direct gene transfer to tissues which are difficult to penetrate such as skin. Although delivery of DNA complexed with lipids or polymers has met with some success with other cell types, epidermal cells have been difficult to transfect efficiently. On the other hand, development of biomaterials for DNA delivery in vivo has met with significant success especially in the area of tissue regeneration and wound healing. A comparison of the main characteristics of viral and non-viral technologies is given in Table 1. For a more detailed discussion on other viral and non-viral technologies see [5].

3.1.2
Routes of Gene Delivery – Short- vs. Long-Term Gene Transfer

Use of gene delivery technologies that result in temporary or permanent genetic modification depends on the requirements of the disease or condition to be treated. In tissue engineering, cells are isolated from the patient or an allogeneic source, genetically modified, expanded in culture and combined with biomaterials to recreate three-dimensional tissues that can be used to restore the lost function. Genetic modification may be used to suppress immune rejection of the transplant or to generate cell-based devices for protein delivery into the systemic circulation. In these cases, **permanent** genetic modification is required to provide long-lasting effects. Consequently, the most suitable vectors are recombinant viruses that can mediate permanent gene transfer such as retrovirus, lentivirus and adeno-associated virus.

On the other hand, genetic modification of the engineered tissue may be used to increase the rate of graft survival by promoting angiogenesis. Alternatively, genes can be delivered in vivo using viral or non-viral technologies to promote wound healing. These applications may require **transient** gene expression until the transplant integrates with the surrounding tissue or until wound healing is complete. Therefore, adenoviruses or non-viral gene transfer technologies may be more appropriate.

3.2
Candidate Disease Conditions for Gene Therapy of the Skin

3.2.1
Genetic Diseases

Identification of genes that are responsible for genetic diseases opens the possibility for treatment by gene therapy approaches. Attempts to correct genetic defects using gene therapy include different forms of epidermolysis bullosa, lamellar ichthyosis and even psoriasis. Epidermolysis bullosa (EB) is a skin blistering disease that is caused by mutations in several genes expressed in the basal cells of the epidermis leading to loss of attachment to the basement membrane. There are three forms of EB that are caused by mutations in different genes. The simplex form of EB is caused by mutations in the keratin genes K5/K14; junctional EB is due to mutations of genes encoding for $\alpha 3$, $\beta 3$ and $\gamma 2$ chains of laminin, integrin $\alpha 6 \beta 4$ or bullous pemphigoid antigen; and dystrophic EB is caused by mutations in type VII collagen [80, 156, 157]. Another genetic disease, ichthyosis is a scaling disorder caused by mutations in genes that regulate the assembly of the cornified envelope. One form of the disease, namely lamellar ichthyosis is the result of mutations in the transglutaminase-1 gene, while another form, X-linked ichthyosis is caused by steroid sulfatase deficiency. Although the etiopathogenesis of these disorders is known, there is no available conventional treatment offering a unique opportunity to develop models for corrective gene delivery.

Several studies have attempted to correct these complex genetic defects using gene therapy. When the $\gamma 2$ or $\beta 3$ integrin subunits were introduced to epidermal keratinocytes from patients with junctional EB using recombinant retrovirus, the modified cells restored expression of laminin-5 and reversed the disease phenotype as evidenced by enhanced cell-substrate adhesion and reduced motility [44, 61, 159]. A clinical trial is under way to test a retroviral-based gene therapy approach for treatment of junctional EB by grafting large skin areas with sheets of gene-corrected epidermal keratinocytes [43]. Another group cloned a truncated form of the collagen VII gene into recombinant retrovirus to deliver the gene to keratinocytes from patients with the dystrophic form of EB. The transduced cells showed enhanced adhesion to matrix proteins, increased proliferative potential and decreased motility leading to reversal of the EB phenotype [30]. Similarly, transduction of dystrophic keratinocytes and fibroblasts with a recombinant lentivirus encoding for the full length collagen VII gene reversed the EB phenotype in vitro and in vivo after transplantation of gene-corrected cells onto immune-deficient mice [29]. Attempts to correct lamellar or X-linked ichthyosis employed recombinant retrovirus to deliver the genes for transglutaminase-1 [33, 34] and steroid sulfatase [59], respectively. When grafted onto athymic mice gene-modified keratinocytes regenerated epidermal tissue with normal morphology and restored barrier function. These studies suggest that gene therapy in combination with tissue engineering may provide a viable approach for the treatment of genetic skin diseases. However, the generalized nature of these disorders may necessitate treatment of large areas or even the entire skin tissue, posing a serious limitation to treatment. In this regard, clinical procedures that allow transplantation of large areas of the body by epidermal sheets may prove useful [43].

Finally, antisense approaches have also been developed for the treatment of localized disorders such as psoriasis. A recent study remonstrated that blocking synthesis of the insulin-like growth factor I receptor (IGF-I) in epidermal keratinocytes resulted in reversal of the hyperproliferative phenotype in vitro and in vivo [166]. Although direct administration of oligonucleotides was enough to elicit a response, multiple doses would be required for a therapeutic effect. Alternatively, small interfering RNA molecules (siRNA) may overcome this drawback as they are more stable and can be cloned into retroviral vectors to achieve permanent inhibition of the target RNA [138].

3.2.2
Wound Healing and Angiogenesis

Gene therapy has the potential to play an important role in wound healing of severe burns or chronic wounds of decubital, vascular or diabetic origin all conditions without adequate conventional treatment. The purpose of gene therapy in this setting is to either promote wound healing or reduce healing complications that lead to scarring, keloid formation or chronic ulceration.

The mode of gene delivery for wound healing would depend on the type and severity of the wound. Burns that cover large areas of the body surface are in need of skin replacement to prevent infection and dehydration. Therefore, tissue-engineered skin composed of genetically modified cells to promote healing and/or prevent scar formation would be the most appropriate treatment. In this case the cells need to be cultured, genetically modified in vitro and used to engineer three dimensional tissues that can be transplanted in the denuded areas. The main issues with this approach may be an acceptable "take" rate of the grafts and efficient gene transfer to have a therapeutic effect. On other hand, smaller wounds and chronic ulcers e.g., diabetic ulcers, may be amenable to in vivo gene delivery using a variety of approaches including direct injection of plasmid DNA, skin electroporation, gene gun and biomaterials that can deliver DNA or recombinant viruses in a controlled way.

In vitro and in vivo gene delivery approaches to date focused on modulating some aspect of the wound healing cascade, such as granulation tissue, vascularization, reepithelialization or long term tissue remodeling (Table 2). The majority of studies have used recombinant retrovirus to modify epidermal keratinocytes in culture, which were then grafted onto athymic mice to study the response of the dermal or epidermal compartments during wound healing. Gene-modified tissue-engineered skin overexpressing PDGF-A reduced wound contraction and increased dermal cell density and blood vessel formation [47, 48]. Epidermal cells were also engineered to express growth factors that are normally expressed in the dermis to study the effects of modulating the autocrine control of keratinocyte proliferation. In one study skin equivalents overexpressing IGF-1 promoted growth of the epidermal compartment [49]. In addition to increased proliferation of basal cells overexpression of KGF induced suprabasal cell proliferation and delayed differentiation without affecting the barrier function of engineered tissues [6]. Interestingly, expression of the secreted isoform of PDGF-B by epidermal keratinocytes increased the density of dermal cells evenly throughout the dermis, while expression of the cell-associated isoform of PDGF-B induced a localized response at the dermo-epidermal interface. This result suggested that the growth factor binding properties may control the spatial organization of cellular events during wound regeneration [51]. Recombinant retrovirus was also used to transfer the gene encoding for VEGF to the cells of the epidermis resulting in increased vascularization of transplanted skin equivalents in vivo [150]. In addition, VEGF expressing grafts showed reduced contraction and altered spatial distribution of blood vessels with more vessels in the upper dermis, which is in close proximity to the modified epidermal cells [149].

A broad armamentarium of technologies has also been developed for in vivo gene delivery to the skin. A single dose of PDGF-B encoding adenovirus significantly enhanced granulation tissue formation, vascularization and extracellular matrix synthesis as compared to topical application of high concentrations of the PDGF-B protein [101]. These results suggest that recombinant adenovirus may provide an efficient treatment for chronic wounds, such as pressure and diabetic ulcers, which showed only modest improvement in clinical trials involving the topical application of PDGF-B

Table 2 Gene therapy for wound healing

Route of Delivery	Target Tissue	Gene(s)	Vehicle/Modified Cells in Engineered Tissue	Refs.
In vitro				
	Reepithelialization	IGF-1	Retrovirus/Keratinocytes in DED-SE	[49]
		KGF	Retrovirus/Keratinocytes in DED-SE	[6]
		HGF	Retrovirus/Keratinocytes in DED-SE	[72]
		EGF	DNA/ Keratinocytes in Fibrin Gels	[8]
[1.5mm]	**Granulation tissue**	PDGF-A	Retrovirus/Keratinocytes in DED-SE	[47, 48]
		PDGF-B	Retrovirus/ Keratinocytes in DED-SE	[51]
		PDGF-B	Retrovirus/Fibroblasts in PGA	[22]
	Vascularization	VEGF	Retrovirus/Keratinocytes in C-SE	[149, 150]
		PDGF-B, VEGF-121	Retrovirus/Fibroblasts in PGA	[21]
		FGF-2	DNA/Encapsulated Myoblasts	[130]
In vivo			**Vehicle/Biomaterial**	
	Reepithelialization	KGF, IGF-1	Liposomes	[84, 85]
	Granulation tissue	PDGF-B	Adenovirus	[101, 102]
		PDGF-B	DNA/PLGA	[142]
		PDGF-A, PDGF-B	DNA/Collagen	[155]
		PDGF-A, PDGF-B	DNA/Collagen Adenovirus/Collagen	[27, 28]
[1.5mm]	**Vascularization**	iNOS	Adenovirus	[168]
		VEGF	DNA	[152]
		Rac1	DNA	[140]
		ORP150	DNA	[121]
	Scar Prevention	TGF-β	Antisense	[35, 89]
		aFGF	DNA	[146]

DED-SE: Skin equivalents containing epidermal keratinocytes grown on the basement membrane of de-epidermized acellular dermis

C-SE: Skin equivalents containing epidermal keratinocytes grown on collagen-embedded fibroblasts

PLGA: Poly(lactic-*co*-glycolic) acid

PGA: polyglycolic acid

protein, despite large doses and long duration of therapy [119, 164]. However, a β-galactosidase encoding adenovirus impaired reepithelialization of excisional wounds possibly as a result of acute inflammatory response to the adenoviral particles, suggesting that improvements in adenoviral vector design and careful scheduling of vector administration may be necessary to avoid adverse effects on the healing response [102].

Physical methods of gene delivery have been employed successfully for gene delivery to the skin. Gene gun or particle-mediated gene transfer employs DNA conjugated to gold microparticles, which are accelerated using a ballistic device to increase penetration through the cell membrane. Although originally developed for plant cells [91], the gene gun has been used to transfect a wide variety of mammalian tissues including skin. In particular microprojectile delivery of EGF [7], TGF-β [17] and PDGF [50] in the wound bed enhanced wound healing rates and increased wound tensile strength. Another approach that may be of practical importance for gene delivery to wounded or unwounded skin is microseeding. This method delivers DNA through a set of solid oscillating microneedles, allowing penetration to various depths of the tissue through multiple sites. Delivery of an EGF-encoding plasmid to partial thickness wounds via microseeding was more efficient than gene delivery by injection and even gene gun. In addition to plasmid DNA this technology can be used to deliver recombinant viruses as well as DNA vaccines [38, 52].

Besides physical methods, chemical methods of gene delivery have been employed for wound healing. Liposomal delivery of FGF-1 in excisional wounds of diabetic mice promoted wound healing in three administrations as compared to 15 administrations required to achieve similar biological effect by delivery of the protein [146]. Similarly, VEGF-encoding DNA complexed with liposomes significantly increased the area of skin flap survival in an ischemic rat model [152]. Other methods, such as topical application of DNA, calcium phosphate precipitation or microinjection of DNA in single cells, are too inefficient for clinical use. Finally, biomaterials have been explored as gene carriers for wound healing and this is discussed in detail in the next section.

In addition to technologies that introduce genes into target cells, antisense oligonucleotides have been used to block unwanted gene function(s). Topical administration of TGF-β antisense oligonucleotides following dermal wounding reduced scar formation in a mouse model [35, 89]. In another study inhibition of the insulin-like growth factor (IGF-1) translation with antisense oligonucleotides reduced hyperproliferation and reversed the psoriatic phenotype [166]. These studies suggest that direct delivery of oligonucleotides to skin may be effective in blocking gene function to prevent scar formation or reverse the course of a disease.

3.2.2.1
Biomaterials for Controlled Gene Delivery at the Wound Site

Biomaterials have enjoyed widespread use in drug delivery for many years because their physicochemical properties can be tailored to achieve controlled delivery and preserve bioactivity of the therapeutic agents. Biomaterials can also be engineered to possess biological activity by decoration with adhesion peptides, enzymatic recognition sites or growth factors [81]. In the context of wound healing bioactive materials can be employed to achieve controlled release of genes locally into the wound bed and at the same time serve as scaffolds to promote tissue regeneration. In addition, biomaterials may protect plasmid DNA or viral particles from the protease-rich environment of the wound [32, 100], reduce immunogenicity and achieve targeted gene transfer only to cells that infiltrate the wound bed.

Natural and synthetic biomaterials have been used to deliver plasmid DNA and recombinant viruses in the wound microenvironment. Collagen-embedded DNA encoding for platelet derived growth factor (PDGF-A or -B) increased granulation tissue, reepithelialization and wound closure in an ischemic rabbit ear model [155]. Poly(lactide-co-glycolide) matrices were also used to deliver the PDGF gene into skin wounds resulting in significantly increased vascularization and granulation tissue formation up to 4 weeks post-wounding [142]. Similarly, PLGA nanoparticles were shown to encapsulate DNA efficiently and exhibited sustained release over a period of four weeks [39]. Although these studies are encouraging, further advances in biomaterial design are required to achieve controlled release for prolonged periods of time, protect DNA from nuclease degradation, and target it to specific cell types of the wound. Strategies that employ liposomes, DNA condensing agents or nuclear localization signals may be integrated into this approach to release the DNA from the endosomes or target it to the nucleus [20, 106, 144, 174].

Bioactive matrix has also been used to deliver recombinant viruses to the site of injury. Encapsulation in gelatin/alginate microspheres protected adenoviral particles from degradation and the release kinetics could be controlled by modulating the composition of the microspheres [88]. Delivery of a PDGF-BB encoding adenovirus increased granulation tissue formation and neo-vascularization of full thickness wounds. Conjugation of adenoviral particles with fibroblast growth factor (FGF2) further increased the potency of the preparation by targeting cellular uptake through the FGF receptors [27, 28]. Additionally, encapsulation of adenovirus in PLGA matrices reduced immunogenicity and decreased inactivation by neutralizing antibodies, thus facilitating repeating virus administrations that may be required for a therapeutic effect [14, 31, 109].

Although temporary genetic modification may be required in most cases to ensure no transgene expression after the healing is complete, permanent genetic modification may be advantageous in the treatment of chronic wounds such as diabetic ulcers [54]. In this context, biomatrix delivery of retrovirus or lentivirus may afford increased viral stability and may be facilitated by the natural propensity of the wound infiltrating cells to divide. More studies are required to establish the efficiency of matrix-assisted retroviral or lentiviral delivery in enhancing tissue regeneration especially in chronic wounds.

3.3
Gene-enhanced Tissue-Engineered Skin:
A Transplantable Bioreactor for Treatment of Systemic Disorders

In general, it has been difficult to maintain full bioactivity of the proteins released from controlled delivery systems mainly due to protein instability. Indeed, clinical trials have resulted in modest improvements in wound healing despite large repetitive doses of growth factors, suggesting that development of alternative means of growth factor delivery is necessary for efficient wound healing [102]. Gene delivery may overcome this problem, as genetically modified cells can produce the therapeutic protein(s) continuously. The skin is an ideal target for gene delivery because it is easily accessible for transplantation or tissue removal if adverse effects occur. Notably, proteins secreted by gene-modified keratinocytes can reach the bloodstream via capillaries (molecular weight < 16 kD) or via the lymphatic return system (molecular weight > 16 kD) [148]. Therefore, gene-modified tissue-engineered skin can be used as a "bioreactor" to deliver proteins to the systemic circulation (Table 3).

One of the first studies employed retrovirus-modified keratinocytes to deliver the gene encoding for human growth hormone (hGH) [116] to human keratinocytes, which expressed the transgenes and secreted functional protein to the systemic circulation of grafted animals. Gene transfer using plasmid DNA transfection was also employed but the efficiency of gene transfer was low and short-lived [83, 139, 153]. Others employed retrovirus-modified keratinocytes to deliver apolipoprotein E (apoE), a protein that is involved in the transport of cholesterol and other lipids [69]. When genetically modified keratinocytes were grafted onto athymic mice, apoE was secreted by basal and suprabasal cells of the differentiated epidermis and was detected in the serum at high concentrations [57].

Other investigators introduced genes into human keratinocytes for the correction of metabolic disorders, in which absence or abnormal function of an enzyme results in systemic accumulation of a toxic substrate (Table 3). Severe combined immunodeficiency syndrome (SCID) is caused by lack of the enzyme adenosine deaminase (ADA) resulting in accumulation of toxic amounts of adenosine and deoxyadenosine. Retroviral gene transfer of the ADA gene into keratinocytes of ADA-deficient patients resulted in complete deamination of deoxyadenosine [56]. Gyrate atrophy is a progressive blinding disorder associated with deficiency of ornithine-delta-aminotransferase

Table 3 Gene-modified skin for treatment of systemic diseases.

Route of delivery and target disease	Vehicle	Gene	RefS.
In vitro			
ADA deficiency	Retrovirus	hGH	[116]
	DNA	hGH	[83, 139, 151, 153]
Hemophilia A	Retrovirus	Factor VIII	[46]
Hemophilia B	Retrovirus	Factor IX	[58, 67, 128, 143]
Emphysema	Retrovirus	α1-antitrypsin	[64]
	Adenovirus	α1-antitrypsin	[141]
Familial Hypercholesterolemia	DNA	ApoE	[57]
Gyrate atrophy	Retrovirus	Ornithine-delta-aminotransferase	[82, 145]
Diabetes	Retrovirus	Proinsulin	[55]
Phenylketonuria	Retrovirus	Phenylalanine hydroxylase and GTP-cyclohydrolase I	[36, 37]
Leptin deficiency (obesity, diabetes, infertility)	Retrovirus	Leptin	[98]
In vivo			
Cancer	Gene gun	IL-6, TNF-α/IFN-γ, IL-2/IFN-γ	[147]
Neutrophil recruitment	Plasmid DNA injection	IL-8	[78]
Contact hypersensitivity	Plasmid DNA injection	IL-10	[114]
Emphysema	Adenovirus	α1-antitrypsin	[141]

(OAT). Overexpression of OAT in retrovirus-modified keratinocytes from a gyrate atrophy patient restored ornithine metabolism and increased the rate of ornithine disappearance from the medium [82, 145]. Interestingly, differentiation of keratinocytes into three-dimensional tissues increased ornithine clearance, suggesting that gene-enhanced differentiated keratinocytes contribute significantly to metabolic function. Retroviral gene transfer was also successful in enhancing the metabolic clearance of phenylalanine by co-transduction of keratinocytes with two genes encoding for phenylalanine hydroxylase (PAH) and GTP cyclohydrolase I (GTP–CH) [37]. PAH catabolizes L-phelyalanine, and GTP–CH participates in the synthesis of BH_4, a co-factor that is required for catalytic activity of PAH. Interestingly, transfer of both genes into the same cell was not necessary since co-cultivation of PAH-expressing with GTP-CH-expressing keratinocytes resulted in similar level of phenylalanine clearance as keratinocytes co-transduced with both genes [37].

Recent studies provided strong evidence on the feasibility and effectiveness of cutaneous gene therapy. The first study demonstrated partial correction of hemophilia A after grafting factor VIII-deficient mice with skin from factor VIII-expressing transgenic animals [53]. The second study employed retroviral gene transfer of the leptin gene into human keratinocytes to examine the effectiveness of gene therapy for obesity, diabetes and infertility associated with leptin deficiency. Tissue-engineered skin with genetically modified keratinocytes was grafted onto transgenic obese (*ob/ob*) mice resulting in significant weight reduction and reversal of the obese phenotype [98]. Notably, skin grafts comprising less than 10% of the body surface area were sufficient for the correction of leptin deficiency. Finally, a recent study indicated that large antibody molecules (150 kDa) can be efficiently secreted by epidermal cells and enter the systemic circulation. Transplantation of genetically modified tissue-engineered epidermis onto SCID mice resulted in sustained production of a monoclonal antibody that crossed the basal layer and entered the bloodstream [120]. Collectively, these studies suggest that cutaneous gene therapy maybe an efficient clinical modality for the treatment of systemic disorders.

Fibroblasts have also been explored as target cells for secretion of protein into the circulation including human growth hormone [139, 151], α1-antitrypsin [64], factor VIII [46], factor IX [128, 143] and insulin (55). However, there are no reports of sustained gene expression by fibroblasts carrying stably integrated transgenes in an in vivo setting [95]. Furthermore, a recent study showed that keratinocytes are transduced more efficiently and have higher metabolic capacity than fibroblasts, suggesting that the epidermis may be a more promising target for gene therapy of systemic disorders [36].

In vivo approaches of gene transfer to the skin have also been tested using viral and non-viral technologies (Table 3). Subcutaneous injection of recombinant adenovirus resulted in significant amounts of α1-antitrypsin in mouse serum [141]. Particle-mediated gene delivery successfully transferred cytokine genes (IL-2, IL-6, IFN-γ, and TNF-α) to the skin resulting in significant reduction in the size of implantable tumors [147]. In addition, naked DNA was used to introduce various genes into the skin including β-galactosidase and IL-8 but the expression levels were low and transient [78, 79]. Significantly, transient expression of the anti-inflammatory cytokine, IL-10 was sufficient to inhibit the effector phase of contact hypersensitivity. Direct injection of IL-10 encoding plasmid DNA into the dorsal skin of DNCB-sensitized hairless rats resulted in protein expression and suppression of swelling of the rat ears [114]. This result is encouraging because it shows that cytokines secreted locally by keratinocytes enter the bloodstream and induce biological effects at distant areas, suggesting that site-directed gene delivery to the skin may result in systemic effects.

3.4
Future Developments for Efficient Gene Transfer

3.4.1
Gene Transfer to Epidermal Stem Cells

Continuously renewing tissues like skin, blood and bone contain cells with different growth potential and at different stages of differentiation: slowly dividing stem cells that continue to proliferate for the lifetime of the tissue; progenitor cells that divide fast but are limited to a finite number of cell divisions before their progeny must commit to differentiate and; cells that are committed to differentiation along a certain lineage and will eventually reach full maturity and die. Although certain gene transfer technologies, such as retrovirus, allow for permanent genetic modification of target cells, differentiation and eventually loss of the transduced cells from the engineered tissue may result in temporary transgene expression. Therefore, to achieve stable long-term gene expression, it is critical that stem cells are transduced with high efficiency. Additionally, recent studies suggest that stem cells do not express molecules that are recognized by the immune system and therefore, they appear to be immuno-privileged [25]. The potential of stem cells for establishing universal donor cells makes them ideal targets for both gene therapy and tissue engineering.

Retrovirus is the most appropriate vector for gene delivery into stem cells due to its ability to integrate into the genome of the target cells and become part of the genome of daughter cells for all future generations. However, retroviral gene transfer depends on cell cycle [3, 115] and the intracellular half-life of retroviruses results in low efficiency of gene transfer to the slowly dividing stem cells [2, 4, 133]. Although lentiviruses can transduce non-dividing stem cells, a recent study that assessed gene transfer to keratinocyte progenitor cells concluded that lentiviruses are not superior to retroviruses at introducing genes into keratinocyte progenitor cells during in vitro culture. Moreover, direct injection of GFP-encoding lentivirus into human skin grafted onto immunocompromised mice, showed low transduction efficiency and provided no evidence for progenitor cell targeting [96]. Therefore, increased efficiency of retroviral and lentiviral gene transfer to stem cells is needed to provide a clinically acceptable means of stem cell gene therapy.

Gene transfer to stem cells can be achieved by isolation and expansion of stem cells that are subsequently genetically modified or by targeting stem cells in cultures containing stem and differentiated cells. Early reports showed that transgene expression in vivo was short-lived suggesting that keratinocyte stem cells had not been transduced [34, 58, 59, 67]. More recently several groups observed sustainable transgene expression from the transplanted tissues indicating successful gene transfer to the stem cell compartment [44, 45, 92, 108, 163]. However, clonal analysis of the transduced cell population showed that the fraction of transduced stem cells was relatively low (< 30%) [92, 108].

Recently, we demonstrated a promising way to transduce epidermal stem cells in cultures containing both stem and differentiated cells. We hypothesized that fibronectin, which is known to inhibit keratinocyte differentiation [1] and to contain heparin binding domains that bind retroviral particles [12, 73, 74, 99], may increase the efficiency of gene transfer to the pool of epidermal stem cells. Results from our laboratory showed that retroviral gene transfer to epidermal keratinocytes is more efficient on fibronectin even in the absence of polycations such as polybrene. The transduction efficiency strongly correlates with the levels of integrin expression ($\alpha5$, $\alpha2$, $\beta1$) and integrin-blocking antibodies decrease the efficiency of gene transfer in a dose dependent way. Notably, cells that adhere rapidly to fibronectin are transduced more efficiently than slowly adherent cells [11]. These findings are novel and potentially important for gene therapy since integrin expression and the rate of adhesion to fibronectin have been associated with epidermal stem cell phenotype [86, 87, 175]. Most important, long-term growth and clonogenic assays showed that transduction on fibronectin promoted gene transfer to epidermal stem cells and prevented loss of clonogenic potential due to exposure of cells to retroviral supernatant [10]. These results are important for cutaneous gene therapy and for biological studies that require efficient and permanent genetic modification.

3.4.2
Regulatable Gene Therapy

The majority of gene transfer vehicles provide constitutive (always on) gene expression that may be advantageous for treatment of genetic diseases. However, tissue-engineering applications may require physiologically regulated gene expression by a subset of cells in the regenerated tissue. Spatial and temporal control of gene expression requires design of a new class of gene delivery vehicles that may be regulated by host mechanisms or by administration of secondary agents.

Tissue-specific transcriptional elements can be used to restrict transgene expression in specific cells. Since they are naturally active in the tissue these promoters may be less prone to methylation and shut-off in vivo. One the first studies used the involucrin promoter to induce suprabasal expression of the β-galactosidase gene in the mouse skin and hair follicles [26]. Similarly, the keratin-14 promoter was used to express human growth hormone in the basal layer of the skin of transgenic mice [161]. Although viral promoters are more active, the keratin-10 promoter was found to induce high levels of transgene expression in human keratinocytes, suggesting that keratin promoters may be useful for cutaneous gene therapy. Indeed, transfer of the VEGF gene under the keratin-5 promoter increased VEGF expression by basal keratinocytes and enhanced neovascularization of transfected skin grafts implanted onto nude mice [131]. Hybrid cassettes have also been constructed to combine the strength of a viral promoter with the specificity of a keratin promoter. One such construct was engineered to contain the cytomegalovirus (CMV) immediate early enhancer/promoter and regulatory elements of the human keratin 5 (hK5) gene and inserted into the backbone of a retroviral vector encoding for Factor IX. When genetically modified keratinocytes were transplanted in vivo the hybrid promoter increased the levels of Factor IX in the plasma of mice by 2–3-fold as compared to a CMV promoter alone [122].

More important for gene therapy is the development of temporally and reversibly regulated transgene expression in primary cells and engineered tissues. Advances in this area have produced synthetic promoters that can be regulated by exogenous agents, which can be administered locally or systemically. Several systems exist including the Cre-loxP recombination, the *lac*-based and the tetracycline (Tet)-based regulatable systems [167]. The most promising of these is the Tet-regulatable system, which contains a tetracycline-dependent promoter constructed by two regulatory elements: the tetracycline resistance *tet* operon embedded in a cytomegalovirus promoter and a hybrid activator protein (tTA). Using this system, it was shown that gene expression could be tightly regulated by addition of tetracycline in a dose-dependent way [13, 60, 70, 90]. Such promoters have been incorporated into retroviral [125] and adenoviral vectors [75, 76] to achieve tetracycline-dependent regulated transgene expression in vitro and in vivo. Improved tetracycline regulatable vectors encode for a transrepressor and a transactivator eliminating low levels of unwanted activity and allowing tight control of gene expression [94, 134]. In the absence of tetracycline gene expression is shut off by the repressor, while in the presence of tetracycline gene expression is induced by the activator (Fig. 2).

This system was used recently to transduce keratinocytes that expressed GFP in two- and three-dimensional culture systems after treatment with doxycycline, a synthetic analog of tetracycline [68]. However, successful doxycycline regulation required excessive selection, which is only possible with immortalized cell lines but not with primary human keratinocytes. Therefore, further development of these genetic switches to control the temporal and spatial expression of the transgene quantitatively and reversibly [90] is necessary to provide physiological control of transgene expression of primary cells and engineered tissues.

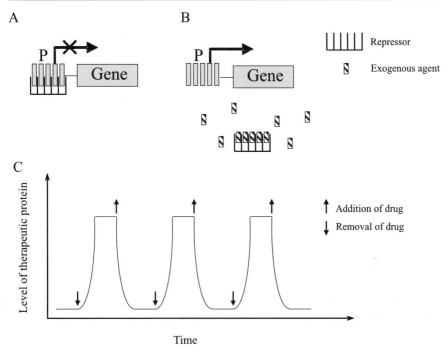

Fig. 2 Schematic of controlled gene expression. **A** In the absence of an exogenous agent the repressor binds to the promoter (P) and prevents gene expression. **B** In the presence of the exogenous agent (drug), binding of the repressor to the promoter is inhibited allowing for expression of the transgene. **C** Kinetic profile of protein secretion upon addition and removal of the drug

4
Gene-Modified Skin Substitutes as Biological Models of Tissue Development and Disease Pathophysiology

In addition to the potential clinical applications of gene therapy and tissue engineering, genetically modified engineered tissues can be employed as model systems for studying tissue development, physiology and disease pathogenesis. They may also be useful as toxicological models for development of new drugs and gene therapeutics. Gene transfer can be used to alter distinct genes in a biosynthetic pathway or express adhesion molecules to alter the interactions of cells with the substrate or with neighboring cells in the tissue. Alternatively, delivery of genes encoding for growth factors can be used to study their effects on tissue growth and differentiation. Although transgenic animal models are currently used to study the effects of genes on tissue development, "transgenic" engineered tissues may be useful for controlled and quantitative studies as they provide in vitro physiological models with morphological, structural and functional similarities to human tissues. Furthermore, while deletion of some genes may be lethal for animal embryos, it may still be possible to study their effects using "transgenic" tissue equivalents.

Several studies have used genetically modified tissues to understand the effects of gene expression on tissue development. We prepared skin equivalents that were genetically modified to express keratinocyte growth factor (KGF), a protein that plays an important role in tissue morphogenesis and wound healing [6]. The modified tissues showed dramatic changes in three-dimensional organization of the epidermis including hyperthickening and flattening of the corrugations of the dermo-epidermal junction (rete-ridges). KGF increased proliferation of basal cells, induced proliferation in the normally quiescent suprabasal cell compartment and delayed differentiation. This study demonstrated that expression of a single growth factor is able to mediate many of the events associated with epidermal growth and differentiation [6].

Engineered tissues were also used to develop in vitro physiological models of wound healing and three-dimensional migration. We found that the response of skin equivalents to wounding mimics that of animal models in terms of the kinetics of healing and the unique phenotype exhibited by the cells as they migrate in a three-dimensional microenvironment. When fibrin was added in the wound bed to mimic the fibrin clot in vivo, wound healing was enhanced mainly due to shortening of the lag phase of keratinocyte activation [65]. More recently we extended this system to develop an in vivo model of wound healing based on tissue-engineered skin [66]. To this end, human skin equivalents were transplanted onto athymic mice. A few weeks later when they had fully integrated with the mouse tissue the transplanted tissues were wounded using a 4-mm biopsy punch. Histology and immuno-histochemistry showed that the kinetics of wound healing and the state of differentiation of the neoepidermis were very similar to those of human skin, suggesting that transplantation of tissue engineered skin may be useful model for studying wound healing of human tissue in an in vivo setting [66]. Using this hybrid wound healing model, we evaluated a novel, cell-controlled delivery system of KGF to promote tissue regeneration in vivo [176].

To understand the molecular determinants of reepithelialization and barrier formation, we used cDNA micro-array technology to identify genes that are expressed upon chemical disruption of the stratum corneum, which results in loss of barrier function [93]. We found that tissue-engineered skin responded to barrier disruption by a two-wave dynamic response. Early on, the cells upregulated signal transducing, stress, proliferation and inflammation genes to protect the tissue and possibly to communicate the damage to the immune system and neighboring tissues. At later times, pro-inflammatory cytokines and some growth-related genes were significantly reduced but enzymes that participate in lipid synthesis increased suggesting that the epidermal cells attempted to restore the lost barrier. Finally, we identified novel genes that were expressed in response to barrier disruption and were not previously known to be expressed by keratinocytes.

We also employed tissue-engineered skin models of complete versus impaired epidermal stratification to discover the genes that may be important in epidermal stratification [177]. Transcriptional profiling at different stages of development showed significant differences in transcription, signaling and metabolism genes that correlated well with functional data on proliferation, expression of adhesion molecules and key metabolic intermediates. Notably, we identified genes that were not previously known to play a role in epidermis and discovered the importance of previously overlooked metabolic pathways in epidermal morphogenesis. Collectively, our work suggests that functional genomics can be used in tissue engineering to study tissue development, response to environmental stimuli and wound regeneration.

Others used recombinant retrovirus to genetically modify keratinocytes in order to follow their fate as they migrate to close the wound. By following the distribution of genetically marked cells in the wound, it was shown that some of the cells that repopulate the wound exhibit a proliferative phenotype, while other cells can migrate long distances without undergoing replication [63]. The same system was also employed to evaluate the role of TGF-β in wound reepithelialization [62]. More recently, skin equivalents from immortalized keratinocytes were used to study the transition of epithelial tumors from a benign to a malignant state. When these cells were genetically modified to overexpress a dominant-negative form of E-cadherin, they developed an invasive phenotype and migrated into the dermis. Transplantation onto athymic mice resulted in the formation of aggressive metastatic tumors in the dermis reminiscent of squamous cell carcinomas [158].

Taken together, these studies demonstrate that engineered tissues can be used to study the cellular and molecular mechanisms of three-dimensional migration and shed light into complex processes such as wound healing and cancer metastasis. Such models may also be useful for evaluating biomaterials as substrates for cell migration and as vehicles for controlled delivery of genes and proteins thus minimizing the number of animal experiments.

5
Summary

Although tissue-engineered skin is the most advanced tissue engineering product, it lacks several functions provided by the natural tissue. Gene therapy can be used to create the next generation of skin equivalents by imparting new properties and enhancing cellular function. Gene-enhanced skin substitutes may be used for the treatment of a variety of disorders ranging from genetic diseases and wound healing to systemic disorders. Ease of accessibility and transplantation make the skin an ideal target for delivery of therapeutics to distant sites. Although previous work has established the feasibility of this approach, several challenges must be overcome before gene therapy can be applied in the clinic. These include development of novel methods for increased gene transfer to the stem cell compartment; biomaterials for efficient and localized gene delivery and construction of advanced regulatable vectors for temporal and spatial control of gene expression in response to physiologic changes in the body or exogenous signals. In addition to disease treatment advances in gene transfer technologies may also facilitate the use of engineered tissues as realistic biological models to understand tissue development and disease pathophysiology.

References

1. Adams JC, Watt FM (1989) Fibronectin inhibits the terminal differentiation of human keratinocytes. Nature 340:307–309
2. Andreadis S, Brott DA, Fuller AO, Palsson BO (1997) Moloney murine leukemia virus-derived retroviral vectors decay intracellularly with a half-life in the range of 5.5 to 7.5 hours. J Virol 71:7541–7548
3. Andreadis S, Fuller AO, Palsson BO (1998) Cell cycle dependence of retroviral transduction: An issue of overlapping time scales. Biotechnol Bioeng 58:272–281
4. Andreadis S, Palsson BO (1996) Kinetics of Retrovirus Mediated Gene Transfer: The Importance of the Intracellular Half-life of Retroviruses. J Theor Biol 182:1–20
5. Andreadis ST (2004) Gene Enhanced Tissue Engineering, Chapter 19. In: Moore J, Zouridakis G (eds) Biomedical Technology and Devices Handbook. CRC Press, Boca Raton
6. Andreadis ST, Hamoen KE, Yarmush ML, Morgan JR (2001) Keratinocyte growth factor induces hyperproliferation and delays differentiation in a skin equivalent model system. Faseb J 15:898–906
7. Andree C, Swain WF, Page CP, Macklin MD, Slama J, Hatzis D, Eriksson E (1994) In vivo transfer and expression of a human epidermal growth factor gene accelerates wound repair. Proc Natl Acad Sci USA 91:12188–12192
8. Andree C, Voigt M, Wenger A, Erichsen T, Bittner K, Schaefer D, Walgenbach KJ, Borges J, Horch RE, Eriksson E, Stark GB (2001) Plasmid gene delivery to human keratinocytes through a fibrin-mediated transfection system. Tissue Eng 7:757–766
9. Asbill C, Kim N, El-Kattan A, Creek K, Wertz P, Michniak B (2000) Evaluation of a human bio-engineered skin equivalent for drug permeation studies. Pharm Res 17:1092–1097
10. Bajaj B, Lei P, Andreadis ST (2005) Efficient gene transfer to human epidermal keratinocytes on fibronectin: in vitro evidence for transduction of epidermal stem cells. Mol Ther 11(6):969–979
11. Bajaj B, Behshad S, Andreadis ST (2002) Retroviral gene transfer to epidermal keratinocytes correlates with integrin expression and is significantly enhanced on fibronectin. Hum Gene Ther 13:1821–1831
12. Bajaj B, Lei P, Andreadis ST (2001) High efficiencies of gene transfer with immobilized recombinant retrovirus: kinetics and optimization. Biotechnol Prog 17:587–596

13. Baron U, Gossen M, Bujard H (1997) Tetracycline-controlled transcription in eukaryotes: novel transactivators with graded transactivation potential. Nucleic Acids Res 25:2723–2729

14. Beer SJ, Matthews CB, Stein CS, Ross BD, Hilfinger JM, Davidson BL (1998) Poly (lactic-glycolic) acid copolymer encapsulation of recombinant adenovirus reduces immunogenicity in vivo. Gene Ther 5:740–746

15. Bell E, Ehrlich HP, Buttle DJ, Nakatsuji T (1981) Living tissue formed in vitro and accepted as skin-equivalent tissue of full thickness. Science 211:1052–1054

16. Bell E, Ehrlich HP, Sher S, Merrill C, Sarber R, Hull B, Nakatsuji T, Church D, Buttle DJ (1981) Development and use of a living skin equivalent. Plast Reconstr Surg 67:386–392

17. Benn SI, Whitsitt JS, Broadley KN, Nanney LB, Perkins D, He L, Patel M, Morgan JR, Swain WF, Davidson JM (1996) Particle-mediated gene transfer with transforming growth factor-beta1 cDNAs enhances wound repair in rat skin. J Clin Invest 98:2894–2902

18. Berthiaume F, Yarmush ML (1995) In: Bronzino GJ (ed) Biomedical Engineering Handbook. CRC Press, New York, Tissue Eng 109:1–12

19. Boyce ST, Medrano EE, Abdel-Malek Z, Supp AP, Dodick JM, Nordlund JJ, Warden GD (1993) Pigmentation and inhibition of wound contraction by cultured skin substitutes with adult melanocytes after transplantation to athymic mice. J Invest Dermatol 100:360–365

20. Branden LJ, Mohamed AJ, Smith CI (1999) A peptide nucleic acid-nuclear localization signal fusion that mediates nuclear transport of DNA. Nat Biotechnol 17:784–787

21. Breitbart AS, Grande DA, Laser J, Barcia M, Porti D, Malhotra S, Kogon A, Grant RT, Mason JM (2001) Treatment of ischemic wounds using cultured dermal fibroblasts transduced retrovirally with PDGF-B and VEGF121 genes. Ann Plast Surg 46:555–61, discussion 561–562

22. Breitbart AS, Mason JM, Urmacher C, Barcia M, Grant RT, Pergolizzi RG, Grande DA (1999) Gene-enhanced tissue engineering: applications for wound healing using cultured dermal fibroblasts transduced retrovirally with the PDGF-B gene. Ann Plast Surg 43:632–639

23. Briscoe DM, Dharnidharka VR, Isaacs C, Downing G, Prosky S, Shaw P, Parenteau NL, Hardin-Young J (1999) The allogeneic response to cultured human skin equivalent in the hu-PBL-SCID mouse model of skin rejection. Transplantation 67:1590–1599

24. Burke JF, Yannas IV, Quinby WC Jr, Bondoc CC, Jung WK (1981) Successful use of a physiologically acceptable artificial skin in the treatment of extensive burn injury. Ann Surg 194:413–428

25. Caplan AI, Bruder SP (2001) Mesenchymal stem cells: building blocks for molecular medicine in the 21st century. Trends Mol Med 7:259–264

26. Carroll JM, Albers KM, Garlick JA, Harrington R, Taichman LB (1993) Tissue- and stratum-specific expression of the human involucrin promoter in transgenic mice. Proc Natl Acad Sci USA 90:10270–10274

27. Chandler LA, Doukas J, Gonzalez AM, Hoganson DK, Gu DL, Ma C, Nesbit M, Crombleholme TM, Herlyn M, Sosnowski BA, Pierce GF (2000) FGF2-Targeted adenovirus encoding platelet-derived growth factor-B enhances de novo tissue formation. Mol Ther 2:153–160

28. Chandler LA, Gu DL, Ma C, Gonzalez MA, Doukas J, Nguyen T, Pierce GF, Phillips ML (2000) Matrix-enabled gene transfer for cutaneous wound repair. Wound Repair Regen 8:473–479

29. Chen M, Kasahara N, Keene DR, Chan L, Hoeffler WK, Finlay D, Barcova M, Cannon PM, Mazurek C, Woodley DT (2002) Restoration of type VII collagen expression and function in dystrophic epidermolysis bullosa. Nat Genet 32:670–675

30. Chen M, O'Toole EA, Muellenhoff M, Medina E, Kasahara N, Woodley DT (2000) Development and characterization of a recombinant truncated type VII collagen minigene. Implication for gene therapy of dystrophic epidermolysis bullosa. J Biol Chem 275:24429–24435

31. Chillon M, Lee JH, Fasbender A, Welsh MJ (1998) Adenovirus complexed with polyethylene glycol and cationic lipid is shielded from neutralizing antibodies in vitro. Gene Ther 5:995–1002

32. Choate KA, Khavari PA (1997) Direct cutaneous gene delivery in a human genetic skin disease. Hum Gene Ther 8:1659–1665

33. Choate KA, Kinsella TM, Williams ML, Nolan GP, Khavari PA (1996) Transglutaminase 1 delivery to lamellar ichthyosis keratinocytes. Hum Gene Ther 7:2247–2253

34. Choate KA, Medalie DA, Morgan JR, Khavari PA (1996) Corrective gene transfer in the human skin disorder lamellar ichthyosis. Nat Med 2:1263–1267

35. Choi BM, Kwak HJ, Jun CD, Park SD, Kim KY, Kim HR, Chung HT (1996) Control of scarring in adult wounds using antisense transforming growth factor-beta 1 oligodeoxynucleotides. Immunol Cell Biol 74:144–150

36. Christensen R, Guttler F, Jensen TG (2002) Comparison of epidermal keratinocytes and dermal fibroblasts as potential target cells for somatic gene therapy of phenylketonuria. Mol Genet Metab 76:313–318

37. Christensen R, Kolvraa S, Blaese RM, Jensen TG (2000) Development of a skin-based metabolic sink for phenylalanine by overexpression of phenylalanine hydroxylase and GTP cyclohydrolase in primary human keratinocytes. Gene Ther 7:1971–1978

38. Ciernik IF, Krayenbuhl BH, Carbone DP (1996) Puncture-mediated gene transfer to the skin. Hum Gene Ther 7:893–899

39. Cohen H, Levy RJ, Gao J, Fishbein I, Kousaev V, Sosnowski S, Slomkowski S, Golomb G (2000) Sustained delivery and expression of DNA encapsulated in polymeric nanoparticles. Gene Ther 7:1896–1905

40. Cuono C, Langdon R, McGuire J (1986) Use of cultured epidermal autografts and dermal allografts as skin replacement after burn injury. Lancet 1:1123–1124

41. Cuono CB, Langdon R, Birchall N, Barttelbort S, McGuire J (1987) Composite autologous-allogeneic skin replacement: development and clinical application. Plast Reconstr Surg 80:626–637

42. De Luca M, Pellegrini G (1997) The importance of epidermal stem cells in keratinocyte-mediated gene therapy [editorial]. Gene Ther 4:381–383

43. Dellambra E, Pellegrini G, Guerra L, Ferrari G, Zambruno G, Mavilio F, De Luca M (2000) Toward epidermal stem cell-mediated ex vivo gene therapy of junctional epidermolysis bullosa. Hum Gene Ther 11:2283–2287

44. Dellambra E, Vailly J, Pellegrini G, Bondanza S, Golisano O, Macchia C, Zambruno G, Meneguzzi G, De Luca M (1998) Corrective transduction of human epidermal stem cells in laminin-5-dependent junctional epidermolysis bullosa. Hum Gene Ther 9:1359–1370

45. Deng H, Lin Q, Khavari PA (1997) Sustainable cutaneous gene delivery. Nat Biotechnol 15:1388–1391

46. Dwarki VJ, Belloni P, Nijjar T, Smith J, Couto L, Rabier M, Clift S, Berns A, Cohen LK (1995) Gene therapy for hemophilia A: production of therapeutic levels of human factor VIII in vivo in mice. Proc Natl Acad Sci USA 92:1023–1027

47. Eming SA, Lee J, Snow RG, Tompkins RG, Yarmush ML, Morgan JR (1995) Genetically modified human epidermis overexpressing PDGF-A directs the development of a cellular and vascular connective tissue stroma when transplanted to athymic mice – implications for the use of genetically modified keratinocytes to modulate dermal regeneration. J Invest Dermatol 105:756–763

48. Eming SA, Medalie DA, Tompkins RG, Yarmush ML, Morgan JR (1998) Genetically modified human keratinocytes overexpressing PDGF-A enhance the performance of a composite skin graft. Hum Gene Ther 9:529–539

49. Eming SA, Snow RG, Yarmush ML, Morgan JR (1996) Targeted expression of insulin-like growth factor to human keratinocytes: modification of the autocrine control of keratinocyte proliferation. J Invest Dermatol 107:113–120

50. Eming SA, Whitsitt JS, He L, Krieg T, Morgan JR, Davidson JM (1999) Particle-mediated gene transfer of PDGF isoforms promotes wound repair. J Invest Dermatol 112:297–302

51. Eming SA, Yarmush ML, Krueger GG, Morgan JR (1999) Regulation of the spatial organization of mesenchymal connective tissue: effects of cell-associated versus released isoforms of platelet-derived growth factor. Am J Pathol 154:281–289

52. Eriksson E, Yao F, Svensjo T, Winkler T, Slama J, Macklin MD, Andree C, McGregor M, Hinshaw V, Swain WF (1998) In vivo gene transfer to skin and wound by microseeding. J Surg Res 78:85–91

53. Fakharzadeh SS, Zhang Y, Sarkar R, Kazazian HH Jr (2000) Correction of the coagulation defect in hemophilia A mice through factor VIII expression in skin. Blood 95:2799–2805

54. Falanga V (1998) Wound healing and chronic wounds. J Cutan Med Surg 3:S1–1–5

55. Falqui L, Martinenghi S, Severini GM, Corbella P, Taglietti MV, Arcelloni C, Sarugeri E, Monti LD, Paroni R, Dozio N, Pozza G, Bordignon C (1999) Reversal of diabetes in mice by implantation of human fibroblasts genetically engineered to release mature human insulin. Hum Gene Ther 10:1753–1762

56. Fenjves ES, Schwartz PM, Blaese RM, Taichman LB (1997) Keratinocyte gene therapy for adenosine deaminase deficiency: a model approach for inherited metabolic disorders. Hum Gene Ther 8:911–927

57. Fenjves ES, Smith J, Zaradic S, Taichman LB (1994) Systemic delivery of secreted protein by grafts of epidermal keratinocytes: prospects for keratinocyte gene therapy. Hum Gene Ther 5:1241–1248

58. Fenjves ES, Yao SN, Kurachi K, Taichman LB (1996) Loss of expression of a retrovirus-transduced gene in human keratinocytes. J Invest Dermatol 106:576–578

59. Freiberg RA, Choate KA, Deng H, Alperin ES, Shapiro LJ, Khavari PA (1997) A model of corrective gene transfer in X-linked ichthyosis. Hum Mol Genet 6:927–933

60. Furth PA, St Onge L, Boger H, Gruss P, Gossen M, Kistner A, Bujard H, Hennighausen L (1994) Temporal control of gene expression in transgenic mice by a tetracycline-responsive promoter. Proc Natl Acad Sci USA 91:9302–9306

61. Gagnoux-Palacios L, Vailly J, Durand-Clement M, Wagner E, Ortonne JP, Meneguzzi G (1996) Functional Re-expression of laminin-5 in laminin-gamma2-deficient human keratinocytes modifies cell morphology, motility, and adhesion. J Biol Chem 271:18437–18444

62. Garlick JA, Taichman LB (1994) Effect of TGF-beta 1 on re-epithelialization of human keratinocytes in vitro: an organotypic model. J Invest Dermatol 103:554–559

63. Garlick JA, Taichman LB (1994) Fate of human keratinocytes during reepithelialization in an organotypic culture model. Lab Invest 70:916–924

64. Garver RI Jr, Chytil A, Courtney M, Crystal RG (1987) Clonal gene therapy: transplanted mouse fibroblast clones express human alpha 1-antitrypsin gene in vivo. Science 237:762–764

65. Geer DJ, Swartz DD, Andreadis ST (2002) Fibrin promotes migration in a three-dimensional in vitro model of wound regeneration. Tissue Eng 8:787–798

66. Geer DJ, Swartz DD, Andreadis ST (2003) In vivo model of wound healing based on transplanted tissue engineered skin. J Invest Dermatol 10(7–8):1006–1017

67. Gerrard AJ, Hudson DL, Brownlee GG, Watt FM (1993) Towards gene therapy for haemophilia B using primary human keratinocytes. Nat Genet 3:180–183

68. Gill PS, Krueger GG, Kohan DE (2002) Doxycycline-inducible retroviral expression of green fluorescent protein in immortalized human keratinocytes. Exp Dermatol 11:266–274

69. Gordon DA, Fenjves ES, Williams DL, Taichman LB (1989) Synthesis and secretion of apolipoprotein E by cultured human keratinocytes. J Invest Dermatol 92:96–99

70. Gossen M, Bujard H (1992) Tight control of gene expression in mammalian cells by tetracycline- responsive promoters. Proc Natl Acad Sci USA 89:5547–5551

71. Griffith LG, Naughton G (2002) Tissue engineering–current challenges and expanding opportunities. Science 295:1009–1014

72. Hamoen KE, Morgan JR (2002) Transient hyperproliferation of a transgenic human epidermis expressing hepatocyte growth factor. Cell Transplant 11:385–395

73. Hanenberg H, Hashino K, Konishi H, Hock RA, Kato I, Williams DA (1997) Optimization of fibronectin-assisted retroviral gene transfer into human CD34+ hematopoietic cells. Hum Gene Ther 8:2193–2206

74. Hanenberg H, Xiao XL, Dilloo D, Hashino K, Kato I, Williams DA (1996) Colocalization of retrovirus and target cells on specific fibronectin fragments increases genetic transduction of mammalian cells. Nat Med 2:876–882

75. Harding TC, Geddes BJ, Murphy D, Knight D, Uney JB (1998) Switching transgene expression in the brain using an adenoviral tetracycline-regulatable system [see comments]. Nat Biotechnol 16:553–555

76. Harding TC, Geddes BJ, Noel JD, Murphy D, Uney JB (1997) Tetracycline-regulated transgene expression in hippocampal neurones following transfection with adenoviral vectors. J Neurochem 69:2620–2623

77. Hayden PJ, Ayehunie S, Jackson GR, Kupfer-Lamore S, Last T, Klausner JM, Kubilus J (2003) In vitro skin equivalent models for toxicity testing, p 225–242. In: Katz S, Salem H (eds) Alternative Toxicological Methods for the New Millennium. CRC Press, Boca Raton

78. Hengge UR, Chan EF, Foster RA, Walker PS, Vogel JC (1995) Cytokine gene expression in epidermis with biological effects following injection of naked DNA. Nat Genet 10:161–166

79. Hengge UR, Walker PS, Vogel JC (1996) Expression of naked DNA in human, pig, and mouse skin. J Clin Invest 97:2911–2916

80. Hilal L, Rochat A, Duquesnoy P, Blanchet-Bardon C, Wechsler J, Martin N, Christiano AM, Barrandon Y, Uitto J, Goossens M et al. (1993) A homozygous insertion-deletion in the type VII collagen gene (COL7A1) in Hallopeau-Siemens dystrophic epidermolysis bullosa. Nat Genet 5:287–293

81. Hubbell JA (1999) Bioactive biomaterials. Curr Opin Biotechnol 10:123–129

82. Jensen TG, Sullivan DM, Morgan RA, Taichman LB, Nussenblatt R, Blaese BRM, Csaky KG (1997) Retrovirus-mediated gene transfer of ornithine-delta-aminotransferase into keratinocytes from gyrate atrophy patients. Hum Gene Ther 8:2125–2132

83. Jensen UB, Jensen TG, Jensen PK, Rygaard J, Hansen BS, Fogh J, Kolvraa S, Bolund L (1994) Gene transfer into cultured human epidermis and its transplantation onto immunodeficient mice: an experimental model for somatic gene therapy. J Invest Dermatol 103:391–394

84. Jeschke MG, Richter G, Herndon DN, Geissler EK, Hartl M, Hofstatter F, Jauch KW, Perez-Polo JR (2001) Therapeutic success and efficacy of nonviral liposomal cDNA gene transfer to the skin in vivo is dose dependent. Gene Ther 8:1777–1784

85. Jeschke MG, Richter G, Hofstadter F, Herndon DN, Perez-Polo JR, Jauch KW (2002) Non-viral liposomal keratinocyte growth factor (KGF) cDNA gene transfer improves dermal and epidermal regeneration through stimulation of epithelial and mesenchymal factors. Gene Ther 9:1065–1074

86. Jones PH, Harper S, Watt FM (1995) Stem cell patterning and fate in human epidermis. Cell 80:83–93

87. Jones PH, Watt FM (1993) Separation of human epidermal stem cells from transit amplifying cells on the basis of differences in integrin function and expression. Cell 73:713–724

88. Kalyanasundaram S, Feinstein S, Nicholson JP, Leong KW, Garver RI Jr (1999) Coacervate microspheres as carriers of recombinant adenoviruses. Cancer Gene Ther 6:107–112

89. Kim HM, Choi DH, Lee YM (1998) Inhibition of wound-induced expression of transforming growth factor-beta 1 mRNA by its antisense oligonucleotides. Pharmacol Res 37:289–293

90. Kistner A, Gossen M, Zimmermann F, Jerecic J, Ullmer C, Lubbert H, Bujard H (1996) Doxycycline-mediated quantitative and tissue-specific control of gene expression in transgenic mice. Proc Natl Acad Sci USA 93:10933–10938

91. Klein RM, Wolf ED, Wu R, Sanford JC (1992) High-velocity microprojectiles for delivering nucleic acids into living cells (1987) Biotechnology 24:384–386

92. Kolodka TM, Garlick JA, Taichman LB (1998) Evidence for keratinocyte stem cells in vitro: long term engraftment and persistence of transgene expression from retrovirus-transduced keratinocytes. Proc Natl Acad Sci USA 95:4356–4361

93. Koria P, Brazeau D, Kirkwood KL, Hayden P, Klausner M, Andreadis ST (2003) Gene Expression Profile of Tissue Engineered Skin Subjected to Acute Barrier Disruption. J Invest Dermatol 121(2):368–382

94. Kringstein AM, Rossi FM, Hofmann A, Blau HM (1998) Graded transcriptional response to different concentrations of a single transactivator. Proc Natl Acad Sci USA 95:13670–13675

95. Krueger GG (2000) Fibroblasts and dermal gene therapy: a minireview. Hum Gene Ther 11:2289–2296

96. Kuhn U, Terunuma A, Pfutzner W, Foster RA, Vogel JC (2002) In vivo assessment of gene delivery to keratinocytes by lentiviral vectors. J Virol 76:1496–1504

97. Langer R, Vacanti JP (1993) Tissue engineering. Science 260:920–926

98. Larcher F, Del Rio M, Serrano F, Segovia JC, Ramirez A, Meana A, Page A, Abad JL, Gonzalez MA, Bueren J, Bernad A, Jorcano JL (2001) A cutaneous gene therapy approach to human leptin deficiencies: correction of the murine ob/ob phenotype using leptin-targeted keratinocyte grafts. Faseb J 15:1529–1538

99. Lei P, Bajaj B, Andreadis ST (2002) Retrovirus-associated heparan sulfate mediates immobilization and gene transfer on recombinant fibronectin. J Virol 76:8722–8728

100. Levy MY, Barron LG, Meyer KB, Szoka FC Jr (1996) Characterization of plasmid DNA transfer into mouse skeletal muscle: evaluation of uptake mechanism, expression and secretion of gene products into blood. Gene Ther 3:201–211

101. Liechty KW, Nesbit M, Herlyn M, Radu A, Adzick NS, Crombleholme TM (1999) Adenoviral-mediated overexpression of platelet-derived growth factor-B corrects ischemic impaired wound healing. J Invest Dermatol 113:375–383

102. Liechty KW, Sablich TJ, Adzick NS, Crombleholme TM (1999) Recombinant adenoviral mediated gene transfer in ischemic impaired wound healing. Wound Repair Regen 7:148–153

103. Limat A, Salomon D, Carraux P, Saurat JH, Hunziker T (1999) Human melanocytes grown in epidermal equivalents transfer their melanin to follicular outer root sheath keratinocytes. Arch Dermatol Res 291:325–332

104. Liu K, Yang Y, Mansbridge J (2000) Comparison of the stress response to cryopreservation in monolayer and three-dimensional human fibroblast cultures: stress proteins, MAP kinases, and growth factor gene expression. Tissue Eng 6:539–554

105. Livesey SA, Herndon DN, Hollyoak MA, Atkinson YH, Nag A (1995) Transplanted acellular allograft dermal matrix. Potential as a template for the reconstruction of viable dermis. Transplantation 60:1–9

106. Madsen S, Mooney DJ (2000) Delivering DNA with polymer matrices: applications in tissue engineering and gene therapy. Pharmaceutical Science and Technology Today 3:381–384

107. Mansbridge J (2002) Tissue-engineered skin substitutes. Expert Opin Biol Ther 2:25–34

108. Mathor MB, Ferrari G, Dellambra E, Cilli M, Mavilio F, Cancedda R, De Luca M (1996) Clonal analysis of stably transduced human epidermal stem cells in culture. Proc Natl Acad Sci USA 93:10371–10376

109. Matthews C, Jenkins G, Hilfinger J, Davidson B (1999) Poly-L-lysine improves gene transfer with adenovirus formulated in PLGA microspheres. Gene Ther 6:1558–1564

110. Meana A, Iglesias J, Del Rio M, Larcher F, Madrigal B, Fresno M, Martin FC, San Roman F, Tevar F (1998) Large surface of cultured human epithelium obtained on a dermal matrix based on live fibroblast-containing fibrin gels. Burns 24:621–630

111. Medalie DA, Eming SA, Collins ME, Tompkins RG, Yarmush ML, Morgan JR (1997) Differences in dermal analogs influence subsequent pigmentation, epidermal differentiation, basement membrane, and rete ridge formation of transplanted composite skin grafts. Transplantation 64:454–465

112. Medalie DA, Eming SA, Tompkins RG, Yarmush ML, Krueger GG, Morgan JR (1996) Evaluation of human skin reconstituted from composite grafts of cultured keratinocytes and human acellular dermis transplanted to athymic mice. J Invest Dermatol 107:121–127

113. Medalie DA, Tompkins RG, Morgan JR (1996) Evaluation of acellular human dermis as a dermal analog in a composite skin graft. J Asaio 42:M455–M462

114. Meng X, Sawamura D, Tamai K, Hanada K, Ishida H, Hashimoto I (1998) Keratinocyte gene therapy for systemic diseases. Circulating interleukin 10 released from gene-transferred keratinocytes inhibits contact hypersensitivity at distant areas of the skin. J Clin Invest 101:1462–1467

115. Miller DG, Adam MA, Miller AD (1990) Gene transfer by retrovirus vectors occurs only in cells that are actively replicating at the time of infection. Mol Cell Biol 10:4239–4242

116. Morgan JR, Barrandon Y, Green H, Mulligan RC (1987) Expression of an exogenous growth hormone gene by transplantable human epidermal cells. Science 237:1476–1479

117. Morgan JR, Yarmush ML (1997) Bioengineered skin substitutes. Science & Medicine. July/August:6–15

118. Morgan JR, Yarmush ML (1998) Gene therapy in tissue engineering, p 278–310. In: Patrick CWJ, Mikos AG, McIntire LV (eds) Frontiers in Tissue Engineering. Pergamon, New York

119. Mustoe TA, Cutler NR, Allman RM, Goode PS, Deuel TF, Prause JA, Bear M, Serdar CM, Pierce GF (1994) A phase II study to evaluate recombinant platelet-derived growth factor-BB in the treatment of stage 3 and 4 pressure ulcers. Arch Surg 129:213–219

120. Noel D, Dazard JE, Pelegrin M, Jacquet C, Piechaczyk M (2002) Skin as a potential organ for ectopic monoclonal antibody production. J Invest Dermatol 118:288–294

121. Ozawa K, Kondo T, Hori O, Kitao Y, Stern DM, Eisenmenger W, Ogawa S, Ohshima T (2001) Expression of the oxygen-regulated protein ORP150 accelerates wound healing by modulating intracellular VEGF transport. J Clin Invest 108:41–50

122. Page SM, Brownlee GG (1998) Differentiation-specific enhancer activity in transduced keratinocytes: a model for epidermal gene therapy. Gene Ther 5:394–402

123. Pandit AS, Wilson DJ, Feldman DS (2000) Fibrin scaffold as an effective vehicle for the delivery of acidic fibroblast growth factor (FGF-1). J Biomater Appl 14:229–242

124. Parenteau NL, Bilbo P, Nolte CJ, Mason VS, Rosenberg M (1992) The organotypic culture of human skin keratinocytes and fibroblasts to achieve form and function. Cytotechnology 9:163–171

125. Paulus W, Baur I, Boyce FM, Breakefield XO, Reeves SA (1996) Self-contained, tetracycline-regulated retroviral vector system for gene delivery to mammalian cells. J Virol 70:62–67

126. Pellegrini G, Ranno R, Stracuzzi G, Bondanza S, Guerra L, Zambruno G, Micali G, De Luca M (1999) The control of epidermal stem cells (holoclones) in the treatment of massive full-thickness burns with autologous keratinocytes cultured on fibrin. Transplantation 68:868–879

127. Pinney E, Liu K, Sheeman B, Mansbridge J (2000) Human three-dimensional fibroblast cultures express angiogenic activity. J Cell Physiol 183:74–82

128. Qiu X, Lu D, Zhou J, Wang J, Yang J, Meng P, Hsueh JL (1996) Implantation of autologous skin fibroblast genetically modified to secrete clotting factor IX partially corrects the hemorrhagic tendencies in two hemophilia B patients. Chin Med J (Engl) 109:832–839

129. Ramos-e-Silva M, Ribeiro de Castro MC (2002) New dressings, including tissue-engineered living skin. Clin Dermatol 20:715–723

130. Rinsch C, Quinodoz P, Pittet B, Alizadeh N, Baetens D, Montandon D, Aebischer P, Pepper MS (2001) Delivery of FGF-2 but not VEGF by encapsulated genetically engineered myoblasts improves survival and vascularization in a model of acute skin flap ischemia. Gene Ther 8:523–533

131. Rio MD, Larcher F, Meana A, Segovia J, Alvarez A, Jorcano J (1999) Nonviral transfer of genes to pig primary keratinocytes. Induction of angiogenesis by composite grafts of modified keratinocytes overexpressing VEGF driven by a keratin promoter. Gene Ther 6:1734–1741

132. Robinson MK, Osborne R, Perkins MA (2000) In vitro and human testing strategies for skin irritation. Ann NY Acad Sci 919:192–204

133. Roe T, Reynolds TC, Yu G, Brown PO (1993) Integration of murine leukemia virus DNA depends on mitosis. Embo J 12:2099–2108

134. Rossi FM, Guicherit OM, Spicher A, Kringstein AM, Fatyol K, Blakely BT, Blau HM (1998) Tetracycline-regulatable factors with distinct dimerization domains allow reversible growth inhibition by p16. Nat Genet 20:389–393

135. Sakiyama SE, Schense JC, Hubbell JA (1999) Incorporation of heparin-binding pep-
     tides into fibrin gels enhances neurite extension: an example of designer matrices in
     tissue engineering. Faseb J 13:2214–2224
136. Sakiyama-Elbert SE, Hubbell JA (2000) Controlled release of nerve growth factor from
     a heparin-containing fibrin-based cell ingrowth matrix. J Control Release 69:149–158
137. Sakiyama-Elbert SE, Hubbell JA (2000) Development of fibrin derivatives for con-
     trolled release of heparin- binding growth factors. J Control Release 65:389–402
138. Scherr M, Morgan MA, Eder M (2003) Gene silencing mediated by small interfering
     RNAs in mammalian cells. Curr Med Chem 10:245–256
139. Selden RF, Skoskiewicz MJ, Howie KB, Russell PS, Goodman HM (1987) Implanta-
     tion of genetically engineered fibroblasts into mice: implications for gene therapy.
     Science 236:714–718
140. Sen CK, Khanna S, Babior BM, Hunt TK, Ellison EC, Roy S (2002) Oxidant-induced
     vascular endothelial growth factor expression in human keratinocytes and cuta-
     neous wound healing. J Biol Chem 277:33284–33290
141. Setoguchi Y, Jaffe HA, Danel C, Crystal RG (1994) Ex vivo and in vivo gene trans-
     fer to the skin using replication- deficient recombinant adenovirus vectors. J Invest
     Dermatol 102:415–421
142. Shea LD, Smiley E, Bonadio J, Mooney DJ (1999) DNA delivery from polymer matri-
     ces for tissue engineering. Nat Biotechnol 17:551–554
143. St Louis D, Verma IM (1988) An alternative approach to somatic cell gene therapy.
     Proc Natl Acad Sci USA 85:3150–3154
144. Subramanian A, Ranganathan P, Diamond SL (1999) Nuclear targeting peptide scaf-
     folds for lipofection of nondividing mammalian cells. Nat Biotechnol 17:873–877
145. Sullivan DM, Jensen TG, Taichman LB, Csaky KG (1997) Ornithine-delta-aminotrans-
     ferase expression and ornithine metabolism in cultured epidermal keratinocytes:
     toward metabolic sink therapy for gyrate atrophy. Gene Ther 4:1036–1044
146. Sun L, Xu L, Chang H, Henry FA, Miller RM, Harmon JM, Nielsen TB (1997) Trans-
     fection with aFGF cDNA improves wound healing. J Invest Dermatol 108:313–318
147. Sun WH, Burkholder JK, Sun J, Culp J, Turner J, Lu XG, Pugh TD, Ershler WB,
     Yang NS (1995) In vivo cytokine gene transfer by gene gun reduces tumor growth
     in mice. Proc Natl Acad Sci USA 92:2889–2893
148. Supersaxo A, Hein WR, Steffen H (1990) Effect of molecular weight on the lymphatic
     absorption of water-soluble compounds following subcutaneous administration.
     Pharm Res 7:167–169
149. Supp DM, Boyce ST (2002) Overexpression of vascular endothelial growth factor ac-
     celerates early vascularization and improves healing of genetically modified cultured
     skin substitutes. J Burn Care Rehabil 23:10–20
150. Supp DM, Supp AP, Bell SM, Boyce ST (2000) Enhanced vascularization of cultured
     skin substitutes genetically modified to overexpress vascular endothelial growth fac-
     tor. J Invest Dermatol 114:5–13
151. Tai IT, Sun AM (1993) Microencapsulation of recombinant cells: a new delivery sys-
     tem for gene therapy. Faseb J 7:1061–1069
152. Taub PJ, Marmur JD, Zhang WX, Senderoff D, Nhat PD, Phelps R, Urken ML, Sil-
     ver L, Weinberg H (1998) Locally administered vascular endothelial growth factor
     cDNA increases survival of ischemic experimental skin flaps. Plast Reconstr Surg
     102:2033–2009
153. Teumer J, Lindahl A, Green H (1990) Human growth hormone in the blood of
     athymic mice grafted with cultures of hormone-secreting human keratinocytes.
     Faseb J 4:3245–3250

154. Todd C, Hewitt SD, Kempenaar J, Noz K, Thody AJ, Ponec M (1993) Co-culture of human melanocytes and keratinocytes in a skin equivalent model: effect of ultraviolet radiation. Arch Dermatol Res 285:455–459
155. Tyrone JW, Mogford JE, Chandler LA, Ma C, Xia Y, Pierce GF, Mustoe TA (2000) Collagen-embedded platelet-derived growth factor DNA plasmid promotes wound healing in a dermal ulcer model. J Surg Res 93:230–236
156. Uitto J, Christiano AM (1994) Molecular basis for the dystrophic forms of epidermolysis bullosa: mutations in the type VII collagen gene. Arch Dermatol Res 287:16–22
157. Uitto J, Pulkkinen L, Christiano AM (1994) Molecular basis of the dystrophic and junctional forms of epidermolysis bullosa: mutations in the type VII collagen and kalinin (laminin 5) genes. J Invest Dermatol 103:39S–46S
158. Vaccariello M, Javaherian A, Wang Y, Fusenig NE, Garlick JA (1999) Cell interactions control the fate of malignant keratinocytes in an organotypic model of early neoplasia. J Invest Dermatol 113:384–391
159. Vailly J, Gagnoux-Palacios L, Dell'Ambra E, Romero C, Pinola M, Zambruno G, De Luca M, Ortonne JP, Meneguzzi G (1998) Corrective gene transfer of keratinocytes from patients with junctional epidermolysis bullosa restores assembly of hemidesmosomes in reconstructed epithelia. Gene Ther 5:1322–1332
160. Wagner H, Kostka KH, Lehr CM, Schaefer UF (2001) Interrelation of permeation and penetration parameters obtained from in vitro experiments with human skin and skin equivalents. J Control Release 75:283–295
161. Wang X, Zinkel S, Polonsky K, Fuchs E (1997) Transgenic studies with a keratin promoter-driven growth hormone transgene: prospects for gene therapy. Proc Natl Acad Sci USA 94:219–226
162. Watt FM, Hogan BL (2000) Out of Eden: stem cells and their niches. Science 287:1427–1430
163. White SJ, Page SM, Margaritis P, Brownlee GG (1998) Long-term expression of human clotting factor IX from retrovirally transduced primary human keratinocytes in vivo. Hum Gene Ther 9:1187–1195
164. Wieman TJ, Smiell JM, Su Y (1998) Efficacy and safety of a topical gel formulation of recombinant human platelet-derived growth factor-BB (becaplermin) in patients with chronic neuropathic diabetic ulcers. A phase III randomized placebo-controlled double-blind study. Diabetes Care 21:822–827
165. Wilkins LM, Watson SR, Prosky SJ, Meunier SF, Parenteau NL (1994) Development of a bilayered living skin equivalent construct for clinical applications. Biotechnol Bioeng 43:747–756
166. Wraight CJ, White PJ, McKean SC, Fogarty RD, Venables DJ, Liepe IJ, Edmondson SR, Werther GA (2000) Reversal of epidermal hyperproliferation in psoriasis by insulin-like growth factor I receptor antisense oligonucleotides. Nat Biotechnol 18:521–526
167. Yamamoto A, Hen R, Dauer WT (2001) The ons and offs of inducible transgenic technology: a review. Neurobiol Dis 8:923–932
168. Yamasaki K, Edington HD, McClosky C, Tzeng E, Lizonova A, Kovesdi I, Steed DL, Billiar TR (1998) Reversal of impaired wound repair in iNOS-deficient mice by topical adenoviral-mediated iNOS gene transfer. J Clin Invest 101:967–971
169. Yannas IV, Burke JF (1980) Design of an artificial skin. Basic I design principles. J Biomed Mater Res 14:65–81
170. Yannas IV, Burke JF, Gordon PL, Huang C, Rubenstein RH (1980) Design of an artificial skin. II. Control of chemical composition. J Biomed Mater Res 14:107–132
171. Yannas IV, Burke JF, Orgill DP, Skrabut EM (1982) Wound tissue can utilize a polymeric template to synthesize a functional extension of skin. Science 215:174–176

172. Yannas IV, Lee E, Orgill DP, Skrabut EM, Murphy GF (1989) Synthesis and characterization of a model extracellular matrix that induces partial regeneration of adult mammalian skin. Proc Natl Acad Sci USA 86:933–937
173. Zandstra PW, Nagy A (2001) Stem cell bioengineering. Annu Rev Biomed Eng 3:275–305
174. Zanta MA, Belguise-Valladier P, Behr JP (1999) Gene delivery: a single nuclear localization signal peptide is sufficient to carry DNA to the cell nucleus. Proc Natl Acad Sci USA 96:91–96
175. Zhu AJ, Haase I, Watt FM (1999) Signaling via beta1 integrins and mitogen-activated protein kinase determines human epidermal stem cell fate in vitro. Proc Natl Acad Sci USA 96:6728–6733
176. Geer DJ, Swartz DD, Andreadis ST (2005) Biomimetic delivery of keratinocyte growth factor upon cellular demand for accelerated wound healing in vitro and in vivo. Am J Pathol 167(6):1575–1586
177. Koria P, Andreadis ST (2006) Epidermal morphogenesis: the transcriptional program of human keratinocytes during stratification. J Invest Dermatol 126(8):1834–1841

Adv Biochem Engin/Biotechnol (2006) 103: 275–308
DOI 10.1007/10_021
© Springer-Verlag Berlin Heidelberg 2006
Published online: 10 June 2006

Nanostructured Biomaterials for Tissue Engineering Bone

Thomas J. Webster[1] (✉) · Edward S. Ahn[2]

[1]Divisions of Engineering and Orthopaedics, Brown University, Providence, RI 02912, USA
Thomas_Webster@Brown.edu

[2]Angstrom Medical Inc., 150 California Street, Newton, MA 02458, USA

Abstract Advances in several critical research fields (processing, catalytic, optical, actuation, electrical, mechanical, etc.) have started to benefit from nanotechnology. Nanotechnology can be broadly defined as the use of materials and systems whose structures and components exhibit novel and significantly changed properties when control is gained at the atomic, molecular, and supramolecular levels. Specifically, such advances have been found for materials when particulate size is decreased to below 100 nm. However, to date, relatively few advantages have been described for biological applications (specifically, those involving bone tissue engineering). This chapter elucidates several promising examples of how nanophase materials can be used to improve orthopedic implant applications. These include mechanical advantages as well as altered cell functions, leading to increased bone tissue regeneration on a wide range of nanophase materials including ceramics, polymers, metals, and composites thereof. Such advances were previously unimaginable with conventional materials possessing large micron-sized particulates.

Keywords Nanostructured materials · Nanotechnology · Bone · Osteoblasts ·
Orthopedic · Tissue engineering

Abbreviations

σ_f	Fracture strength
K_{1c}	Mode 1 fracture toughness (under tension)
c	Size of the largest defect in the material
Y	Geometric factor related to the location of the defect
E	Young's modulus

1
Introduction

1.1
Nanotechnology and Bone Tissue Engineering

Over the past nine decades of administering implants to humans, most synthetic orthopedic prostheses consist of material particles and/or grain sizes
with conventional dimensions (of the order 1 to 10^4 μm) [2]. The majority of
these early implants were made out of the following: vanadium steel (in the
1920s), stainless steel, cobalt alloys, titanium (in the 1930s), gold and amal-
gams (metal alloys containing mercury) [2]. However, the lack of sufficient
bonding of synthetic implants to surrounding bone and the inability to obtain mechanical characteristics (such as flexural strength, bending strength,
modulus of elasticity, toughness, and ductility) as well as electrical characteristics (such as resistivity and, even, piezoelectricity) that simulate their
human tissue equivalents have, in recent years, led to the investigations of
novel materials [3–13]. Several different approaches (such as altering ceramic
coating chemistry, utilizing biodegradable polymers or composites of natural
and synthetic materials) are currently under heavy investigation to find this
next generation of improved bone biomaterials.

Although some of these methods are showing promise, this entry seeks
to describe an alternative approach to developing the next generation of orthopedic implant that lies at the intersection between nanotechnology and
medicine; such an approach has not received much attention to date. Nanotechnology can be broadly defined as the methods to produce and use materials whose components exhibit novel and significantly changed properties
when control is gained at the atomic, molecular, and supramolecular levels [14]. Several critical research fields (such as catalysis, optics, electronics,
MEMS, etc.) have started to benefit from new technological advancements
in the area of nanotechnology, particularly through the use of nanomaterials [15–24]. Specifically, such advances have been found for materials when
particulate size is decreased to below 100 nm. For that reason, although var-

ious definitions have been attached to the word "nanomaterial" by different experts [23], the commonly accepted concept refers to those materials with basic structural units in the range 1–100 nm (nanostructured), crystalline solids with grain sizes between 1 and 100 nm (nanocrystals), individual layers or multilayer surface coatings in the range 1–100 nm (nanocoatings), extremely fine powders with an average particle size in the range 1–100 nm (nanopowders), and fibers with a diameter in the range 1–100 nm (nanofibers).

Although showing promise for traditional science and engineering applications, to date, relatively few advantages of nanophase materials have been described for biological applications (specifically, those involving bone tissue engineering applications). Yet, since nanotechnology embraces a system whose core of materials is in the range of nanometers (10^{-9} m), there are many similarities between nanophase materials and biological organs, particularly bone (Fig. 1). Living systems are clearly governed by molecular behavior at nanometer scales [25, 26]. The molecular building blocks of life – proteins, nucleic acids, lipids, carbohydrates, and their non-biological mimics – are examples of materials that possess unique properties determined by the size, folding, and patterns at the nanoscale. More specifically, constituents in natural bone also possess dimensions in the nanometer regime. Specifically, hydroxyapatite plates are approximately between 25 nm in width and 35 nm in length while Type I collagen is a triple helix 300 nm in length, 0.5 nm in width, and has a periodicity of 67 nm [2]. It is because of this that it stands to reason that bone cells are naturally accustomed to interacting with sur-

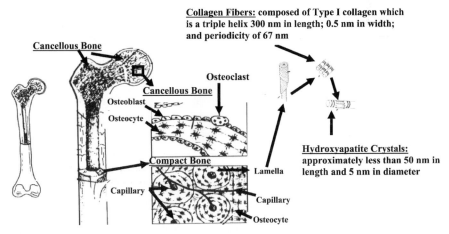

Fig. 1 Nanostructure of bone. The fundamental nanosize characteristics of the inorganic (specifically, HA) and organic (such as Type I collagen) phases of bone have yet to be accurately duplicated in materials proposed as bone implants. Redrawn and adapted from [2]

faces with a large degree of nanometer roughness. This is in contrast to the surfaces provided by traditional orthopedic implants. Specifically, materials being used and proposed as bone prosthetics have constituent grain or particle sizes that provide topographies that are rough at the micron scale yet smooth at the nanoscale.

Because of this intrigue, this chapter elucidates: (i) exciting bulk properties (such as composition, microstructure, and mechanical strength) and (ii) novel surface properties (such as protein and cellular interactions) of nanophase materials pertinent for orthopedic applications. Moreover, several promising processing techniques for forming nanostructured bone constructs will be discussed throughout this chapter. Such advances were previously unimaginable with conventional materials that possess large, non-biologically inspired, micron-sized constituent particulates and subsequent topographies.

1.2
Bone: A Nanostructured Biomaterial

Bone is an organic–inorganic nanocomposite composed of (1) collagen (35 dry wt %) for flexibility and toughness; (2) carbonated apatite (65 dry wt %) for structural reinforcement, stiffness and mineral homeostasis; and (3) other non-collagenous proteins for support of cellular functions. Its microstructure is organized three-dimensionally along multiple length scales so that maximum strength and toughness are along the lines of the applied stress (Fig. 2). At the macroscopic level, bone consists of a dense shell of cortical

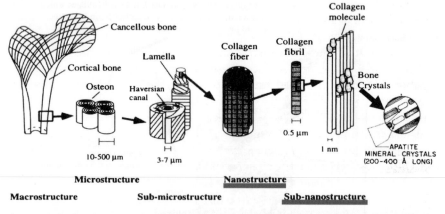

Fig. 2 Hierarchical structure of bone. Bone is a highly organized nanoscale hierarchical structure. Thus, it stands to reason that bone cells are accustomed to interacting with nanoscale features; this is in contrast to the materials that are being implanted today, which have micron-scale topographies smooth at the nanoscale (Redrawn and adapted from [98])

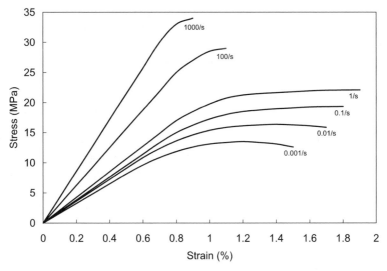

Fig. 3 Representative stress–strain curve of human compact bone, illustrating strain rate sensitivity. *X-axis* is in natural log (ln)

bone along the length and a porous trabecular bone at the proximal and distal ends to optimize weight transfer and minimize friction at the articulating joints, as is the case in long bones. At the microscopic level, the dense cortical bone and trabecular struts are composed of mineralized collagen fibrils that are stacked in parallel to form layers, and then assembled in a ±45° fashion. Each of the collagen fibrils consists of bundles of microfibrils that in turn contain bundles of collagen molecules. The collagen molecules are arranged with a periodicity of 67 nm; where two ends meet, there is a small gap of 40 nm, or hole zone, where most of the 35 × 25 nm plates of bone mineral are located.

Because of this hierarchical structure, bone possesses unique mechanical and biological properties. As the microstructure of bone involves the coordination of highly dispersed bone mineral nanocrystals within collagen, it behaves as a tough material at low strain rates (0.001/s) but as a brittle material at high strain rates (1000/s) (see Fig. 3). Furthermore, this microstructure also acts as a scaffold for cell regulation. The surface features that these cells encounter are mostly nanometers in dimension, suggesting that this length scale is critical in regulating cell behavior.

1.3
Clinical Need for Better Orthopedic Implant Materials

Though bone possesses a unique capability for continuous renewal, like most of the body's other components, the human skeleton represents a variety of challenges to medicine including: fracture, degeneration, congenital abnormality, and disease. At some time in many peoples' lives, part of the skeleton may require treatment, repair, or ultimately replacement. During surgery, the orthopedic surgeon repairs or replaces the damaged bone using various implants to restore function, and bone grafts to regenerate bone. The acute injuries associated with trauma often require orthopedic intervention and are the second most common reason why patients visit the emergency room; approximately 7.7 million visits to the emergency room will require immediate orthopedic surgery each year; an additional 700 000 patients will require a total joint replacement; and approximately 450 000 spinal fusion procedures will be performed this year in the USA to alleviate back pain [27–29]. Integrated into these high numbers of orthopedic surgeries are revision surgeries. This is because, to date, no optimal material exists for orthopedic implantation. This is highlighted by the fact that the average lifetime of an implant is only 15 years [1].

A clear reason why a perfect orthopedic implant material does not currently exist is because bone is a structurally and biologically complex multifunctional tissue. This dynamic tissue responds to biochemical, mechanical, and electrical signals by remodeling itself so that maximum strength and toughness are along the lines of the greatest applied stress. Current orthopedic practice decouples mechanical and biological function. Metallic implants are often used as structural elements whereas bone grafts are used to restore biological function. The ideal orthopedic biomaterial should simultaneously restore mechanical and biological function by mimicking the mechanical properties of bone, sustaining a cell population on its surface, and responding to various stimuli. The challenge is to develop an orthopedic biomaterial that emulates the micro- and nanostructural elements and compositions of bone to obtain these desired characteristics; nanophase materials may be such materials.

1.3.1
Metallic Implants:
Mechanical Stabilization During Skeletal Reconstruction

Typically, screws, plates, rods, nails, and pins are used to immobilize portions of the spine and fractured bones; such materials are primarily constructed of metal or, occasionally, of polyester polymers. The healing response immediately following implantation determines the long-term functionality of the device and involves the recruitment of a variety of body fluids, proteins, and unwanted cell types to the tissue–implant interface. Healing characteristics common of unsuccessful bone prosthetic wound-healing include fibrous encapsulation and chronic inflammation. Fibrous encapsulation (or callus formation) at the interface or at the implant–body tissue site decreases the effectiveness of the implant to bond to bone and often results in clinical failures. This slows down rapid attainment of the physical strength in the region that would be juxtaposed to that of normal tissue. Most importantly, to date, these failures (common to conventional orthopedic implant materials) often lead to insufficient bonding with juxtaposed bone; this event is necessary to stabilize the implant for subsequent physiological loading. That is, select, timely, and desirable responses from surrounding bone cells are required to enhance deposition of the mineralized matrix at the tissue–implant interface, which provides crucial mechanical stability to implants. Therefore, modulating the healing response may improve biomaterial efforts to increase new bone regeneration and decrease unwanted fibrous tissue formation.

However, bioactive properties that would promote such a positive sequence of healing such as bone bonding and osteoconduction have been sacrificed for mechanical strength in material selection for load-bearing orthopedic implant applications in joint reconstruction, spinal fusion, internal fixation, and craniomaxillofacial implants. Consequently, conventional implant materials do not biologically integrate with the natural bone, result in non-uniform healing, elicit foreign body reactions, or contribute to loss of surrounding bone or re-injury. Because of these shortcomings, orthopedic medicine continues to pursue procedures that are increasingly less invasive and implants that are better adapted to restoring both bone function and structure.

Although metallic implants have been clinically successful in the short term (< 15 years), there are major problems associated with stress shielding, corrosion, implant–tissue interfacial stability, and postsurgical recovery of the surrounding bone tissue. For all these applications, metals have several short- and long-term disadvantages as an implant material, such as:

- While adequate to meet short-term mechanical goals, metal implants raise a number of longer-term clinical concerns including protuberance over the skin, non-uniform healing, and loosening, all of which may lead to a second surgery to remove the implant [30–35]. In such cases, patients may undergo a second surgery to remove the implant due to pain or, in the case of small digit applications, erosion of tissue over the implant. In patients or procedures for which this is not possible, the potential risks of loosening, stress shielding, implant degeneration, re-injury and pain remain during the life of the patient.
- Although metals provide structural support to the area during healing, bone cannot directly integrate with, or bond to, metallic surfaces. Lack of firm attachment can result in non-uniform healing, implant migration, pain, or infection at the implantation site, as well as other foreign body complications.
- The presence of metal limits the use of postoperative MRI because the metallic objects distort the MRI image. Postoperative MRI is highly desired in spinal fusion.

Most importantly, as previously mentioned, there is now overwhelming clinical evidence of the poor long-term performance of metals as hip, internal fixation devices, and spinal fusion bridges to name a few. First, as a result of these shortcomings, 12.8% of the total hip arthroplasties are revised [1]. The fact that such a high percentage of hip replacements performed every year are revision surgeries is not surprising when considering the life expectancy of the implant versus that of the patient receiving the implant. Consistently, over 30% of those requiring total hip replacements have been below the age of 65, yet, the longevity of orthopedic implants ranges from only about 12–15 years. For this reason, the majority of those that receive an implant at age 65 or below will require at least one revision surgery.

Second, of the roughly six million fractures reconstructed by internal fixation each year in the USA, an estimated 5–10% have impaired or delayed healing. This may be due to inadequate immobilization of the fracture, fracture fragments that may be distracted by fixation devices or traction, excessive early motion of a fracture, or excessive damage to other soft tissue during surgery.

Third, while the spine fusion procedure has been reasonably successful, in a number of cases the spine may not properly align due to movement of the stabilizing implant. There are also reports of patients suffering from foreign body complications associated with the implant or particles of material generated from the implant. Titanium cages also have been used as an implant to maintain interbody separation because they are capable of providing sufficient strength to carry the loads imposed on the spinal processes. Unfortunately, the durability of the material often results in subsidence into the adjacent vertebral structures. When this occurs, a second surgery must be used to remove the cage – a difficult and time-consuming procedure. Collectively such data suggests that the life expectancy of orthopedic prostheses is a recurring problem that has to be dealt with since, clearly, current approaches fail with metals.

1.3.2
Autograft and Allograft:
Bone Regeneration During Skeletal Reconstruction

Another approach to regenerating bone in small defects (such as in the dissection of osteosarcomas) or fusion in extraskeletal sites (such as in spine) include autografts and allografts. Bone replacement and regeneration technologies include a wide variety of grafting materials to regenerate or replace bone and are often required in cases ranging from trauma, congenital and degenerative diseases, cancer, and total joint replacements to cosmetics. Of the estimated 550 000 bone graft procedures performed annually, the vast majority of allografts are in the spine [36].

These bone graft materials are the most implanted materials, second only to transfused blood and are categorized into three groups: autografts, allografts, and synthetic bone substitutes. The key differentiating factor between the grafts is the degree to which they are osteoconductive (serve as a scaffold for bone to grow around, but do not trigger new bone growth) and osteoinductive (recruit *and* trigger cells to differentiate into active osteoblasts, which promote new bone). However, these grafts are far from ideal and have many associated problems.

Autografts are considered the "gold standard" because it is a living tissue and, consequently, yields the best clinical results. However, autografts are harvested from the iliac crest and are clearly limited in size and quantity. Consequently, procedures requiring an autograft require an additional painful and traumatic surgery, create an additional defect, and are limited by the availability of bone. As a result, operating room time and potential for complications (infection, chronic pain, etc.) are increased. More importantly, patients find the second surgery to be more traumatic than the actual procedure. There is an 8–10% surgical morbidity associated with bone graft harvesting, including donor site pain, prolonged anesthesia time, and infection. In general, autografts are not used where the mechanical requirements are demanding and, thus, are only used only to fill small voids and defects.

As a substitute to autografts, allografts from human or animal cadavers have been used and are available in a variety of fresh, frozen, or freeze dried forms (e.g., chips, paste, blocks, gels, and putties). Though not a living tissue, allografts can still retain many of the biological factors allowing them to be osteoinductive; however, clinical results can be unpredictable [37–43]. Though surgeons spend considerable time in preparing the surgical site to receive the allograft, the clinical result ultimately depends on the quality of the graft and not on surgical skill. For example, if the allograft is obtained from an older donor, the clinical result maybe poor whereas younger donors may provide more fecund grafts. However, the limitations in size and strength are preventing broader applications of allografts. While allografts have found wide utility as interbody fusion devices in spine and, consequently, acceptance as a structural implant, size and strength limitations prevent the further development into other implant geometries such as allograft plates and screws. Allograft cages have met with good market acceptance – they provide structural support, can resorb into the healing tissue and thus provide a scaffold for spinal fusion, but still possess a failure rate approaching 30% [44–53]. Finally, transmission of viruses and contaminating agents is a concern. While allograft bone is available from numerous bone banks, there is the potential for implant rejection as well as the transmission of infectious agents. In addition, supply is an issue given the limited availability of cadavers coupled with the storage, processing, and donor screening required. Because of these problems, materials other than autografts and allografts are needed.

1.4
Nanostructured Tissue Engineered Synthetic Bone

The deficiencies in conventional metallic implants, autografts, and allografts have resulted in an aggressive effort to develop a tissue-engineered bone possessing strength, reliable and consistent bioactivity, and unrestricted size and geometry. This engineered bone must possess:

1. Mechanical stability at the injured site for the required duration to allow adequate healing
2. Biocompatibility with the surrounding host tissue
3. Osseointegration with the host bone
4. Elimination of aseptic inflammation

Compared to metallic and polymeric implants, engineered bone would possess characteristics vital for the effective functioning of orthopedic implants in the host tissue, including lack of any toxicity tendencies, resistance to corrosion by surrounding tissue fluids, sufficient strength to support normal physical motion or loading, resistance to bodily fatigue forces, and the ability to promote appropriate cellular adhesion and function leading to juxtaposed bone tissue regeneration [2].

The classifications of nanostructured tissue engineered bone proposed in this chapter could improve upon the autograft "gold standard" by eliminating the need for the expensive and painful harvesting procedure, and by reducing surgery time, trauma, and cost. More importantly, it could be produced in implant geometries with structural properties (for example, a bone screw). Nanostructured tissue engineered synthetic bone can also be delivered in more varied geometries and sizes than either allograft or autograft so that it can be utilized by the physician as a conventional metallic or polymeric implant. Furthermore, its improved mechanical properties can provide for a greater number of clinical indications than traditional grafts and can protect the regenerating tissue construct during ex vivo expansion and implantation. Finally, nanostructured matrices can confer enhanced bioactivity leading to better cell attachment, adhesion, and proliferation during ex vivo expansion to minimize incubation time, and in vivo to rapidly integrate with surrounding tissue.

2
Properties of Nanostructured Tissue Engineered Synthetic Bone

2.1
The Promise

Perhaps the most extensively studied substrate for tissue engineering has been hydroxyapatite, HA ($Ca_{10}(PO_4)_6(OH)_2$) because its composition closely resembles native bone mineral ($P6_3/m$) and it is inherently osteoconductive. HA has generated great interest as an advanced orthopedic and dental implant material as it elicits a favorable biological response and forms a bond with the surrounding tissues. However, applications of HA are currently limited to powders, coatings, porous bodies, and non-load-bearing implants due to processing difficulties and the resulting poor mechanical properties. By designing constituent components of such materials in the nanometer regime, some of these problems may be solved. For example, calcium phosphate biomaterials can uniquely benefit from nanostructured materials processing and subsequently can result in unique properties necessary for tissue-engineered bone applications.

The use of nanostructured HA as the foundation for bone tissue engineering offers several advantages. Emerging evidence indicates that nanophase (< 100 nm) substrates enhance the adhesion and subsequent functions (including deposition of calcium) of osteoblasts. The ceramic nature of nanocrystalline HA (nano-HA) also affords chemical and structural stability and therefore confers the advantage of long-term shape integrity. Finally, nano-HA is mechanically robust and can support and protect the bone tissue construct during the implantation procedure and within the body. These enabling properties make nano-HA an attractive platform for tissue engineering bone in vivo or ex vivo.

Recent reports also suggest that this enhanced bioactivity is not exclusive to HA and applies to a wide range of nanostructured surfaces as a group. Specifically, unique enhanced bone-regeneration has been observed on other nanophase ceramics (like alumina and titania), polymers like poly(lactic-co-glycolic acid) and polyurethane, carbon nanofibers, metals (like Ti, Ti6Al4V, and CoCrMo), and composites thereof, presenting a wide range of potentially exciting nanomaterials for bone applications.

2.2
Mechanical Properties

Conventional HA's low strength and brittleness can be attributed to defects introduced by conventional synthesis and processing techniques and its relatively low fracture toughness. According to the following equation:

$$\sigma_f = \frac{K_{1c}}{Y\sqrt{c}},$$

where σ_f is the fracture strength; K_{1c} is the mode 1 fracture toughness (under tension); c is the size of the largest defect in the material; and Y is a geometric factor related to the location of the defect ($Y = 0.73$ for a half-circular surface crack, $= 1.12$ for a through-thickness surface crack, and $= 1.00$ for an interior crack) [54–56]. Mechanical failure in HA occurs when the tensile stress concentration at a critical flaw exceeds the fracture toughness of the material.

Hence, the strength ultimately depends on the fracture toughness, defect size distribution, and stress field applied to the HA. Consequently, reducing HA's functional mechanical performance can be increased by reducing the size and number of defects.

2.2.1
Effect of Defect Size

A fully dense, phase-pure, transparent, nano-HA can be achieved through nanostructured processing, as shown in Fig. 4a. The transparency of this nano-HA indicates that the fundamental microstructural size is very small and less than the wavelength of light; microstructural analysis by scanning electron microscopy (SEM) in Fig. 4b illustrates that the fundamental microstructural size of nano-HA is about 100 nm. Consequently, 100 nm becomes the fundamental lower limit for a tolerable defect size in the material, which in turn, governs the fundamental upper limit of mechanical strength. By achieving full density, defects (such as residual porosity) related to the processing of HA powders into dense implants via powder compaction and sintering have been removed or minimized in size (i.e., below the wavelength of light). Furthermore, nanostructured processing of HA results in phase stability during the sintering process, removing the negative contribution of decomposition due to mechanical strength.

Given that the Young's modulus (E) of nano-HA has been measured to be approximately 150 GPa (Van Vliet et al. unpublished results), the ideal strength of HA is between 18.75 GPa (12.5% of E) to 10 GPa (6.67% of E), based on the maximum interatomic force per bond prior to bond rupture. Relating the ideal strength to the fracture toughness of nano-HA in the above equation can establish the dependence of fracture strength to defect size. The

a   b

Fig. 4 a Fully dense, transparent nanocrystalline HA implants. **b** SEM of a nanocrystalline, phase-pure HA implant. Nominal HA crystal size of this implant is less than 100 nm. *Scale bar* represents 100 nm

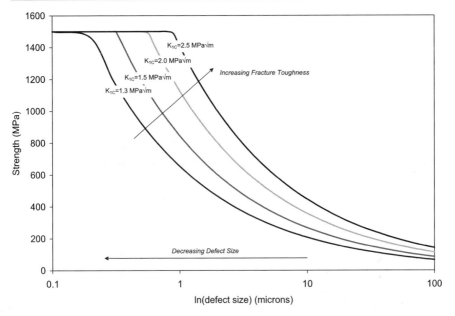

Fig. 5 Effect of defect size and fracture toughness on the strength of nano-HA

fracture toughness of nano-HA has been measured to be 1.3 MPa$\sqrt{m}$ (Ahn et al. unpublished results); fracture strength of nano-HA as a function of defect size is illustrated in Fig. 5. For a nano-HA tissue engineered bone construct, the ideal strength can be achieved if the largest defect is reduced to the smallest fundamental microstructural unit (i.e., the grain size) of 100 nm. Consequently, this material can satisfy the load requirements of most orthopedic applications if processed appropriately.

However, the strength of HA, nano- or otherwise, has yet to approach this ideal strength. Jarcho et al. reported an average three point bending strength of 196 MPa [57], which is among the highest strengths observed for HA. Assuming a fracture toughness of 1.0 MPa$\sqrt{m}$, Jarcho's HA is likely to have possessed defects between 5 and 10 μm. Equibiaxial strengths (ASTM C 1499) of 240 MPa have been observed for nano-HA by the authors of this chapter; this equibiaxial strength corresponds to a pure bending strength of 300 MPa for a Weibull modulus of m= 5, which is the highest observed bending strength for HA to date. With a measured fracture toughness of 1.3 MPa$\sqrt{m}$, nano-HA is likely to possess defects from 4 to 8 μm in size. While both Jarcho's and the authors' HA were fully dense, phase-pure HA and the authors' nano-HA possessed a higher strength due to the presence of smaller defects and higher fracture toughness. However, the strength of both systems is sensitive to defects, primarily introduced during the processing of the materials. Consequently, improved processing to remove defects will be the key to substantially increasing strength; if defects are limited to $1\,\mu m$, strength can increase beyond 600 MPa!

2.2.2
The Effect of Fracture Toughness

Nanocomposite engineering can be used to manipulate the microstructure of HA to increase its fracture toughness and overall strength. For example, as shown again in Fig. 5, the strength can be increased if the fracture toughness is increased. If the fracture toughness of nano-HA is increased to $2.0\,MPa\sqrt{m}$, the strength can be increased to over 600 MPa. Consequently, a higher fracture toughness can increase the strength for a given defect size. Given this, HA nanocomposites with increased fracture toughness can be used for the most mechanically demanding clinical indications. By incorporating structural reinforcing agents into the HA matrix, the matrix can be toughened through crack deflection, crack bridging, or phase transformation mechanisms. In all the cases, either the energy required to propagate a crack through the matrix is increased or the energy of a propagating crack is more effectively dissipated, preventing crack propagation and thereby increasing fracture toughness.

To increase toughness by crack deflection toughening, the structural reinforcing agent is typically anisotropic (i.e., platelets, whiskers, or fibers) and is usually harder, tougher, and stronger than HA. In this mechanism, crack energy is dissipated when a crack is forced to propagate around the agent. Common reinforcing agents of this type are alumina (K_{1c} from 1.4 to $2.5\,MPa\sqrt{m}$) [58–60], carbon [61], iron chromium alloys (K_{1c} from 6.0 to $7.4\,MPa\sqrt{m}$) [62], silicon carbide (K_{1c} from $2.11\,MPa\sqrt{m}$) [63–65], or silicon nitride (K_{1c} from 2.5 to $3.2\,MPa\sqrt{m}$) [66–69]. While these reinforcing agents were effectively able to increase the fracture toughness, the reported flexural strengths were still below that of the strength observed for nano-HA. In many cases a substantial volume of the secondary phase (from 5 to 60 vol %) is required to increase the fracture toughness. Consequently, these HA composites have been more difficult to process, resulting in large amounts of defects.

To increase toughness by crack bridging toughening, the structural reinforcing agent is also typically anisotropic but is more plastic than HA. Crack energy is dissipated when the crack energy is absorbed by the plastic deformation of reinforcing agents bridging the crack path. Common crack bridging reinforcing agents are relatively ductile metals such as gold, silver, or platinum. Zhang et al. reported an increased fracture toughness of $2.45\,MPa\sqrt{m}$ at a 30 vol % silver [70]; however, 8 vol % residual porosity remained, limiting strength to a 80 MPa four point bending strength. Similar to crack deflection toughened HA matrices, difficulty in processing has limited the strength despite the higher fracture toughness. Alternatively, nanocomposite engineering can achieve a three point bending strength of 210 MPa at 5 vol % of silver [71]. These preliminary results have been achieved by highly dispersing silver within the HA matrix. The size and vol% of the crack-bridging dispersoids should be tailored to further optimize the mechanical properties of systems of this type.

To increase toughness by a phase transformation mechanism, the structural reinforcing agent must undergo a phase transformation when interacting with a crack that results in a specific volume expansion. This specific volume expansion results in a compressive force at the crack tip, closing the tip and preventing further crack propagation. The most common reinforcing agent of this type is a partially stabilized tetragonal zirconia and has resulted in a fracture toughness from 1.73 to 2.3 MPa$\sqrt{m}$ [72–75]. Similar to the previously described mechanisms, a high vol% of zirconia reinforcing agents (20 to 50 vol %) were required to achieve these fracture toughness improvements. Consequently, microstructures with extensive defects were formed and bending strengths did not exceed 300 MPa despite fracture toughness improvements. Alternatively, a HA–zirconia nanocomposite achieved a three point bending strength of 243 MPa and a fracture toughness of 2.0 MPa$\sqrt{m}$ at 1.5 vol % zirconia (Ahn et al. unpublished results)!

Although composite engineering can increase the strength and toughness of ceramics, the processing of conventional HA composites have been challenging. These composites often suffer from microstructural heterogeneity and poor phase dispersion, requiring a substantial volume of the reinforcing phase to increase the fracture toughness and to strengthen the HA matrix. Consequently, the presence of these poorly dispersed secondary phases introduces large flaws and interfacial stability problems and decreases the bioactivity of these systems. HA nanocomposites, alternatively, can result in similar levels of mechanical reinforcement at substantially reduced volume loadings, thereby persevering HA's bioactivity.

2.3
Bioactivity

As the previous sections have discussed, nanophase materials (particularly nanophase HA) hold much promise for optimizing the mechanical properties necessary for orthopedic applications. Equally as intriguing towards using nanophase materials in orthopedic applications, is the overwhelming evidence of optimal cell interactions leading to increased bone tissue regeneration on nanophase materials compared to conventional ones. This section will emphasize such promising findings.

2.3.1
Effect of Nanostructured Surfaces on Protein Interactions

Since nanophase materials are composed of grains or particulates of the same atoms but fewer (less than tens of thousands) and smaller (less than 100 nm in diameter) than conventional forms (which contain several billions of atoms and have grain or particulate sizes of microns to millimeters in diameter), it is clear that their surface properties such as topography, energy, wettability, etc. are unique [17]. To demonstrate such differences in topography, representative atomic force micrographs of nanophase and conventional titania are presented in Fig. 6. Compared to conventional materials, examination of Fig. 6 illustrates much higher surface roughness at the nanoscale, higher surface areas, greater numbers of atoms at the surface (Fig. 7), and increased portions of material defects at the surface (such as edge/corner site and particulate/grain boundaries; Fig. 8) for nanophase materials.

In this respect, nanophase materials, by their very nature, have special surface properties for interactions with charged and uncharged species. To date, the increased surface reactivity of nanomaterials has been utilized for catalytic applications almost exclusively [21–23]. For example, studies have shown that compared to conventional grain size magnesium oxide (MgO), nanophase (i.e., 4 nm average grain size) MgO possessed increased numbers

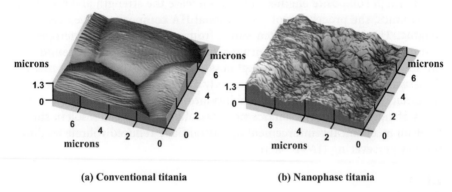

(a) Conventional titania   (b) Nanophase titania

Fig. 6 Representative atomic force micrographs of **a** conventional and **b** nanophase materials (specifically, titania is depicted here). Clearly, unique biologically inspired surface properties exist for nanophase materials [19]

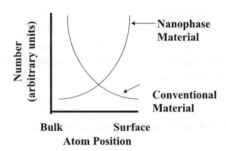

Fig. 7 Greater numbers of atoms at the surface of nanophase compared to conventional materials provide for unique surface properties

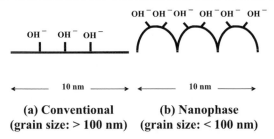

(a) Conventional
(grain size: > 100 nm)

(b) Nanophase
(grain size: < 100 nm)

Fig. 8 Increased material defects at the surface of nanophase compared to conventional materials provide for unique surface properties. Compared to **a** conventional materials, **b** nanophase materials possess higher surface areas, increased numbers of surface defects, and altered electron delocalization. For example, less acidic OH⁻ groups (due to an increase in electron delocalization) would exist for nanomaterials with hydroxide layers. Redrawn and adapted from [23]

of atoms at the surface (Fig. 7), higher surface areas ($100-160$ m^2/g compared to $200-500$ m^2/g, respectively), and less acidic OH⁻ groups (due to a much higher proportion of edge sites for nanophase MgO to cause delocalization of electrons; Fig. 8), which increased adsorption of acidic species and increased destructive adsorption of organophosphorous and of chlorocarbons [21–23]. However, for control of cell function leading to increased tissue regeneration, it is the immediate interaction of other charged species (i.e., proteins) with material surfaces that are important. Since such promising results have been found when utilizing nanophase materials in catalytic applications, it is intriguing to ponder what promise these materials may have in applications involving proteins that mediate bone cell function leading to bone tissue regeneration.

2.3.1.1
Protein Adsorption on Nanomaterial Surfaces

A great deal of attention must center on protein structure in aqueous media since soluble proteins present in biological fluids (for example, plasma or serum) are the type of proteins that are involved in immediate adsorption to surfaces [76]. In contrast, insoluble proteins that comprise tissues (like collagen and elastin) are not normally free to diffuse to a solid surface; these proteins may, however, appear on solid surfaces of implantable devices due to synthesis and deposition by cells [77]. In as short a time as can be measured (i.e., less than 1 s), soluble proteins from fluids adsorb to surfaces [25]. This is true for implants (such as orthopedic, vascular, etc.), bioseparation devices, immunoassays, catheters, marine fouling surfaces, biosensors, or for any device involved in protein contact from a liquid to a solid surface. In seconds to minutes, a monolayer of adsorbed protein will form on solid surfaces [25]. Typical values for protein adsorption on solid surfaces are in the range of 1 µg/cm^2 and exhibit a plateau with respect to initial protein concentration [76]. That is, protein adsorption will reach a maximum and will not be influenced by higher bulk protein concentrations. Moreover, the surface concentration of proteins adsorbed on a material surface is often 1000 times more concentrated than in the bulk phase [25].

It is apparent that since there is a limited amount of space on a material surface, extreme competition exists for protein adsorption to solid surfaces. Depending on the two major driving forces for adsorption (specifically, the relative bulk concentration of each protein and the properties of the surface that control reactivity), the outcome of this competitive adsorption process is that surfaces will be rich in some proteins contained in the supernatant solution while lacking in others. Because proteins are composed of amino acids that have vastly different properties (specifically, chemical composition and subsequent charge, hydrophilicity/hydrophobicity, etc.) and thus reaction to surfaces [76], adsorption will clearly be different depending on material properties such as chemistry, wettability, roughness, charge, etc.

It is also imperative to note that with some exceptions, protein adsorption to material surfaces is irreversible and, thus, leads to "immobilization" of specific proteins since they are for the most part not free to diffuse away [25]. Harsh treatments like using detergents (such as sodium dodecyl sulfide or SDS) are usually required to remove adsorbed proteins from a material surface [20]. Clearly, due to this efficiency, proteins have an inherent tendency to adsorb on surfaces as a tightly bound adsorbate, and removing unwanted protein adsorption (or fouling) from surfaces has become an active area of research [78, 79]. More importantly, this type of immediate, almost irreversible, interaction of proteins with material surfaces suggests the importance of altering surface properties to control such events.

2.3.1.2
Protein Orientation on Nanomaterial Surfaces

As previously mentioned, protein adsorption is only one manner in which proteins may interact with solid surfaces. The orientation (or bioactivity) of proteins in the adsorbed monolayer must also be considered since this interaction leads to extreme consequences for the ultimate function of a device [20, 25]. For example, for implants, biosensors, or immunoassay applications, cells, antibodies, or other agents that drastically influence device function will bond to specific exposed amino acids in adsorbed proteins. An example of the importance of protein orientation for the adhesion of cells is illustrated in Fig. 9 [26]. Protein orientation will alter from surface to surface since proteins are not uniform in properties or structure on the exterior. The existence of regions that are largely acidic/basic, hydrophobic/hydrophilic, or with select amino acids exposed to the supernatant liquid solution will greatly influence how that protein will adsorb and, thus, its orientation. As previ-

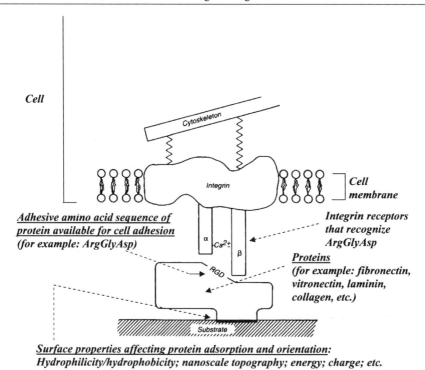

Cell

Adhesive amino acid sequence of
protein available for cell adhesion
(for example: ArgGlyAsp)

Cell
membrane

Integrin receptors
that recognize
ArgGlyAsp

Proteins
(for example: fibronectin,
vitronectin, laminin,
collagen, etc.)

Surface properties affecting protein adsorption and orientation:
Hydrophilicity/hydrophobicity; nanoscale topography; energy; charge; etc.

Fig. 9 Interaction of cells with proteins adsorbed on materials such as an implant. Illustrated here is the importance of the orientation of the adsorbed protein in exposing the adhesive peptide sequence (RGD) to a cell to promote adhesion. Altering material surface properties will influence the exposure of amino acids in adsorbed proteins to cell integrin receptors. Redrawn and adapted from [26]

ously mentioned, proteins are not normally free to rotate once adsorbed due to multiple bonding mechanisms. Thus, immediately upon adsorption, proteins are usually fixed in a preferred orientation or bioactivity to the bulk media [25]. Again, this points to the importance of designing a material surface to manipulate protein orientation important for specific device function.

As an aside, it is important to note that under certain extreme conditions (for example, conditions that are outside of the physiological range or outside the range of 0 to 45 °C, pH 5 to 8, and in aqueous solutions of about 0.15 M ionic strength), proteins may lose their normal structure [80]. In other words, under such conditions, the spherical or globular tertiary structure most soluble proteins assume in aqueous media will unfold or denature. The structure of denatured proteins has been described as a random coil similar to those found in synthetic polymers [80]. Since the structure of the protein has changed from that of a hydrophilic/hydrophobic exterior/interior to a more random arrangement, denatured proteins often lose their solubility, become less dense (folded protein structures have densities of approximately $1.4\,g/cm^3$), and lose their bioactivity [25]. Although there have been many examples of protein denaturation under non-physiological conditions in solution, in general, there have been few reports of full protein denaturation on material surfaces [25]. That is, generally, proteins adsorbed at the solid-liquid interface are not fully denatured and retain some degree of structure and bioactivity.

2.3.1.3
Nanosurface Properties that Influence Protein Interactions

Not only do the properties of proteins determine the degree of protein interactions with surfaces, but the properties of the solution in which they reside and of the surface (specifically, wettability, surface energy, chemistry, roughness, etc.) also influence the degree of protein interactions [25]. This is true because proteins are relatively large in size and have correspondingly large numbers of charged amino acid residues of varying acidity/basicity well-distributed on the exterior of the protein. Therefore, the inherent polyelectrolytic behavior of proteins becomes increasingly important, depending on alterations in pH and ionic strength of the protein-containing solution. The polyelectrolytic property of proteins provides for exciting design criteria in surfaces to maximize or minimize protein interactions. Not surprisingly, at a neutral or slightly charged surface and at a pH in which the net charge on the protein is minimal (that is, near the isoelectric pH), most proteins will exhibit maximum adsorption [25]. For surfaces with a large net charge, initial protein interactions will be dominated by the degree of the opposite charge on the surface [25]. It is important to note that these are general trends, and exceptions do exist. However, the key consideration in the design of successful devices is to manipulate surface properties to interact with select proteins important for device function.

Simple consideration of the spatial organization of amino acids can also be used in the design of surfaces to enhance protein interactions [20]. As previously discussed, hydrophilic and hydrophobic amino acids are present primarily on the exterior and interior of soluble proteins, respectively. This spatial arrangement has a direct consequence on the initial interaction of proteins with surfaces. For example, for the most part, a surface that initiates interactions with the exterior hydrophilic amino acid residues can promote adsorption. In contrast, for the interior hydrophobic amino acid residues to interact with a material surface, the soluble protein would have to unfold or lose some degree of tertiary structure. Generally, it is the amino acids in the interior of globular proteins that mediate antibody, cellular, or other agent attachment. It is examination of these specific amino acid sequences in proteins that gives promise for the design of materials with specific properties at the nanometer level. In this manner, extreme excitement exists for nanostructured materials because it is only these materials that can be manipulated at the same dimensions as amino acids to control specific protein interactions for device function.

2.3.1.4
Experimental Evidence

Indeed, reports in the literature have demonstrated that proteins from bodily fluids interact differently with nanostructured than with conventional surfaces. Of first importance, cumulative adsorption of proteins contained in serum was significantly higher on smaller, nanometer grain size materials [20, 81]. In particular, the interaction of two proteins (fibronectin and vitronectin – both proteins known to enhance osteoblast or bone-forming cell function [81]) is consistently altered on a wide range of nanophase compared to conventional materials [20]. For example, by just decreasing grain size to below 100 nm, select competitive vitronectin adsorption increased 10% on alumina formulations. Since these proteins have calcium binding sites, studies have also demonstrated that initial enhanced calcium adsorption on nanophase materials was a key mechanism for the observed increased protein interactions [20, 81].

It has also been shown that calcium-mediated protein adsorption affected the orientation of the proteins adsorbed on nanophase materials [20]. Specifically, a novel adaptation of the standard surface-enhanced Raman scattering (SERS) technique provided evidence of increased unfolding of fibronectin and vitronectin on nanophase compared to conventional grain size ceramics [20]. Unfolding of these proteins promoted availability of specific cell-adhesive epitopes (such as the amino acid sequence Arg-Gly-Asp or RGD and those that recognize Lys-Arg-Ser-Arg or KRSR) that undoubtedly increased bone cell adhesion and function (again this interaction is depicted in Fig. 9). Evidence supporting this claim was further provided by these investigators through competitive cell adhesion inhibition studies in which cell integrin membrane receptors were blocked with RGD and KRSR and less osteoblast adhesion on protein-coated nanostructured than on conventional surfaces was observed [20].

These studies were the first to demonstrate that by decreasing grain or particle size (or, in effect, decreasing the size of surface features) to below 100 nm, protein interactions were altered in ways previously unobserved on conventional formulations. To date, studies have reported enhanced interactions of vitronectin and fibronectin on nanophase compared to conventional spherical particle size ceramics (specifically, alumina (Fig. 6), titania, and HA [18–20]), fibrous particle size alumina (Fig. 10) [82], titanium and Ti6Al4V (Fig. 11) [83], poly(lactic-co-glycolic acid) or PLGA (Fig. 12) [84], carbon fibers (Fig. 13) [85], and PLGA–titania composites (Fig. 14) [86]. In this manner, when comparing all of these studies, interactions (specifically,

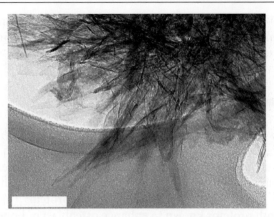

Fig. 10 Representative TEM of individual alumina nanodimensional fibers. Interactions of proteins (specifically, fibronectin and vitronectin) and subsequent osteoblast function are enhanced on nanophase compared to conventional alumina. Further enhancement of osteoblast function can be obtained through the use of nanofiber (as opposed to nanospherical) materials. *Scale bar* represents 100 nm [82]

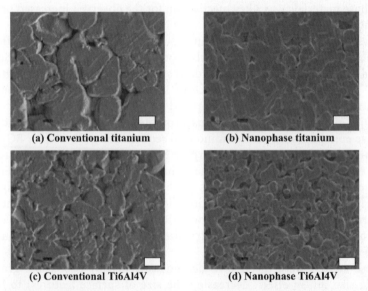

Fig. 11 SEM images depicting surface roughness of **a** conventional and **b** nanostructured Ti as well as **c** conventional and **d** nanostructured Ti6Al4V. Interactions of proteins (specifically, fibronectin and vitronectin) and subsequent osteoblast functions are enhanced on nanophase compared to conventional Ti and Ti6Al4V. *Scale bar* represents 10 μm [83]

adsorption and bioactivity) of fibronectin and vitronectin seem to be more uniformly promoted on nanophase than on conventional ceramics, metals, polymers and composites thereof [20, 81–86].

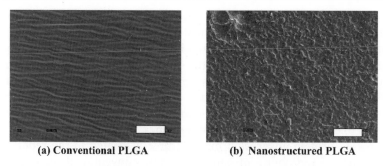

(a) Conventional PLGA   (b) Nanostructured PLGA

Fig. 12 SEM images depicting surface roughness of **a** conventional and **b** nanostructured PLGA scaffolds. Interactions of proteins (specifically, fibronectin and vitronectin) and subsequent osteoblast functions are enhanced on nanophase compared to conventional PLGA. *Scale bar* represents 100 μm [84]

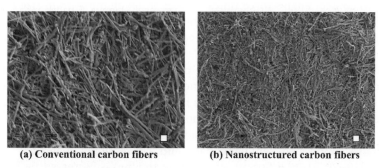

(a) Conventional carbon fibers   (b) Nanostructured carbon fibers

Fig. 13 SEM images depicting **a** conventional carbon fibers (0.125 μm diameter) and **b** nanophase carbon fibers (60 nm diameter). Interactions of proteins (specifically, fibronectin and vitronectin) and subsequent osteoblast functions are enhanced on nanophase compared to conventional carbon fibers. *Scale bar* represents 1 μm [85, 88]

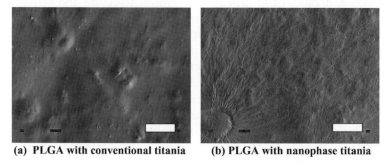

(a) PLGA with conventional titania   (b) PLGA with nanophase titania

Fig. 14 SEM images depicting PLGA composites containing **a** conventional titania (4.120 μm grain size) and **b** nanophase titania (32 nm grain size); 70/30 wt % PLGA/titania. Interactions of proteins (specifically, fibronectin and vitronectin) and subsequent osteoblast functions are enhanced on PLGA containing nanophase compared to conventional titania. *Scale bar* represents 100 μm [86]

It is interesting to consider what properties of nanophase materials manipulate protein adsorption and orientation to benefit orthopedic implant applications. Select adsorption of proteins important for bone cell function was enhanced by simply decreasing the grain or particle size of the materials into the nanometer regime. Equally as important, the tertiary structure of the proteins (namely the positioning of the RGD amino acid sequence in the interior of some of these proteins) was manipulated as the protein was partially unfolded to expose these cell-adhesive epitopes.

Increased protein adsorption may be directly related to the reported increased surface wettability of nanophase over conventional materials [18–20, 81–86]. It was previously suspected that nanophase materials have much greater surface reactivity due to increased numbers of atoms (Fig. 7) and defect boundaries at the surface (such as edge/corner sites, grain boundaries, etc.; Fig. 8). This was confirmed by studies that provided evidence of aqueous contact angles three times smaller when alumina grain size was reduced from 167 to 24 nm [19]. As previously discussed, since proteins assume a tertiary structure with mostly hydrophilic amino acid residues on the exterior of globular shapes, a material surface with increased wettability (or hydrophilic) properties may increase protein adsorption and may even enhance protein unfolding. Since fibronectin and vitronectin have numerous exterior hydrophilic amino acids, this is one hypothesis for why interactions of these proteins are enhanced on nanostructured surfaces.

Large topographic differences resulting from grain or particulate sizes were also found between nanophase and conventional materials, as expected [18–20, 81–86]. Specifically, surface roughness increased by 35–50% on nanophase compared to conventional grain size alumina, titania, and HA [18–20]. These investigators hypothesized that for the first time, due to the ability to create surface topographies with features that approximate the size of proteins, extreme control over protein interactions can result. For example, because of protein stereochemical structure and ceramic pore size dimensions (in Angstroms for nanophase compared to microns for conventional alumina, titania, and HA), many of these nanostructured proteins preferentially adsorbed to the smaller pores sizes in nanophase ceramics [18–20, 81]. In contrast, some of the larger proteins contained in blood plasma (such as albumin) were preferentially sterically excluded from the Angstrom-sized pores of nanostructured ceramics [20, 81]. In addition, variations in ceramic surface topography, on the same order of magnitude as the size of proteins, influenced adsorbed protein orientation and, thus, availability of select amino acid sequences for promoting bone-forming cell function [20, 81]. Due to protein dimensions in the nanometer regime, through the use of nanostructured topographies, scientists can now modify a surface to control and manipulate adsorbed protein conformation; this has, most likely, the largest unexplored and most promising potential for controlling select protein interactions with nanophase materials.

2.3.2
The Effect of Nanostructured Surfaces on Cellular Interactions

Initial protein interactions with implantable materials are important. However, without subsequent cellular adhesion leading to appropriate extracellular matrix development, a successful tissue-engineering scaffold will never be found. As previously mentioned, the type of proteins that initially adsorb, and the bioactivity of the proteins once adsorbed, will determine cellular adhesion. For example, osteoblasts (as well as other cells) adhere to select amino acid sequences (such as RGD) in adsorbed proteins as illustrated in Fig. 9. Whether this and other amino acid sequences are exposed to cells after proteins adsorb will be influential in whether cells adhere or not. Thus, it can be argued that protein interactions are the imperative step to controlling subsequent cell function and eventual implant success/failure. For these reasons, modification of surface properties such as topography, chemistry, wettability, charge, etc. will influence protein interactions important for cell adhesion.

As discussed, many surface properties are altered for nanostructured as compared to conventional substrate formulations. Specifically, as the surface grain or particulate size decreases, the number of grain boundaries (or other material defects) at the surface increases, the topography changes, surface area increases, the number of atoms at the surface compared to bulk increases, etc. Such material properties will influence surface energetics important for mediating initial protein interactions essential for cell adhesion and, subsequently, cell function. In addition, since the dimensions of proteins that mediate cell adhesion are at the nanometer level, a surface with a nanometer topography can influence the availability of amino acids for cell adhesion to a greater extent than conventional surfaces which are smooth at the nanoscale. Evidence has already been provided and discussed demonstrating altered interactions of such proteins with nanostructured as compared to conventionally structured materials. It is with these events in mind, that the following experimental evidence of increased bone cell function on nanophase compared to conventional materials is discussed.

2.3.2.1
Increased Osteoblast Functions on Nanophase Materials

The first reports correlating increased bone cell function with decreased material grain or particulate size into the nanometer regime date back to 1998 [87]. Some of the best examples of how nanophase, compared to conventional, materials alter interactions with proteins are outlined in investigations of the potential use of nanostructured materials as the next generation of bone prosthetics [18–20, 81–83, 85–89]. Reports in the literature have determined that in vitro osteoblast adhesion, proliferation, differentiation (as measured by intracellular and extracellular matrix protein synthesis such as alkaline phosphatase), and calcium deposition is enhanced on materials with particulate or grain sizes less than 100 nm [18–20]. Specifically, this has been demonstrated for a wide range of materials and chemistries. Ceramics investigated to date include titania (Fig. 6), alumina (Fig. 10), and HA [18–20]. Metals investigated to date include titanium (Fig. 11), Ti6Al4V (Fig. 11), and CoCrMo [83]. Polymers investigated to date include poly(lactic-co-glycolic acid) (PLGA; Fig. 12), polyurethane, and polycaprolactone [84, 86]. Osteoblast function has also been increased on nanophase compared to conventionally dimensioned carbon fibers (Fig. 13) [85]. Lastly, composites investigated to date include PLGA/nanophase alumina, titania (Fig. 14), and HA (70/30 wt % PLGA/ceramic) [86].

For example, four, three, and two times the amount of calcium-mineral deposition was observed when osteoblasts were cultured on nanophase rather than on conventional alumina, titania, and HA, respectively [89]. Up to three times more osteoblasts adhered to PLGA when it contained nanophase rather than conventional titania particles [86]. Moreover, Elias et al. showed significantly greater osteoblast function leading to new bone synthesis on bulk compacts containing carbon fibers with nanometer compared to conventional dimensions [88]. Such novel properties have also been seen when carbon nanofibers were incorporated into polymer composites. Specifically, increased osteoblast adhesion was observed in polyurethane with increasing weight percentages of nanometer rather than conventionally dimensioned carbon fibers [85]. It is important to note that for each respective nanophase and conventional material example of these trends, similar chemistry and material phases were studied. That is to say, only nanometer surface topography was altered, no other material properties changed between the nanophase and conventional material formulations.

Equally as interesting, a step-function increase in osteoblast function has been reported at distinct ceramic (specifically, alumina and titania) grain sizes; specifically at ceramic spherical grain sizes below 60 nm [19]. This is intriguing since when creating alumina or titania ceramics with average grain sizes below 60 nm, a drastic increase in osteoblast function is observed when compared to ceramics with grain sizes just 10 nm higher (i.e., those with average grain sizes of 70 nm). Although an explanation as to why this occurs is not known to date, it is believed that the importance of this specific grain size in improving osteoblast function is connected with interactions of vitronectin (a protein known to increased osteoblast adhesion with linear protein dimensions remarkably similar to 60 nm) [90]. This critical grain size for improving osteoblast function is also interesting since, as previously mentioned, numerous special properties (such as mechanical, electrical, catalytic, etc.) of nanophase materials exist when grain or particulate size is reduced specifically to below 100 nm [15–24]. With this information, evidence has been provided to show for the first time that the ability of nanophase materials to increase bone cell function is indeed limited to grain sizes below 100 nm, specifically those below 60 nm [19]. Thus, another novel property of nanos-tructured materials has been elucidated.

2.3.2.2
Increased Osteoclast Functions on Nanophase Materials

In addition to studies highlighting enhanced osteoblast function on nano-phase materials, increased functions of osteoclasts (bone-resorbing cells) have been reported on nanospherical compared to larger grain size ce-ramics [91]. Specifically, osteoclast synthesis of tartrate-resistant acid phos-phatase (TRAP) and subsequent formation of resorption pits was up to two times greater on nanophase compared to conventional ceramics such as HA. Coordinated functions of osteoblasts and osteoclasts are imperative for the formation and maintenance of healthy new bone juxtaposed to an orthope-dic implant (Fig. 15). Frequently newly formed bone juxtaposed to implants is not remodeled by osteoclasts and thus becomes unhealthy or necrotic [2]. At this time, the exact mechanism of increased functions of osteoclasts on nanophase ceramics is not known, but it may be tied to the well-documented

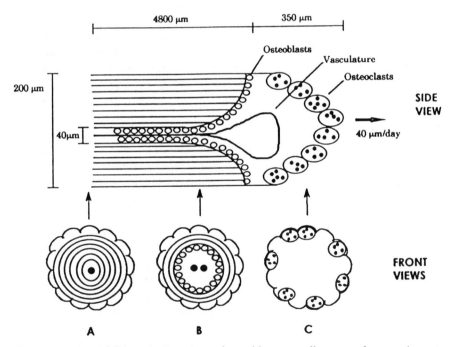

Fig. 15 Bone remodeling unit. Functions of osteoblasts as well as osteoclasts are import-ant to form and maintain healthy juxtaposed bone tissue. Functions of osteoblasts and osteoclasts are enhanced on nanophase compared to conventional ceramics. Adapted and redrawn from [99]

increased solubility properties of nanophase compared to conventional materials [17, 22, 23]. In other words, due to increased numbers of grain boundaries at the surface of smaller grain size materials, increased diffusion of chemicals (such as TRAP) may occur, subsequently resulting in the formation of more resorption pits.

2.3.2.3
Decreased Competitive Cell Functions on Nanophase Materials

Importantly, it has also been shown that competitive cells do not respond in the same way to nanophase ceramics as do osteoblasts and osteoclasts [20]. In fact, decreased functions of fibroblasts (cells that contribute to fibrous encapsulation and callus formation events that may lead to implant loosening and failure [2]) and of endothelial cells (cells that line the vasculature of the body) have been observed on nanophase compared to conventional ceramics, metals, and polymers [20, 85, 86, 92]. Previously, such selectively in bone cell function on materials has only been observed through delicate surface chemistry (such as through immobilization of peptide sequences KRSR) [93]. It has been argued that immobilized delicate surface chemistries may be compromised once implanted due to macromolecular interactions that render such epitopes non-functional in vivo. For these reason, it is important to note that studies demonstrating select enhanced osteoblast and osteoclast functions on nanophase compared to conventional materials have been conducted on surfaces that have not been chemically modified (by immobilization of proteins, amino acids, peptides, etc.). It is the unmodified, raw material surface that results in unique altered cell responses.

2.3.2.4
Increased Osteoblast Functions on Nanophase Materials
that Simulate the Aspect Ratio of Constituent Components of Bone

Recently, researchers have further modified nanophase ceramics to simulate not only the nanometer dimension but also the aspect ratio of proteins and HA found in the extracellular matrix of bone (Figs. 10 and 13) [82, 85, 88]. For example, consolidated substrates formulated from nanofibrous alumina (diameter 2 nm, length > 50 nm; Fig. 10) increased osteoblast functions in comparison with similar alumina substrates formulated from the aforementioned nanospherical particles [82]. Specifically, Price et al. determined a twofold increased osteoblast cell adhesion density on nanofiber versus conventional nanospherical alumina substrates, following only a 2-h culture [82]. Thus, perhaps not only is the nanometer grain size of components of bone important to mimic in materials, but the aspect ratio may also be key to simulate in synthetic materials in order to optimize bone cell response. Similar results of the importance of nanofibrous materials have been reported for carbon (Fig. 13) [85, 88]. These findings consistently testify to the unprecedented and excellent ability to create nanomaterials to mimic constituent components of physiological bone to promote new bone formation.

3
Unassesed Risks of Using Nanophase Particles as Implantable Materials

Since the research on and the use of nanophase materials is still in its infancy, risks to human health and environment must not be overlooked. Many issues relating to safe and healthy fabrication of nanophase materials still need to be addressed. For example, the health effects of small nanoparticles entering the human body through pores and accumulating in the cells of the respiratory or other organ systems (for example, when becoming dislodged through wear debris) are still largely unknown. This would happen during commercial scale processing of the nanoparticles as well as through the use of these materials as implants.

Although some studies have begun to examine the influence of nanoparticulate wear debris on cell (particularly, osteoblast) viability, there is a need for continuous, more comprehensive monitoring of the potential effects of newly designed and fabricated nanomaterials under in vivo situations. For example, even though nanophase materials have increased wear fatigue properties [16], debris may form from articulating components of orthopedic implants composed of nanophase ceramics when subjected to physiological loading properties. Conventionally sized wear debris particles induce bone loss, which leads to implant loosening and sometimes results in clinical failure of bone prostheses [2]. In recent reports, however, results of a more well-spread morphology, increased osteoblast viability, and increased osteoblast proliferation in the presence of nanophase compared to conventional wear debris demonstrated a less adverse influence of nanoparticulates on bone cell function than of larger conventional ceramic wear particles [94, 95]. For this reason, nanophase wear debris may not be as harmful as its conventional counterparts, although more testing is clearly necessary.

Clearly, much more testing (especially, in vivo) is needed before nanomaterials can be widely considered for improving orthopedic applications.

4
Conclusions

According to the US government's research agenda, the current and future broad interests in nanobiomedical activity can be categorized as given in Table 1 [96, 97]. Marsch [96] further grouped the entire activity in three broad related fronts:

1. Development of pharmaceuticals for inside-the-body applications such as drugs for anticancer and gene therapy
2. Development of diagnostic sensors and lab-on-a-chip techniques for outside-the-body applications such as biosensors to identify bacteriological infections in biowarfare
3. Development of prostheses and implants for inside-the-body uses

Whereas the European governments emphasize commercial applications in all three fronts above (according to Marsch), the US government (as can be seen in Table 1) tends to gear towards fundamental research on biomedical implants and biodefense, leaving commercial applications to industry. Both classifications identify nanobiomedical tissue engineering applications (item 8 in Table 1) as potential interests. The biological and biomimetic nanostructures to be used as an implant involve some sort of an assembly in which smaller materials later on assume the shape of a body part, such as hipbone. These final biomimetic, bulk nanostructures can start with a pre-defined nanochemical (like an array of large reactive molecules attached to a surface) or nanophysical (like a small crystal) structure. It is believed that by using these fundamental nanostructured building blocks as seed molecules or crystals, a larger bulk material will self-assemble or keep growing by itself.

Clearly, the nanomaterials mentioned in this chapter are in their infancy and much more testing must be conducted before their full potential is realized. It is also important to note that the scientific developments reported above do not exhaust the current global beehive research efforts on the biological potentials of nanoparticulates as implants. It is believed, however, that following the trends of these impressive application properties of nanomaterials in the biomedical domain, there exists a bright future for therapies and treatments though prosthetic implantation.

In summary, nanostructured materials provide alternatives not yet fully explored for controlling interactions with proteins and cells. This is in add-

Table 1 US government current and future broad interests in nanobiomedical research activity [96, 97]

US government research interests for nanobiomedical research

1	Synthesis and use of nanostructures
2	Applications of nanotechnology in therapy
3	Biomimetic nanostructures
4	Biological nanostructures
5	Electronic–biological interface
6	Devices for early detection of diseases
7	Instruments for studying individual molecules
8	Nanotechnology for tissue engineering

ition to their promising bulk mechanical properties. Since technologies now exist to manipulate materials at the atomic, molecular, and supramolecular level, bulk materials and surfaces can be designed at a similar dimension to that of constituent components of bone: the nanometer level. Moreover, in conjunction with this promise, the inherent increased surface area, higher proportions of edge sites, and greater number of material defects (such as grain boundaries for ceramics) at the surface, give nanophase materials special surface properties for regenerating bone. When these concepts are fully appreciated and realized, it will be possible to design highly engineered surfaces for improving implantable devices as well as other applications (including, but not limited to, filtration systems, bioseparations, bioMEMS, etc.). Such advances were previously unimaginable with conventional materials. As the disciplines of protein/cell biology and nanophase material science continue to develop and mature, the design criteria mentioned in this chapter for simulating the mechanical properties of bone and simultaneously controlling bone cell interactions will be expanded and refined. Undoubtedly, nanophase materials have the potential to become the next generation of choice proactive materials for innovative biotechnology and biomedical applications and could have profound impact in many diverse fields including bone tissue engineering.

References

1. http://www.aaos.org/wordhtml/press/arthropl.htm
2. Kaplan FS, Hayes WC, Keaveny TM, Boskey A, Einhorn TA, Iannotti JP (1994) In: Simon SP (ed) Orthopedic basic science. American Academy of Orthopedic Surgeons, Columbus, Ohio, p 127
3. Baraton MI, Chen X, Gonsalves KE (1997) Nanostructured Mater 8:435
4. Bohn R, Haubold R, Birringer R, Gleiter H (1991) Scripta Metal Mater 25:811
5. Carry C, Mocellin A (1987) Ceramics Int 13:89
6. Catledge S, Vohra Y (1999) J App Phys 86:698
7. Ciftcioglu M, Mayo MJ (1990) Processing of nanocrystalline ceramics. In: Mayo MJ, Kobayashi M, Wadsworth J (eds) Proceedings symposium on superplasticity in metals, ceramics, and intermetallics. Materials Research Society, Pittsburgh, 1990, p 77
8. Cui Z, Hahn H (1992) Nanostructured Mater 1:419
9. Nieman GW, Weertman JR, Siegel RW (1991) Mechanical behavior of nanocrystalline Cu and Pd. In: Van Aken DC (ed) Microcomposites and nanophase materials. TMS, Warrendale, PA, p 15
10. Mayo M, Siegel RW, Liao YX, Nix WD (1992) J Mat Res 7:973
11. Mayo M, Siegel RW, Narayanasamy A, Nix WD (1990) J Mat Res 5:1073
12. Nieman GW, Weertman JR, Siegel RW (1989) Scripta Mettallurgica 23:2013
13. Nieman GW, Weertman JR, Siegel RW (1991) J Mat Res 6:1012
14. Roco MS, Williams RS, Alivisatos P (1999) Nano-technology research directions: IWGN Workshop Report

15. Siegel RW, Fougere GE (1994) Mechanical properties of nanophase materials. In: Hadjipanayis GC, Siegel RW (eds) Nanophase materials: synthesis-properties-applications. Kulwer, Dordrecht, p 233
16. Siegel RW, Fougere GE (1995) Nanostructured Mater 6:205
17. Siegel RW (1996) Sci Amer 275:42
18. Webster TJ, Siegel RW, Bizios R (1999) Nanostructured Mater 12:983
19. Webster TJ, Siegel RW, Bizios R (1999) Biomaterials 20:1221
20. Webster TJ, Schadler LS, Siegel RW, Bizios R (2001) Tissue Eng 7:291
21. Siegel RW, Hu E, Roco MC (1999) In: Nano-structure science and technology. Kluwer, Boston, MA
22. Baraton MI, Chen X, Gonsalves KE (1999) Nanostructured Mater 8:435
23. Klabunde KJ, Strak J, Koper O, Mohs C, Park D, Decker S, Jiang Y, Lagadic I, Zhang D (1996) J Phys Chem 100:12141
24. Wu SJ, DeJong LC, Rahaman MN (1996) J Am Ceram Soc 79:2207
25. Horbett TA (1996) Proteins: structure, properties and adsorption to surfaces. In: Ratner BD, Hoffman AS, Schoen AS, Lemmons JE (eds) Biomaterials science: an introduction to materials in medicine. Academic, New York, p 133
26. Schakenraad JM (1996) Cell: their surfaces and interactions with materials. In: Ratner BD, Hoffman AS, Schoen AS, Lemmons JE (eds) Biomaterials Science: an introduction to materials in medicine. Academic, New York, p 141
27. Martinelli KA, Lematire DT, Lee TJ, Jungling M (2003) Merrill Lynch, December 12
28. Gibbonsm W, Chazan A (1999) The orthopedic industry: continued profitable growth. William Blair, Boston, MA
29. Davidson SR, Wise FA (1998) Orthopedics: a group for all seasons. Bear Stearns, Boston, MA
30. Gautier E, Perren SM, Cordey J (2000) Injury 31:C14-20
31. Kumar A, Jasani V, Butt MS (2000) Injury 31:169
32. Oohashi Y (2001) Arthroscopy 17:1007
33. Lang GJ, Cohen BE, Bosse MJ, Kellam JF (1995) Clin Orthop 315:64
34. Hawkins LG (1970) J Bone Joint Surg Am 52:991
35. Freedman EL, Johnson EE (1995) Clin Orthop 315:25
36. Martinelli KA, Lematire DT, Lee TJ, Jungling M (2003) Orthopedic industry: cruise control, Merrill Lynch, 12 December
37. Patel JC, Watson K, Joseph E, Garcia J, Wollstein R (2003) J Hand Surg (Am) 28:784
38. Fox EJ, Hau MA, Gebhardtm MC, Hornicek FJ, Tomford WW, Mankin HJ (2002) Paper presented to American Academy of Orthopeadic Surgeons, Dallas, Texas, 12–17 February, 2002
39. Norman-Taylor FH, Santori N, Villar RN (1997) BMJ 315:498
40. Springfield DS (1997) Semin Surg Oncol 13:11
41. Kerry RM, Masri BA, Garbuz DS, Czitrom A, Duncan CP (1999) Instr Course Lect 48:645
42. Berrey BH Jr, Lord CF, Gebhardt MC, Mankin HJ (1990) J Bone Joint Surg Am 72:825
43. Dick HM, Strauch RJ (1994) Clin Orthop 306:46
44. Finkenberg J, Banta C, Cross GL, Dawson E, Gutzman D, Highland T, Kucharzyk D, Lenderman L, Murphy J, Neely W, Rogozinski A, Rogozinski C (2001) Spine J 1:102
45. Price CT, Connolly JF, Carantzas AC, Ilyas I (2003) Spine 28:793
46. Allen MJ, Hai Y, Ordway NR, Park CK, Bai B, Yuan HA (2002) Spine J 2:261
47. Barnes B, Rodts GE Jr, Haid RW Jr, Subach BR, McLaughlin MR (2002) Neurosurgery 51:1191

48. Lofgren H, Johannsson V, Olsson T, Ryd L, Levander B (2000) Spine 25:1908
49. van Limbeek J, Jacobs WC, Anderson PG, Pavlov PW (2000) Eur Spine J 9:129
50. Cohen DB, Chotivichit A, Fujita T, Wong TH, Huckell CB, Sieber AN, Kostuik JP, Lawson HC (2000) Clin Orthop 371:46
51. Ehrler DM, Vaccaro AR (2000) Clin Orthop 371:38
52. Finkelstein JA, Chapman JR, Mirza S (1999) J Spinal Discord 12:424
53. Martin GJ Jr, Haid RW Jr, MacMillan M, Rodts GE Jr, Berkman R (1999) Spine 24:852
54. Griffith AA (1920) Philos Trans R Soc London A221:163
55. Irwin GR (1957) J Appl Mech 24:361
56. Irwin GR (1964) Appl Mech Res 3:65
57. Jarcho M, Bolen CH, Thomas MB, Bobick J, Kay JF, Doremus RH (1976) J Mater Sci 11:2027
58. Noma T, Shoji N, Wada S, Suzuki T (1993) J Ceram Soc Japan 101:923
59. Gautier S, Champion E, Bernache-Assollant D (1995) In: Ravaglioli A (ed) Proceedings of fourth conference on euroceramics, vol 8 (Biomaterials), p 201
60. Li J, Fartash B, Hermansson L (1990) Interceram 39:20
61. Slosarczyk A, Klisch M, Blazewicz M, Piekarczyk J, Stobierski L, Rapacz-Kmita A (2000) J Eur Ceram Soc 20:1397
62. DeWith G, Corbijn AJ (1989) J Mater Sci, 24:3411
63. Noma T, Shoji N, Wada S, Suzuki T (1992) J Ceram Soc Jpn 100:1175
64. Tian J, Yi SZ, Shao Y, Shan H (1995) 97th Annual meeting of the ACerS, Cincinnati, OH, April 30–May 3, 1995
65. Tian JM, Zhang S, Shao Y, Shan H (1996) Ceramic Trans 63 (Bioceramics: Materials and Applications II)
66. Nonami T (1990) Multifunctional materials. In: Buckely AJ, Gallagher-Daggitt G, Karasz FE, Ulrich DR (eds) Mater Res Soc Symp Proc 175, Pittsburgh, PA p 71
67. Nonami T, Satoh J (1995) J Ceram Soc Japan 103:804
68. Ioku K, Noma T, Ishizawa N, Yoshimura M (1990) J Ceram Soc Japan Int Ed 98:1348
69. Fang Y, Roy DM, Cheng J, Roy R, Agrawal DK (1993) Ceram Trans 36:397
70. Zhang X, Gubbels GHM, Terpstra RA, Metselaar R (1997) J Mater Sci 32:235
71. Ahn ES (2001) PhD Thesis, Massachusetts Institute of Technology
72. Ioku K, Yoshimura M, Somiya S (1990) Biomater 11:57–61
73. Kim HK, Koh YH, Yoon BH, Kim HE (2002) J Am Ceram Soc 85:1634
74. Takagi M, Mochida M, Uchida N, Saito K, Uematsu K (1992) J Mater Sci Mater Med 3:199
75. Delgados JA, Morejon L, Martinez S, Ginebra MP, Carlsson N, Fernandez E, Planell JA, Clavaguera-Mora MT, Rodriguez-Viejo J (1999) J Mater Sci Mater Med 10:715
76. Horbett TA (1986) Techniques for protein adsorption studies. In: Williams DF (ed) Techniques of biocompatibility testing. CRC, Boca Raton, FL, p 183
77. Hench LL, Ethridge EC (1975) Adv Biomed Eng 5:35
78. Norde W, Lyklema J (1991) J Biomater Sci Polymer Ed 2:183
79. Yutani K, Ogasahara K, Tsujita T, Sugino Y (1987) Proc Natl Acad Sci USA 84:4441
80. Peters T (1985) Serum albumin. In: Anfinsen CB, Edsall JT, Richards FM (eds) Advances in protein chemistry, vol 37. Academic, New York, p 161
81. Webster TJ, Ergun C, Doremus RH, Siegel RW, Bizios R (2000) J Biomed Mater Res 51:475
82. Price RL, Ellison K, Haberstroh KM, Webster TJ (2004) J Biomed Mater Res 70A:129
83. Webster TJ, Ejiofor JU (2004) Biomaterials 25:4731
84. Thapa A, Webster TJ, Haberstroh KM (2003) Biomaterials 24:2915

85. Price RL, Waid MC, Haberstroh KM, Webster TJ (2003) Biomaterials 24:1877
86. Kay S, Thapa A, Haberstroh KM, Webster TJ (2002) Tissue Engineering 8:753
87. Webster TJ, Siegel RW, Bizios R (1998) In: LeGeros RZ, LeGeros JP (eds) Bioceramics 11: Proceedings of the 11th international symposium on ceramics in medicine. World Scientific, New York, p 273
88. Elias KL, Price RL, Webster TJ (2002) Biomaterials 23:3279
89. Webster TJ, Siegel RW, Bizios R (2000) Biomaterials 21:1803
90. Ayad A (1994) The extracellular matrix factsbook. Elsevier, NY
91. Webster TJ, Ergun C, Doremus RH, Siegel RW, Bizios R (2001) Biomaterials 22:1327
92. Vance RJ, Miller DC, Thapa A, Haberstroh K, Webster TJ (2004) Biomaterials 25:2095
93. Dee KC, Andersen TT, Rueger DC, Bizios R (1996) Biomaterials 17:209
94. Price RL, Haberstroh KM, Webster TJ (2004) Nanotechnology 15:892
95. Gutwein LG, Webster TJ (2002) J Nanoparticle Res 4:231
96. Malsch I (2001) Lecture for COST and NanoSTAG conference, Leuven, 29 Oct 2001
97. Malsch I (2002) The Industrial Physicist June/July:15
98. Cowin J (1987) Handbook of bioengineering. McGraw Hill, New York
99. Martin BR and Burr DB (1989) Structure, function and adaptation of compact bone. Raven, New York

Adv Biochem Engin/Biotechnol (2006) 103: 309–329
DOI 10.1007/10_029
© Springer-Verlag Berlin Heidelberg 2006
Published online: 12 September 2006

Integration of Technologies for Hepatic Tissue Engineering

Yaakov Nahmias · Francois Berthiaume · Martin L. Yarmush (✉)

Center for Engineering in Medicine/Department of Surgery,
Massachusetts General Hospital, Shriners Burns Hospital,
Harvard Medical School, 51 Blossom St, Boston, MA 02114, USA
ireis@sbi.org

Abstract The liver is the largest internal organ in the body, responsible for over 500 metabolic, regulatory, and immune functions. Loss of liver function leads to liver failure which causes over 25 000 deaths/year in the United States. Efforts in the field of hepatic tissue engineering include the design of bioartificial liver systems to prolong patient's lives during liver failure, for drug toxicity screening and for the study of liver regeneration, ischemia/reperfusion injury, fibrosis, viral infection, and inflammation. This chapter will overview the current state-of-the-art in hepatology including isolated perfused liver, culture of liver slices and tissue explants, hepatocyte culture on collagen "sandwich" and spheroids, coculture of hepatocytes with non-parenchymal cells, and the integration of these culture techniques with microfluidics and reactor design. This work will discuss the role of oxygen and medium composition in hepatocyte culture and present promising new technologies for hepatocyte proliferation and function. We will also discuss liver development, architecture, and function as they relate to these culture techniques. Finally, we will review current opportunities and major challenges in integrating cell culture, bioreactor design, and microtechnology to develop new systems for novel applications.

1
Introduction

The liver is the largest internal organ in the body, accounting for 2% of the weight of an adult ($\sim$ 1.5 kg) and around 5% of the weight in a neonate [1, 2]. More than 500 functions are ascribed to the liver, which serves an endocrine function secreting factors such as albumin and urea into the blood, an exocrine function secreting bile to the intestine, and a storage function for glycogen. The liver also serves as a hub for carbohydrate, lipid, and amino acid regulation. These hepatic functions are an integral part of the body's metabolic homeostasis mechanism [3]. Furthermore, the liver actively participates in systemic responses to trauma or injury by producing acute phase proteins and assuming a hypermetabolic state [4–6].

In addition to its metabolic activity, the liver is one of the body's first lines of defense, inactivating toxins and xenobiotics absorbed by the intestine and clearing foreign particles from the blood [7–10]. Finally, the liver is unique in its capacity to regenerate from even the most massive injuries, restoring to its original mass even if less than 20% of the cells remain undamaged [3, 11, 12].

Loss of liver function causes 25 000 deaths/year and is the tenth most frequent cause of death in the United States [13]. The only known cure for liver failure is orthotopic liver transplantation, but unfortunately less than 7000 organs are available per year in the United States. This organ shortage has spurred research in hepatocyte-based bioartificial liver systems (BAL), which have recently entered the market and promise to extend the life of those awaiting transplantation [14, 15]. Similar hepatocyte culture technology is also used in the pharmaceutical industry, screening the effects of new drugs and toxins prior to in vivo animal and human studies [9, 10]. Hepatocyte cultures are also used for other basic or applied purposes including the study of liver development [16], regeneration [12], ischemia/reperfusion injury [17], fibrosis [18], viral infection [19], and inflammation [20].

This report will provide an overview of the current state-of-the-art of in vitro liver models, detailing major advantages and disadvantages of each model. We will discuss liver development, architecture, and function as they relate to these culture techniques. Finally, we will review current opportunities and major challenges in integrating cell culture, bioreactor design, and microtechnology to develop new systems for novel applications.

2
Liver Development and Biology

The liver arises as a bud from the thickening of the endodermal epithelium lining of the foregut. As the bud grows it invades the mesenchyme of the septum transversum and separates into three parts, one of which will form the liver [21–23]. The proliferating endodermal cells of the hepatic portion migrate forward in the form of cords or strands into a network of capillaries emanating from the mesodermal umbilical and vitelline veins. In doing so, the hepatic cords subdivide the capillaries, giving rise to the development of parenchymal plates and liver sinusoids [1, 24]. Recent studies suggest that early endothelial cells promote hepatic differentiation and migration in both liver and pancreas [16, 25]. This interaction between mesenchymal cells and hepatocytes may prove to be essential for liver tissue reconstruction.

One of the liver's basic structural unit is the hepatic sinusoid, a specialized capillary of fenestrated liver sinusoidal endothelial cells (LSEC) averaging 10 μm in diameter and 275 μm in length (Fig. 1) [2, 26]. Blood flows through the sinusoid at an average flow rate of 144 μm/sec delivering approximately 2000 nmol/mL of oxygen to the surrounding shell of hepatocytes [26]. Between the hepatocytes and the LSEC is the, 1.4 μm wide, space of Disse, a thin reticular basement membrane, composed of fibronectin, laminin, proteoglycan, collagen IV, and collagen I [2, 27]. The hepatocytes surrounding the sinusoid extend numerous microvilli into the space of Disse and beyond and are therefore in direct contact with the blood [1], providing the cells ample access to oxygen, growth factors, and hormones. It is thought that the gradient of oxygen, growth factors, and hormones that forms along the sinusoid is responsible for the zonation of hepatic function (Table 1) [2, 28].

One example of the zonation of hepatic function is ammonia detoxification [29, 30]. Ammonia is removed by the liver by urea synthesis, which occurs along the sinusoid length, except for the last few cells near the central vein, which convert ammonia to glutamine [29]. Glutamine synthesis occurs at much lower ammonia concentrations than urea synthesis, allowing the perivenous hepatocyte to removes the remaining molecules of ammonia that could not be converted to urea in time. Therefore, perivenous hepatocytes act as a scrubber that reduces ammonia levels to well below toxic values in the

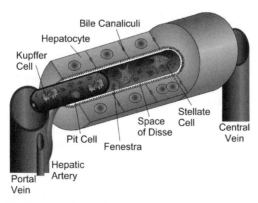

Fig. 1 Schematic of the hepatic sinusoid

Table 1 Metabolic zonation of the liver

	Periportal	Perivenus
Oxygen	60–70 mmHg	25–35 mmHg
Hormones	High	Low
Glucose	Gluconeogenesis	Glycolysis
		Detoxification
Nitrogen	Ureagenesis	Glutamine synthesis
Lipid	β-Oxidation	Liponeogenesis

exiting blood. Glutamine eventually returns to the liver via the systemic circulation. At this point, periportal hepatocytes expressing glutaminase release ammonia from glutamine. This results in enhanced levels of free ammonia in the periportal region of the sinusoid, which increases the rate of urea synthesis due to mass action effects. Thus, the combination of a high capacity urea synthetic ability in the periportal region with a high affinity ammonia removal system in the perivenous region enables a high rate of ammonia detoxification, while keeping circulating levels of ammonia very low [29].

Sinusoid zonation is not limited to the metabolic function of hepatocytes. The sinusoid architecture and cellular composition also varies from the periportal to the perivenous region [30]. Variations include the composition of the extracellular matrix along the sinusoid [31], cell morphology, cellular concentrations, and activity [32]. For example, liver sinusoidal cells have larger fenestrae in the periportal region, but smaller and more numerous fenestrae close to the central vein [33]. In addition, Kuppfer cells are more

Table 2 Cellular composition of rat liver, adapted from Morin et al. (1988)

Cell type	Diameter (μm)	Volume (% of total)	Number (% of total)	Type	Density (g/mL)
Parenchymal cells					
Hepatocytes	20–25	77.8	65	Epithelial	1.10–15
Non-parenchymal cells					
Sinusoidal endothelial cells	6.5–11	2.8	16	Endothelial	1.061–80
Kupffer cells	10–13	2.1	12	Macrophages	1.076
Stellate (Ito) cells	10.7–11.5	1.4	8	Fibroblasts	1.05
Pit cells		Minor	Minor		
Extracellular spaces					
Sinusoidal lumina		10.6			
Space of Disse		4.9			
Bile canaliculi		0.43			

concentrated near the portal triads and differ metabolically with their lobular locations [34]. Hepatocytes are also known to vary in size along the sinusoid. The cellular composition of the liver is listed in Table 2.

The liver parenchymal cells, the hepatocytes, are large cuboidal epithelial cells 20–25 μm in diameter [2]. Although hepatocytes are epithelial cells, they do not form a surface, but rather form a tubular space between adjacent hepatocytes, which is delimited by tight junctions. These apical spaces fuse along the hepatic plate, forming bile canaliculi, which drain secreted bile into the bile duct [2]. The hepatocyte basal surface faces the space of Disse and the blood, and is the site of albumin secretion and lipoprotein uptake [2]. Hepatic architecture and sinusoidal organization are essential for proper liver function (Fig. 1). Loss of liver architecture due to trauma or disease, such as fibrosis, leads to loss of tissue function.

3
Maintenance of Liver Tissue Ex Vivo

The need to preserve relevant tissue organization and function has led many groups to develop in situ models of liver function. One example is the isolated perfused liver, in which the organ remains intact and is nourished by perfusing the portal vein with media following ligation of the hepatic artery [35]. This technique preserves the heterotypic cell–cell interaction, as well as the organ's three-dimensional architecture and function [35]. Isolated liver perfusion has been used to study metabolic changes during injury and disease [36], protein turnover [37], toxic response [38, 39], and liver synthetic function [40, 41]. However, isolated livers perfused in this fashion can only be maintained for a few hours, and only a single experiment can be carried out on each liver, leading to a significant variance in the results [35].

In an effort to perform multiple experiments on each isolated liver, several groups developed methods to maintain precision-cut liver slices in culture [42–44]. Such slices can be created down to 200 μm in thickness (∼ 8 cell layers) and cultured for up to 3 days in a high-oxygen environment [43, 45]. Liver slices have been successfully used to study drug-induced toxicity and hepatic metabolism [42, 45–47]. Their main advantage is that they preserve the heterotypic interactions and architecture of the liver, while enabling a series of experiments to be run on the liver of a single animal, greatly reducing the assay variability [44]. However, the hyper-physiological oxygen concentration required to maintain the liver slices in culture might significantly alter hepatic metabolism, influencing the response of cultured liver slices to various stimuli.

The oxygen requirement of the liver is less stringent during development, when hepatic differentiation is not yet complete. This allows for the in vitro culture of embryonic tissue explant in which specific regions of the developing endoderm (i.e., ventral foregut, liver bud) are dissected and cultured on a cell culture insert or plastic dish [16, 23]. Embryonic tissue explants have been used to elucidate important mechanisms of liver development, such as the role of the cardiac mesoderm [48], endothelial cells [16], fibroblast growth factor (FGF) [21], and hepatocyte growth factor (HGF) [22].

4
Hepatocyte Culture Techniques

The most commonly used hepatocyte isolation technique was introduced by Seglen in 1976 [49, 50]. The isolation relies on a two-step in situ collagenase perfusion, followed by mechanical segregation of the tissue, and a purification step based on cell density [49]. This technique allows for high viability (> 90%) and, due to the high density of hepatocytes, a relatively pure population (> 95%). Between 100 to 400 million hepatocytes can be routinely isolated from a single rat liver [50].

Although purified hepatocytes have been available for the last three decades, reliable culture techniques have only become available in the last two decades (Fig. 2) [51, 52]. The most significant problem with hepatocyte culture is that the cells rapidly lose their differentiated structure and function following isolation. Isolated hepatocytes cultured under standard tissue culture conditions lose their cuboidal morphology and their liver-specific functions, accumulate actin stress fibers on the ventral surface in contact with the substrate (become fibroblast-like), lack bile canaliculi, and die within 1–3 days [51, 53]. In order to study hepatpcytes in vitro, several groups have developed advanced culture techniques that preserve hepatocyte function and structure for a period of time ranging from a week to several months.

The most common technique for culturing primary hepatocytes is to seed the cells on a single layer of collagen gel in conditioned medium. Standard hepatic culture medium contains fetal bovine serum, corticosteroids (either hydrocortisone or dexamethasone), insulin, and epidermal growth factor (EGF) [54]. Several serum-free media formulations are also available commercially and are detailed in the literature [55]. Hepatocytes cultured on a single layer of collagen gel in conditioned media secrete albumin and urea, and show cytochrome P450 activity [54]. However, these liver-specific functions steadily decline within the first week of culture, suggesting that significant survival cues are missing from this single gel culture [56].

Interestingly, the addition of a second layer of collagen on top of the cultured hepatocytes can maintain and even rescue hepatocyte polarity and function (Fig. 2b) [56]. When primary hepatocytes are cultured on a single collagen layer and allowed to "de-differentiate" for a few days, the addition of a second collagen layer on top of the cells causes a dramatic reorganization and expression of cytoskeletal proteins [51, 56, 57]. This collagen "sandwich" configuration induces the formation of distinct apical and lateral membrane

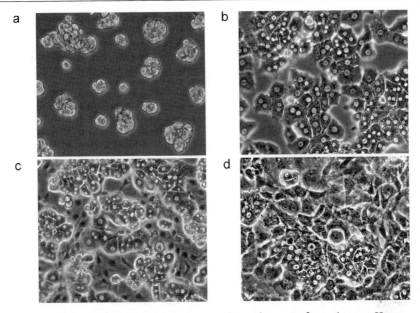

Fig. 2 Phase images of hepatocyte culture and coculture configurations. **a** Hepatocyte spheroids formed on Matrigel exhibit significant synthetic and enzymatic activity. Spheroids were shown to express E-cadherin [58], form extensive bile canaliculi [53], and show sinusoidal surface markers at the interface between cells in the spheroid [53]. **b** Hepatocytes cultured in a collagen double gel configuration form plate-like structures and stabilize synthetic and enzymatic activity a week after isolation. Cells in double gel were shown to exhibit native cell–cell contacts such as E-cadherin and bile canaliculi [58], but do not express sinusoidal receptors such as EGF-R [58] and LDL-R [81]. **c** Hepatocytes cocultured with LSEC show both traditional polarity markers [74, 75], and express a high level of sinusoidal receptors (EGF-R, LDL-R) at the interface between the hepatocytes and the LSEC. At least part of the interaction between hepatocytes and LSEC has been shown to be mediated by growth factors [76, 132]. **d** Hepatocytes cocultured with 3T3 fibroblasts grow in distinct clusters and exhibit hepatic cell–cell contacts such as connexin-32 [133] and bile canaliculi [134], but also do not express the EGF-R and LDL-R. At least part of the interaction between hepatocytes and 3T3 fibroblasts was shown to be mediated by N-cadherin and decorin [73]

domains, each expressing specific surface markers [58]. Actin filaments concentrate in regions of contact with neighboring cells, forming the sheathing of a functional bile canalicular network. Most importantly, the collagen sandwich induces a dramatic increase in the expression of liver-specific functions, which stabilize following one week of culture. Primary hepatocytes can be maintained in a collagen double gel for several months [51, 56, 57, 59]. The addition of heparan sulfate proteoglycan to the top collagen layer allows the expression of basal surface markers and gap-junction proteins, which are not observed in the standard collagen sandwich configuration [58–60].

Another stable culture technique involves the formation of hepatic spheroids on soft or non-adhesive extracellular matrix (Fig. 2a) [61]. When primary hepatocytes are cultured on a single layer of Matrigel, they form spherical aggregates during the first 48 h of culture [52, 60, 62, 63]. These hepatic spheroids also form when the cells are cultured on other protein membranes or relatively non-adhesive plastic such as Primaria [63]. Hepatocyte spheroids maintain a high level of hepatic structural polarity forming distinct apical, lateral, and basal domains [53]. Liver-specific function of hepatic spheroids is also enhanced showing high levels of albumin production [52], urea secretion, and cytochrome P450 activity [61, 64]. A practical limitation of hepatocyte spheroids is that the methods used to create them typically lack precise control over their size, which causes variability in the transport of metabolites in and out of the aggregates. In addition, the sheer size of the spheroids can lead to the formation of necrotic cores due to internal mass transfer limitations.

Although both the collagen "sandwich" and the spheroid culture configuration show similar enhancement and maintenance of function, the morphology of the cultured hepatocytes is distinctly different [60, 61]. Hepatocytes trapped between the layers of collagen form distinct plate-like structures with aligned bile canaliculi closely resembling the in vivo organization of the liver [59]. Hepatocyte spheroids, however, are round and form closely associated cellular aggregates, which are not found in the mature liver. One possibility is that hepatic spheroids resemble the organization of liver during regeneration, in which hepatocytes dissociate from the sinusoid and form large proliferating clusters [3, 12]. However, hepatocytes cultured in both configurations show little to no proliferation capacity, and therefore cannot serve as platforms for generating the large quantities of cells that may be required for clinical applications such as hepatocyte transplantation or the development of a bioartificial liver [54].

5
Hepatic Heterotypic Interactions

In an effort to more faithfully reproduce liver-like organization and hepatic proliferation in vitro, several groups have cocultured hepatocytes with the non-parenchymal cells of the liver (stellate, Kupffer, and endothelial cells) (Table 3) [65]. Small hepatocytes, a progenitor cell population characterized by Mitaka [66], were shown to proliferate in culture, forming large aggregates supported by epithelial cells and fibroblasts [66, 67]. These hepatic organoids were later formed on a collagen mesh and show remarkable liver-like structure and function [66, 68]. Hepatocyte plate-like structures emanated from the organoids in response to Matrigel [67]. Likewise, Michalopoulos found that hepatocytes proliferate and form plates when parenchymal and non-

Table 3 Liver sinusoidal endothelial cells (LSEC)

Liver sinusoidal endothelial cells (LSEC)
Liver sinusoids are composed of a highly specialized type of microvascular endothelial cells which play an important role in lipid metabolism, coagulation, cellular growth, differentiation, immune, and inflammatory response. LSEC contain numerous open pores in their membrane, called fenestra, which range from 100 to 1000 nm in diameter and are thought to act as a sieve for blood-borne particles. LSEC are highly phagocytic scavenger cells, which express MHC class I and II. LSEC are usually purified from the nonparenchymal fraction of the liver by centrifugal elutriation or a two-step Percoll gradient separation. L-SIGN is a specific marker for LSEC. Although LSEC do not proliferate in culture, a tumor cell line is available (SK-HEP-1) as well as a reversably immortalized cell line (TMNK-1).

Kupffer cells
Kupffer cells are the liver resident macrophage, thought to originate from the bone marrow. Kupffer cells ingest and degrade old erythrocytes, bacteria, various endotoxins, and play an important role in iron metabolism. Due to their similar density and size it is difficult to separate Kupffer cells from LSEC. However, relatively pure populations can be purified using centrifugal elutriation. The ED2 antibody is specific for Kupffer cells. Kupffer cells do not proliferate, and rapidly activate in culture. A mouse Kupffer cell line is available (KC13-2).

Hepatic stellate cells (HSC)
Stellate cells (Ito cells, fat-storing cells) are vitamin A-storing pericytes that decorate the liver's sinusoids. HSC are the main matrix-producing cell in the liver and play an important role in regeneration, differentiation, and inflammation. HSC are thought to become activated during liver fibrosis, increasing collagen and DNA synthesis and acquiring a myofibroblast-like phenotype. The low density of the HSC allows for simple purification using density centrifugation. CD95 and Desmin II are specific markers for HSC in the liver. HSC become rapidly activated in culture. An immortalized rat liver stellate cell line (HSC-T6) is available. Activated stellate cells proliferate in culture.

---

parenchymal cells are cultured on Matrigel [69]. In another study, the same group demonstrated that parenchymal and non-parenchymal cells self-organize in roller bottles to form simple epithelial structures consisting of a superficial layer of biliary epithelial cells, a middle layer of hepatocytes and connective tissue, and an inner layer of endothelial cells [70]. Although the coculture of hepatocytes with the non-parenchymal cells of the liver presents many opportunities for liver research, the role of the individual cell fractions has not been elucidated [65].

In order to study the mechanism by which non-parenchymal cells support hepatic function, our group has created micropatterns of hepatocytes and fibroblasts (3T3-J2) using photolithography [71]. These micropatterning studies showed that the fibroblasts maintain hepatocyte liver-specific function (albumin, urea) through a combination of cell–cell contact and short-acting diffusible substances [72]. Hepatocytes cultured several cell layers away from the fibroblasts lost albumin expression after a few days in culture, while those in close contact remained albumin positive. Gene expression studies carried out using different fibroblast populations implicated the adhesion molecule N-cadherin as well as the secreted proteoglycan decorin in the elevated hepatic function [73]. Most importantly, recent results from our group have shown that rat hepatocytes seeded at low densities on a feeder-layer of growth-arrested fibroblasts are able to proliferate in vitro for over 20 cell doublings (Fig. 3), Cho et al. (submitted for publication). These data suggest that fibroblasts are responsible for hepatic proliferation in coculture, enabling several clinical applications to utilize laboratory-grown hepatocytes.

In addition to fibroblasts, liver sinusoidal endothelial cells have also been shown to support hepatocyte function during in vitro culture (Fig. 2c). Pioneering work by Morin and Normand has shown that liver sinusoidal endothelial cells stabilize hepatic urea and albumin secretion for up to a month in culture [74, 75]. Others have shown that endothelial cells also play an important role in liver development prior to vascular development by stimulating hepatic migration and differentiation in vivo [16]. In addition, liver sinusoidal endothelial cells were shown to secrete hepatocyte growth factor (HGF) and interleukin-6 (IL-6) in response to vascular endothelial growth factor (VEGF) signaling from hepatocytes, protecting hepatocytes from toxic or ischemic damage [76]. A similar interaction between endothelial cells and hepatocytes was recently shown to occur in vitro. Hepatocytes cultured on Matrigel specifically migrate toward endothelial capillaries in response to endothelial-secreted HGF [77]. Microscale patterning of the endothelial cells using laser guidance [78, 79], allowed the researchers to align hepatocytes

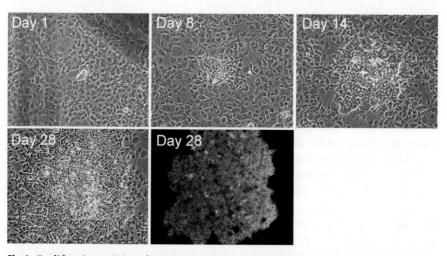

Fig. 3 Proliferative activity of primary rat hepatocytes on growth-arrested 3T3-J2 fibroblasts. Phase images are shown on day 1, 8, 14, and 28 followed by immunofluorescence staining for albumin on day 28

in a sinusoid-like structure, suggesting that endothelial cells might impart structural information to the developing and regenerating liver [80]. More recently, our group has shown that liver sinusoidal endothelial cells, but not fibroblasts, induce the hepatic expression of the basal receptors LDL-R and EGF-R and the uptake of HCV-like particles [81]. These receptors and the associated low density lipoprotein (LDL) uptake are a major function of hepatocytes in vivo, but are expressed at very low levels during in vitro culture. The inclusion of liver sinusoidal endothelial cells in culture induced the emergence of basal surface markers in hepatocytes, further enhancing their in vitro function [81].

Another important heterotypic cell–cell interaction occurs between hepatocytes and Kupffer cells, the liver's resident macrophages. Kupffer cells are important scavengers, which ingest and degrade aged erythrocytes, bacteria, various endogenous toxins, and play a role in iron metabolism [1, 8]. Kupffer cells have been implicated in drug toxicity and ischemic injury, becoming activated in response to a foreign stimulus and secreting active oxygen and nitrogen species, which cause hepatic damage [46, 82, 83]. Cocultures of hepatocytes and Kupffer cells, either in direct contact [84] or separated using a culture insert [85], have been shown to mimic in vivo-like damage due to ischemia and drug toxicity [10].

6
Role of Oxygen in Hepatocyte Culture

Oxygen is an important component of the hepatic microenvironment, mediating cellular metabolism, differentiation, and growth [28, 86]. Energy production in primary hepatocytes is highly dependent on oxidative phosphorylation, as each cell contains over 1500 mitochondria which consume oxygen at a rate of 0.3 to 0.9 nmol/sec/10^6 cells [87, 88]. In order to supply hepatocytes with this amount of oxygen in vivo, the liver is connected to the highly-oxygenated arterial circulation in addition to the portal circulation [89]. During in vitro culture, oxygen is supplied by diffusion from the air–liquid interface (Fig. 4). This oxygen diffusion limits the density of hepatocytes that can be seeded in culture or grown in a bioreactor [14].

In order to better understand the impact of oxygen supply in culture, hepatocytes have been cultured by several groups under varying oxygen tensions. Albumin and urea secretion were shown to increase when hepatocytes were cultured under high oxygen tension [90]. Conversely, hepatic cytochrome P450 gene expression and activity was show to increase under hypoxic conditions, suggesting that oxygen gradients could contribute to the establishment of metabolic zonation observed in vivo [28]. However, high oxygen tensions were also shown to stimulate the production of free radicals, possibly damaging the cultured cells [43, 91].

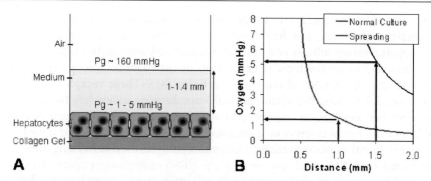

Fig. 4 Oxygen supply in standard culture configuration. **a** Schematic of hepatoctyes cultured on collagen gel in a standard tissue culture well and conditions. **b** Assuming Michaelis–Menten uptake of oxygen by hepatocytes at rates of 0.3–0.9 nmol/sec/10^6 cells for normal and spreading cultures, respectively [87, 88]. The *curves* show the partial pressure of oxygen at the cell surface. Hepatocytes were assumed to be seeded at standard densities of 100 000 cells/cm^2. Oxygen partial pressure is below 5 mmHg

Another technique recently developed by our group involves increasing the oxygen concentration in the system by adding an oxygen carrier to the extracellular matrix. Hepatocytes cultured in a "sandwich" of oxygen-carrying collagen showed a significant increase in albumin and urea secretion (Nahmias et al. 2006, accepted for publication). In addition, cytochrome P450 activity was shown to be dramatically increased during the first 24 h of culture, possibly due to the role of molecular oxygen in the enzymatic reaction (Nahmias et al. 2006, accepted for publication). Oxygen carriers such as emulsified fluorocarbon and red blood cells have also been used in a number of bioartificial liver studies and were shown to similarly increase hepatic survival and function [92–94].

Oxygen supply is especially important during the initial phase of cellular spreading, when the oxygen uptake rate is 40–300% higher than the value observed during the stable phase of culture [87, 95]. This initial demand for high oxygen concentration makes seeding of hepatocytes in various bioreactors and microfluidic devices very challenging [54, 88]. The small liquid volume in which cells are seeded stores very little oxygen, while flow cannot be used to deliver more oxygen until the cells have adhered. Therefore, many devices and reactors are seeded in static open configuration, which is hermetically closed and perfused only after cellular adhesion [86].

7
Impact of Culture Medium Formulation

Typical hepatocyte culture media contain high levels of hormones compared to physiological values. For example, insulin is used at levels approximately

10^4 times physiological. These formulations were developed in the early days of hepatocyte culture and have not received as much scrutiny as other aspects of hepatic tissue culture [96]. It is likely that some of the requirements for hormonal and other supplements in the culture medium may be relaxed in the newer hepatocyte culture systems that provide more in vivo cues from complex ECM and cell–cell interactions. Literature data suggest that collagen-sandwiched hepatocytes can be placed in media containing physiologically relevant hormone levels – at least for a few days – to observe metabolic responses to stress hormones [97]. Supraphysiological levels of hormones can also lead to paradoxical responses when cultured hepatocytes are placed in animal or human plasma, clearly a more physiologically relevant fluid than culture medium. Prior studies show that rat hepatocytes become severely fatty and lose hepatic functions when transferred from culture medium to plasma [98], but that plasma-induced intracellular lipid accumulation can be eliminated if culture medium containing low insulin levels is used prior to exposure to plasma [99].

8
Dynamic Flow Cultures

Microfabrication and microfluidics are relatively new technologies that allow for the control of the cellular microenvironment at the micron scale [100]. Cells and cellular complexes cultured in a microfluidic device can be addressed by a variety of soluble (growth factors) and mechanical factors (shear). The technology allows for the study of cellular response to stimuli that cannot be created in static culture, such as cellular interaction with leukocytes following ischemia-reperfusion injury, or hepatocyte metabolic differentiation in response to a gradient of oxygen or hormones generated in vitro (zonation).

The flat-plate bioreactor is a simple design with established flow geometry that can be machined without the need for microfabrication (Fig. 5a) [86]. Flat-plate bioreactors have been used to study hepatocyte function and differentiation by several groups, including ours [28, 86, 101]. Using this model we and others have shown that the hepatocyte metabolic function is significantly reduced when the cells are exposed to high shear rates (> 5 dyn/cm^2) [86]. While reducing the shear flow would reduce the putative mechanical damage, it would also reduce the delivery of oxygen to the hepatocytes [86]. Two strategies have been developed to decouple the oxygen supply from flow in the bioreactor. One technique is to increase the oxygen concentration in the bioreactor by incorporating a membrane oxygenator that allows for diffusion of oxygen across the reactor wall [86]. Another strategy is to reduce the exposure of hepatocytes to shear by seeding the cells in microfabricated groves perpendicular to the flow direction [102] or microwells [103].

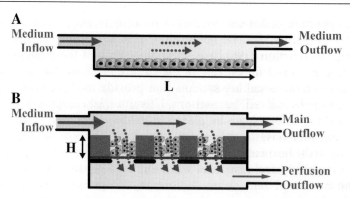

Fig. 5 Hepatic flow bioreactors. **a** Flat-plate bioreactor design. *L*: perfusion length of 2–3 cm allowing for oxygen and hormone gradients to develop along the reactor. The two-dimensional layout allows for clear optical imaging of cells in the reactor. **b** Packed-bed bioreactor design. *H*: perfusion height of 200–300 μm, allowing for an in vivo-like microenvironment in a physiological perfusion length

The flat-plate bioreactor system has also been used to culture hepatocytes under a stable oxygen and hormone gradient in vitro. The cultured hepatocytes showed aspects of zonal differentiation including the localization of phosphoenolpyruvate carboxykinase in the upstream oxygen-rich region, and cytochrome P450 2B in the downstream oxygen-poor region, which is consistent with the in vivo zonation [28]. This system was more recently used to study the effects of acetaminophen toxicity on metabolically zonated hepatocytes [104].

A different version of a hepatic bioreactor is the packed-bed reactor in which hepatocyte aggregates are perfused in an environment that allows for three-dimensional organization [105, 106]. Designs include hepatocytes cultured on polyvinyl formal resin [107], entrapped in alginate particles [108], or hepatocyte aggregates packed between silica beads [109]. One interesting design, introduced by investigators at MIT, is a microfabricated array bioreactor (Fig. 5b) [110]. The bioreactor's heart is a silicon scaffold perforated with a regular array of square holes and seated atop a microporous filter. Wells are 300 μm wide and 235 μm in height, designed for physiological shear [110, 111]. Hepatocyte aggregates seeded in packed-bed reactors such as the one described maintain albumin and urea secretion as well as cytochrome P450 activity for weeks in culture [107–109, 112]. The microfabricated array bioreactor was more recently used to study drug toxicity and hepatitis B virus infection [113]. Although these packed-bed reactors allow for the three-dimensional organization of tissue-like structures under physiological shear, they do so at the cost of losing control over cellular architecture and optical clarity. Tissue-like structures form randomly in each well and cannot be faithfully reproduced. In addition, the three-dimensional nature of the reactors makes optical imaging of the hepatic aggregates difficult.

The integration of heterotypic cell–cell interactions in perfused hepato-cyte cultures is an additional level of complexity that will be required for capturing the function and characteristics of the in vivo liver. Our group has cocultured 3T3-J2 fibroblasts with hepatocytes in the flat-plate bioreac-tor and showed increased hepatic function [86, 114]. Others have integrated the non-parenchymal cell fraction of the liver with hepatocytes, with simi-lar results [115]. Gerlach's group, at Humboldt University, showed remarkable organization of mouse liver fetal cells in a hollow fiber bioreactor [116]. When fetal cells were cultured in the bioreactor for several weeks, the cells formed liver-like tissue structures including hepatic, endothelial, and stel-late components and demonstrated albumin secretion and cytochrome P450 activity [116]. Another group showed similar organization of hepatic, en-dothelial, and stellate cell lines in a radial-flow bioreactor [117].

9
Current Challenges and Opportunities

The last decade has seen immense growth in knowledge about hepato-cyte survival, differentiation, and function both in vivo and in vitro. Par-allel growth in microfabrication and imaging technologies allows for mi-croscale control and observation of the cellular microenvironment. These ad-vances create opportunities for the development of hepatic tissue engineered models that more closely mimic the in vivo physiology and pathology of the liver [106, 111]. Liver fibrosis, for example, is a global health problem associ-ated with a gross disruption of liver architecture, impaired hepatic function, portal hypertension, and significant resultant morbidity and mortality [118]. There is overwhelming evidence that hepatic stellate cells become activated following chronic liver injury, producing a wide variety of collagenous and non-collagenous extracellular matrix proteins [119, 120]. The accumulation of these extracellular matrix proteins is thought to alter the phenotype of the sinusoidal endothelial cells and hepatocytes, causing sinusoidal capilariza-tion and loss of hepatic function [18]. Several antifibrotic therapies have been shown to inhibit stellate cell activation in vitro and in vivo, but the next gen-eration of treatments will attempt to reverse fibrosis and restore normal liver architecture and function [121, 122]. Tissue engineered liver models, such as those described above, can be constructed from human hepatic, endothelial, and stellate cells and used to screen vast arrays of potential antifibrotic ther-apies prior to clinical trials. In addition to drug development, such studies would potentially discover the basic biological mechanism of sinusoid differ-entiation and hepatic function, which is lost and regained in the process.

Another emerging opportunity is the study of hepatitis C virus infec-tion. The hepatitis C virus infects over 3% of the world population and it is currently the leading cause of liver failure in the United States [123, 124].

Several groups have recently developed hepatoma cell lines, which can be transfected with the viral RNA and produce high titers of infectious viral particles [125, 126]. However, blood-borne hepatitis C virus is not able to stably infect hepatocytes in vitro although it is able to efficiently infect hepatocytes in vivo [124]. This discrepancy between hepatocytes in vitro and in vivo suggests a phenotypical difference caused by culture conditions or improper cell–cell or cell–matrix interactions. Loss of liver-specific receptors, cellular polarization, or liver function could be at fault. Therefore, stable hepatitis C virus infection is an excellent litmus test of whether a specific liver model faithfully reproduces the in vivo phenotype.

One limitation of current liver culture technologies is the absence of a separate compartment to collect secreted bile, unlike in the native liver [54]. Bile secreted by hepatocytes in vivo flows slowly within the hepatic canaliculi, generally in the opposite direction to the blood flow in the sinusoid, and discharges into the bile duct [2, 127]. Bile is a complex fluid containing detergent-like bile acids, excess cholesterol, and bilirubin, which is a toxic breakdown product of hemoglobin [127]. Lacking a functional clearance mechanism for bile, current in vitro culture models and bioreactors have these toxic products accumulate and mix with the basal medium, potentially causing significant damage over time to cultured hepatocytes [128]. A major challenge of hepatic tissue engineering is to integrate a bile canalicular collection system for hepatocyte cultures [129]. Figure 6 suggests one microfabrication strategy to achieve that goal. Hepatocytes would be seeded in microfabricated channels set 50 μm apart, limiting cellular organization to two aligned rows of hepatocytes. The hepatocytes would then be layered with collagen and a monolayer of endothelial cells exposed to shear flow. A small negative pressure could be applied to ∼ 10 μm high microchannels,

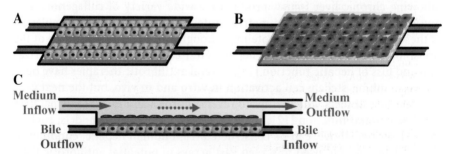

Fig. 6 Microfabrication design of bile collection in a flat-plate bioreactor. **a** Hepatocytes seeded in microfabricated channels set 50 μm apart, limiting cellular organization to two aligned rows of hepatocytes. **b** The hepatocytes would then be layered with collagen and a monolayer of endothelial cells exposed to shear flow. A small negative pressure could be applied to ∼ 10 μm high microchannels, allowing secreted bile to be cleared from the underlying hepatocytes. **c** Cross-section of a layered hepatocyte endothelial culture in a flat-plate bioreactor with integrated bile-collection system

allowing secreted bile to be cleared from the underlying hepatocytes. This design would also allow studies of bile acid composition in response to various stimuli.

Finally, the integration of techniques from cellular and molecular biology, tissue engineering, and microelectromechanical systems (MEMS) will spawn new designs of systems with many tissue-engineered hepatic units enabling massively parallel screening strategies [130]. For example, fluorescecent reporter genes can be introduced into cells in a model liver sinusoid such as those described above [130, 131]. Addressed by various stimuli (drug, toxin, cytokine) using microfluidics, the cells would respond by emitting a fluorescent signal following reporter gene activation (i.e., NFκ-b). Signal detection could be computerized, allowing the collection of thousands of data points in real time [130, 131]. Such liver-on-chip devices could be used to screen the hepatotoxic effects of drugs or environmental toxins, screen potential growth factors for hepatic differentiation of stem cells, or track the response of tumors to various anticancer therapies.

Acknowledgements This work was supported by grants from the National Institutes of Health (R01DK43371 and P41ED002503) and the Shriners Hospitals for Children.

References

1. Zakim D, Boyer T (1996) Hepatology: A textbook of liver disease. W.B. Saunders, Philadelphia
2. Desmet VJ (2001) In: Arias IM (ed) The liver, biology and pathobiology. Lippincott Williams & Wilkins, Philadelphia, p 3
3. Taub R (2004) Nat Rev Mol Cell Biol 5:836
4. Dahn MS, Lange MP, Berberoglu ED (1996) Shock 6:52
5. Lee K, Berthiaume F, Stephanopoulos GN, Yarmush ML (2003) Biotechnol Bioeng 83:400
6. Ramadori G, Christ B (1999) Semin Liver Dis 19:141
7. Nedredal GI, Elvevold KH, Ytrebo LM, Olsen R, Revhaug A, Smedsrod B (2003) Comp Hepatol 2:1
8. Willekens FLA, Werre JM, Kruijt JK, Roerdinkholder-Stoelwinder B, Groenen-Döpp YAM, Bos AGvd, Bosman GJCGM, Berkel TJCv (2005) Blood 105:2141
9. Behnia K, Bhatia S, Jastromb N, Balis U, Sullivan S, Yarmush ML, Toner M (2000) Tissue Eng 6:467
10. Grattagliano I, Portincasa P, Palmieri VO, Palasciano G (2002) Annal Hepatol 1:162
11. Ankoma-Sey V (1999) News Physiol Sci 14:149
12. Michalopoulos GK, DeFrances MC (1997) Science 276:60
13. Popovic JR, Kozak LJ (2000) Vital Health Stat 13
14. Yarmush ML, Toner M, Dunn JCY, Rotem A, Hubel A, Tompkins RG (1992) Annal NY Acad Sci 665:238
15. Tsiaoussis J, Newsome PN, Nelson LJ, Hayes PC, Plevris JN (2001) Liver Tansplant 7:2
16. Matsumoto K, Yoshitomi H, Rossant J, Zaret KS (2001) Science 297:559

17. Fondevila C, Busuttil RW, Kupiec-Weglinski JW (2003) Exp Mol Pathol 74:86
18. Neubauer K, Saile B, Ramadori G (2001) Can J Gastroenterol 15:187
19. Gardner JP, Durso RJ, Arrigale RR, Donovan GP, Maddon PJ, Dragic T, Olson WC (2003) PNAS 100:4498
20. Jayaraman A, Yarmush ML, Roth CM (2005) Tissue Eng 11:50
21. Jung J, Zheng M, Goldfarb M, Zaret KS (1999) Science 284:1998
22. Schmidt C, Bladt F, Goedecke S, Brinkmann V, Zschiesche W, Sharpe M, Gherardi E, Birchmeler C (1995) Nature 373:699
23. Lemaigre F, Zaret KS (2004) Curr Opin Genet Dev 14:582
24. Sosa-Pineda B, Wigle JT, Oliver G (2000) Nat Genet 25:254
25. Lammert E, Cleaver O, Melton D (2001) Science 294:564
26. McCuskey RS, Ekataksin W, LeBouton AV, Nishida J, McCuskey MK, McDonnell D, Williams C, Bethea NW, Dvorak B, Koldovsky O (2003) Anatom Rec Part A 275A:1019–1030
27. Martinez-Hernandez A, Amenta PS (1995) FASEB J 9:1401
28. Allen JW, Bhatia SN (2003) Biotechnol Bioeng 82:253
29. Haussinger D, Lamers WH, Moorman AF (1992) Enzyme 46:72
30. Jungermann K (1992) Diabete Metab 18:81
31. Reid LM, Fiorino AS, Sigal SH, Brill S, Holst PA (1992) Hepatology 15:1198
32. Bouwens L, Bleser PD, Vanderkerken K, Geerts B, Wisse E (1992) Enzyme 46:155
33. Braet F, Wisse E (2002) Comparat Hepatol 1:1
34. Morin O, Goulet F, Normand C (1988) Rev Sobre Biol Cel 15:1
35. Gores GJ, Kost LJ, LaRusso NF (1986) Hepatology 6:511
36. Lee K, Berthiaume F, Stephanopoulos GN, Yarmush DM, Yarmush ML (2000) Metab Eng 2:312
37. Tavill AS, East AG, Black EG, Nadkarni D, Hoffenberg E (1972) Ciba Found Symp 9:155
38. Palmen NG, Evelo CT, Borm PJ, Henderson PT (1993) Hum Exp Toxicol 12:127
39. McKindley DS, Chichester C, Raymond R (1999) Shock 12:468
40. Gordon AH, Humphrey JH (1960) Biochem J 75:240
41. Burke WT (1960) Biochem Biophys Res Commun 3:525
42. Zhao P, Kalhorn TF, Slattery JT (2002) Hepatology 36:326
43. Martin H, Sarsat JP, Lerche-Langrand C, Housset C, Balladur P, Toutain H, Al-baladejo V (2002) Cell Biol Toxicol 18:73
44. Fisher RL, Ulreich JB, Nakazato PZ, Brendel K (2001) Toxicol Method 11:59
45. Lerche-Langrand C, Toutain HJ (2000) Toxicology 153:221
46. Neyrinck A, Eeckhoudt SL, Meunier CJ, Pampfer S, Taper HS, Verbeeck RK, Delzenne N (1999) Life Sci 65:2851
47. Olinga P, Hof IH, Merema MT, Smit M, Jager MHd, Swart PJ, Slooff MJ, Meijer DK, Groothuis GM (2001) J Pharmacol Toxicol Method 45:55
48. Gualdi R, Bossard P, Zheng M, Hamada Y, Coleman JR, Zaret KS (1996) Gene Dev 10:1670
49. Seglen PO (1976) Method Cell Biol 13:29
50. Strain AJ (1994) Gut 35:433
51. Dunn JCY, Tompkins RG, Yarmush ML (1991) Biotechnol Prog 7:237
52. Koide N, Shinji T, Tanabe T, Asano K, Kawaguchi M, Sakaguchi K, Koide Y, Mori M, Tsuji T (1989) Biochem Biophys Res Commun 161:385
53. Abu-Absi SF, Friend JR, Hansen LK, Hu W-S (2002) Exp Cell Res 274:56
54. Yarmush ML, Toner M, Dunn JC, Rotem A, Hubel A, Tompkins RG (1992) Ann NY Acad Sci 665:238

55. Hamilton GA, Westmorel C, George AE (2001) In Vitro Cell Dev Biol Animal 37:656
56. Dunn JC, Yarmush ML, Koebe HG, Tompkins RG (1989) FASEB J 3:174
57. Dunn JC, Tompkins RG, Yarmush ML (1992) J Cell Biol 116:1043
58. Moghe PV, Berthiaume F, Ezzell RM, Toner M, Tompkins RG, Yarmush ML (1996) Biomaterials 17:373
59. Berthiaume F, Moghe PV, Toner M, Yarmush ML (1996) FASEB J 10:1471
60. Moghe PV, Coger RN, Toner M, Yarmush ML (1997) Biotechnol Bioeng 56:706
61. Landry J, Bernier D, Ouellet C, Goyette R, Marceau N (1985) J Cell Biol 101:914
62. Schuetz EG, Li D, Omiecinski CJ, Muller-Eberhard U, Kleinman HK, Elswick B, Guzelian PS (1988) J Cell Physiol 134:309
63. Peshwa MV, Wu FJ, Sharp HL, Cerra FB, Hu W-S (1996) In Vitro Cell Dev Biol Animal 32:197
64. Wu FJ, Friend JR, Remmel RP, Cerra FB, Hu W-S (1999) Cell Transplant 8:233
65. Strain AJ (1999) Hepatology 29:288
66. Mitaka T, Sato F, Mizuguchi T, Yokono T, Mochizuki Y (1999) Hepatology 29:111
67. Mitaka T (2002) J Hepatobiliary Pancreat Surg 9:697–703
68. Harada K, Mitaka T, Miyamoto S, Sugimoto S, Ikeda S, Takeda H, Mochizuki Y, Hirata K (2003) J Hepatol 39:716–723
69. Michalopoulos GK, Bowen WC, Zajac VF, Stolz DB, Watkins S, Kostrubsky V, Strom SC (1999) Hepatology 29:90
70. Michalopoulos GK, Bowen WC, Mule K, Stolz DB (2001) Am J Pathol 159:1877–1887
71. Bhatia SN, Yarmush ML, Toner M (1997) J Biomed Mater Res 34:189
72. Bhatia SN, Balis UJ, Yarmush ML, Toner M (1999) FASEB J 13:1883
73. Khetani SR, Szulgit G, Rio JAD, Barlow C, Bhatia SN (2004) Hepatology 40:545
74. Morin O, Normand C (1986) J Cell Physiol 129:103
75. Goulet F, Normand C, Morin O (1988) Hepatology 8:1010
76. LeCouter J, Moritz DR, Li B, Phillips GL, Liang XH, Gerber H-P, Hillan KJ, Ferrara N (2003) Science 299:890
77. Nahmias Y, Schwartz RE, Wei-Shou H, Verfaillie CM, Odde DJ (2006) Tissue Eng 12:1627
78. Nahmias YK, Odde DJ (2002) IEEE J Quantum Electron 38:131
79. Nahmias YK, Gao BZ, Odde DJ (2004) Appl Opt 43:3999
80. Nahmias Y, Schwartz RE, Verfaillie CM, Odde DJ (2005) Biotechnol Bioeng 92:129
81. Nahmias Y, Casali M, Barbe L, Berthiaume F, Yarmush ML (2006) Hepatology 43:257
82. Jaeschke H, Farhood A (1991) Am J Physiol 260:G355
83. James LP, Mayeux PR, Hinson JA (2003) Drug Metabol Disp 31:1499
84. Hoebe KH, Witkamp RF, Fink-Gremmels J, Miert ASV, Monshouwer M (2001) Am J Physiol Gastrointest Liver Physiol 280:G720
85. Milosevic N, Schawalder H, Maier P (1999) Eur J Pharmacol 368:75
86. Tilles AW, Baskaran H, Roy P, Yarmush ML, Toner M (2001) Biotechnol Bioeng 73:379
87. Balis UJ, Behnia K, Dwarakanath B, Bhatia SN, Sullivan SJ, Yarmush ML, Toner M (1999) Metabol Eng 1:49
88. Foy BD, Rotem A, Toner M, Tompkins RG, Yarmush ML (1994) Cell Transplant 3:515
89. Lemasters JJ (2001) In: Arias IM (ed) The liver: biology and pathobiology. Lippincott Williams & Wilkins, Philadelphia, p 257
90. Bhatia SN, Toner M, Foy BD, Rotem A, O'Neil KM, Tompkins RG, Yarmush ML (1996) J Cell Eng 1:125
91. Fariss MW (1990) Free Radic Biol Med 9:333
92. King AT, Mulligan BJ, Lowe KC (1989) Nat Biotechnol 7:1037

93. Gordon JE, Dare MR, Palmer AF (2005) Biotechnol Prog 21:1700
94. Rappaport C, Rensch Y, Abbasi M, Kempe M, Rocaboy C, Gladysz J, Trujillo EM (2002) Biotechniques 32:142
95. Rotem A, Toner M, Bhatia S, Foy BD, Tompkins RG, Yarmush ML (1994) Biotechnol Bioeng 43:654
96. Williams GM, Bermudez E, Scaramuzzino D (1977) In Vitro 13:809
97. Zupke CA, Stefanovich P, Berthiaume F, Yarmush ML (1998) Biotechnol Bioeng 58:222
98. Matthew HWT, Sternberg J, Stefanovich P, Morgan JR, Toner M, Tompkins RG, Yarmush ML (1996) Biotechnol Bioeng 51:100
99. Chan C, Berthiaume F, Washizu J, Toner M, Yarmush ML (2002) Biotechnol Bioeng 78:753
100. Andersson H, Berg Avd (2004) Lab Chip 4:98
101. Kataropoulou M, Henderson C, Grant MH (2005) Tissue Engineering 11:1263
102. Park J, Berthiaume F, Toner M, Yarmush ML, Tilles AW (2005) Biotechnol Bioeng 90:632
103. Khademhosseini A, Yeh J, Eng G, Karp J, Kaji H, Borenstein J, Farokhzad OC, Langer R (2005) Lab Chip 5:1380
104. Allen JW, Khetani SR, Bhatia SN (2005) Toxicol Sci 84:110
105. Strain AJ, Neuberger JM (2002) Science 295:1005
106. Griffith LG, Naughton G (2002) Science 295:1009
107. Ohshima N, Yanagi K, Miyoshi H (1997) Artif Organs 21:1169
108. Murtas S, Capuani G, Dentini M, Manetti C, Masci G, Massimi M, Miccheli A, Crescenzi V (2005) J Biomater Sci Polym Ed 16:829
109. Li AP, Barker G, Beck D, Colburn S, Monsell R, Pellegrin C (1993) In Vitro Cell Dev Biol 29A:249
110. Powers MJ, Domansky K, Kaazempur-Mofrad MR, Kalezi A, Capitano A, Upadhyaya A, Kurzawski P, Wack KE, Stolz DB, Kamm R, Griffith LG (2002) Biotechnol Bioeng 78:257
111. Griffith LG, Swartz MA (2006) Nat Rev Mol Cell Biol 7:211
112. Powers MJ, Janigian DM, Wack KE, Baker CS, Stolz DB, Griffith LG (2002) Tissue Eng 8:499
113. Sivaraman A, Leach JK, Townsend S, Iida T, Hogan BJ, Stolz DB, Fry R, Samson LD, Tannenbaum SR, Griffith LG (2005) Curr Drug Metab 6:569
114. Bhatia SN, Balis UJ, Yarmush ML, Toner M (1998) J Biomater Sci Polym Ed 9:1137
115. Kan P, Miyoshi H, Yanagi K, Ohshima N (1998) ASAIO 44: M441
116. Monga SPS, Hout MS, Baun MJ, Micsenyi A, Muller P, Tummalapalli L, Ranade AR, Luo J-H, Strom SC, Gerlach JC (2005) Am J Pathol 167:1279
117. Saito M, Matsuura T, Masaki T, Maehashi H, Shimizu K, Hataba Y, Iwahori T, Suzuki T, Braet F (2006) World J Gastroenterol 12:1881
118. Benyon RC, Iredale JP (2000) Gut 46:443
119. Bataller R, Brenner DA (2005) J Clin Invest 115:209
120. Friedman SL, Bansal MB (2006) Hepatology 43:S82
121. Albanis E, Friedman SL (2006) Am J Transplant 6:12
122. Prosser CC, Yen RD, Wu J (2006) World J Gastroenterol 12:509
123. Penin F, Dubuisson J, Rey FA, Moradpour D, Pawlotsky J-M (2004) Hepatology 39:5–19
124. Guidotti LG, Chisari FV (2006) Ann Rev Pathol Mech Dis 1:23–61
125. Wakita T, Pietschmann T, Kato T, Date T, Miyamoto M, Zhao Z, Murthy K, Habermann A, Krausslich HG, Mizokami M, Bartenschlager R, Liang TJ (2005) Nat Med 11:791

126. Lindenbach BD, Evans MJ, Syder AJ, Wolk B, Tellinghuisen TL, Liu CC, Maruyama T, Hynes RO, Burton DR, McKeating JA, Rice CM (2005) Science 309:623
127. Ujhazy P, Kipp H, Misra S, Wakabayashi Y, Arias IM (2001) In: Arias IM (ed) The Liver, Biology and Pathobiology. Lippincott Williams & Wilkins, Philadelphia, p 361
128. Clouzeau-Girard H, Guyot C, Combe C, Moronvalle-Halley V, Housset C, Lamireau T, Rosenbaum J, Desmoulière A (2006) Lab Invest 86:275
129. Tilles AW, Berthiaume F, Yarmush ML, Tompkins RG, Toner M (2002) J Hepatobiliary Pancreat Surg 9:686
130. Thompson DM, King KR, Wieder KJ, Toner M, Yarmush ML, Jayaraman A (2004) Anal Chem 76:4098
131. Wieder KJ, King KR, Thompson DM, Zia C, Yarmush ML, Jayaraman A (2005) Biomed Microdev 7:213
132. Davidson AJ, Zon LI (2003) Science 299:835
133. Sugimachi K, Sosef MN, Baust JM, Fowler A, Tompkins RG, Toner M (2004) Cell Transplant 13:187
134. Bhandari RN, Riccalton LA, Lewis AL, Fry JR, Hammond AH, Tendler SJ, Shakesheff KM (2001) Tissue Eng 7:345

Author Index Volumes 101–103

Subject Index

Printing: Krips bv, Meppel
Binding: Stürtz, Würzburg